Military Ethics and Emerging Technologies

This volume looks at current and emerging technologies of war and some of the ethical issues surrounding their use. Although the nature and politics of war never change, the weapons and technologies used in war do change and are always undergoing development. Because of that, the arsenal of weapons for twenty-first century conflict is different from previous centuries. Weapons in today's world include an array of instruments of war that include, robotics, cyber war capabilities, human performance enhancement for warriors, and the proliferation of an entire spectrum of unmanned weapons systems and platforms. Tactical weapons now have the potential of strategic results and have changed the understanding of the battle space creating ethical, legal, and political issues unknown in the pre-9/11 world. What do these technologies mean for things such as contemporary international relations, the just-war tradition, and civil-military relations?

Directed at readers in the academic, scientific, military, and public policy communities, this volume offers current thought on ethics and emerging technologies from internationally-recognized scholars addressing the full spectrum of issues in present warfare technology. It includes current and ongoing topics of multi-discipline and international interest, such as ethics, law, international relations, war studies, public policy, science and technology.

The chapters in this book were originally published in various issues and volumes of the *Journal of Military Ethics.*

Timothy J. Demy, Th.D., Ph.D. is a Professor of Military Ethics at the US Naval War College in Newport, Rhode Island.

George R. Lucas, Jr., Ph.D. is a Professor of Ethics & Public Policy, US Naval Postgraduate School and Distinguished Chair of Ethics (Retired), U.S. Naval Academy.

Bradley J. Strawser, Ph.D. is an Assistant Professor of Philosophy in the Defense Analysis Department at the US Naval Postgraduate School in Monterey, California.

Military Ethics and Emerging Technologies

Edited by

Timothy J. Demy, George R. Lucas, Jr., and Bradley J. Strawser

Routledge
Taylor & Francis Group
LONDON AND NEW YORK

First published 2014
by Routledge
2 Park Square, Milton Park, Abingdon, Oxon, OX14 4RN, UK

and by Routledge
711 Third Avenue, New York, NY 10017, USA

Routledge is an imprint of the Taylor & Francis Group, an informa business

British Library Cataloguing in Publication Data
A catalogue record for this book is available from the British Library

ISBN 13: 978-0-415-73710-4

Typeset in Times New Roman
by Taylor & Francis Books

Publisher's Note
The publisher accepts responsibility for any inconsistencies that may have arisen during the conversion of this book from journal articles to book chapters, namely the possible inclusion of journal terminology.

Disclaimer
Every effort has been made to contact copyright holders for their permission to reprint material in this book. The publishers would be grateful to hear from any copyright holder who is not here acknowledged and will undertake to rectify any errors or omissions in future editions of this book.

Editors' Disclaimer
The views expressed in this volume are solely those of the editors and authors and do not represent or reflect the position or endorsement of any governmental agency or organization, military or otherwise.

Contents

Citation Information

The following chapters were originally published in the *Journal of Military Ethics*, from various issues and volumes. When citing this material, please use the original page numbering for each article, as follows:

Chapter 2
The Ethics of Killer Applications: Why Is It So Hard To Talk About Morality When It Comes to New Military Technology?
P.W. Singer
Journal of Military Ethics, volume 9, issue 4 (2010) pp. 299–312

Chapter 3
With Fear and Trembling: An Ethical Framework for Non-Lethal Weapons
Pauline Kaurin
Journal of Military Ethics, volume 9, issue 1 (2010) pp. 100–114

Chapter 4
Do Non-Lethal Capabilities License to 'Silence'?
Sjef Orbons
Journal of Military Ethics, volume 9, issue 1 (2010) pp. 78–99

Chapter 5
Ethical Blowback from Emerging Technologies
Patrick Lin
Journal of Military Ethics, volume 9, issue 4 (2010) pp. 313–331

Chapter 6
The Case for Ethical Autonomy in Unmanned Systems
Ronald C. Arkin
Journal of Military Ethics, volume 9, issue 4 (2010) pp. 332–341

Chapter 7
The Case Against Robotic Warfare: A Response To Arkin
Ryan Tonkens
Journal of Military Ethics, volume 11, issue 2 (2012) pp. 149–168

Chapter 8
Moral Predators: The Duty to Employ Uninhabited Aerial Vehicles
Bradley Jay Strawser
Journal of Military Ethics, volume 9, issue 4 (2010) pp. 342–368

Chapter 9
Saying 'No!' to Lethal Autonomous Targeting
Noel Sharkey
Journal of Military Ethics, volume 9, issue 4 (2010) pp. 369–383

Chapter 10
Additive Manufacturing and its Implications for Military Ethics
John Mark Mattox
Journal of Military Ethics, volume 12, issue 3 (2013) pp. 225–234

Chapter 11
The Ethics of Cyberwarfare
Randall R. Dipert
Journal of Military Ethics, volume 9, issue 4 (2010) pp. 384–410

Chapter 12
'Cyberation' and Just War Doctrine: A Response to Randall Dipert
Colonel James Cook
Journal of Military Ethics, volume 9, issue 4 (2010) pp. 411–423

Chapter 13
Warfare in a New Domain: The Ethics of Military Cyber-Operations
Edward T. Barrett
Journal of Military Ethics, volume 12, issue 1 (2013) pp. 4–17

Chapter 14
The Moral Permissibility of Automated Responses During Cyberwarfare
David Danks and Joseph H. Danks
Journal of Military Ethics, volume 12, issue 1 (2013) pp. 18–33

Chapter 15
Other-Than-Internet (OTI) Cyberwarfare: Challenges for Ethics, Law, and Policy
Randall R. Dipert
Journal of Military Ethics, volume 12, issue 1 (2013) pp. 34–53

Chapter 16
Just War and Cyberwar
Christopher J. Eberle
Journal of Military Ethics, volume 12, issue 1 (2013) pp. 54–67

Chapter 17
Is Stuxnet Physical? Does it Matter?
Ryan Jenkins
Journal of Military Ethics, volume 12, issue 1 (2013) pp. 68–79

Chapter 18
Twenty Years of Cyberwar
John Arquilla
Journal of Military Ethics, volume 12, issue 1 (2013) pp. 80–87

Please direct any queries you may have about the citations to clsuk.permissions@cengage.com

Foreword

Henrik Syse

In hindsight, it is relatively easy to see which events or changes, in technology or otherwise, represent watersheds in human history. But as they actually happen, it is much harder to see. The Zeppelin Airship, to mention one example, was seen by many in the early 1900s as representing the future of air transport and a milestone in technological development. However, it never reached its expected potential and even today remains elusive, even with contemporary experimentation. Along with quadrophonic records and fax machines, it seemed there and then to represent profound change; later, we could see that it was either quite insignificant, or a quickly outdated mode of technology, or simply part of something much larger.

Another complex question, often answerable only with the benefit of hindsight, is the actual newness of that which is new. The development of the ballpoint pen was important, but hardly a revolution as compared to already existing writing materials. The invention of the printing press by Johann Gutenberg, on the other hand, clearly represented a revolution. To say that it merely sped up writing and copying, so that what had previously been done with the aid of pens and ardent copy-writers was now done with greater speed, is disingenuous, to say the least. Although it carried on an activity as old as humanity itself – the spreading of thoughts and ideas through writing – it revolutionized the way in which that happened to an extent that deeply changed the activity itself, and in turn the whole world.

These days, new military or dual-use civilian-military technologies are appearing at breakneck speed. Moral optimists regarding these technologies, and moral pessimists regarding the same, both claim that they represent developments that are truly new, for good or ill. However, among both groups, but especially among the optimists, there are those who will insist that the moral challenges essentially remain the same. In war, weapons are still being fired, and targets are still being chosen. The trigger may be pulled from elsewhere or by different actors, and the flying machines may move differently and possibly even without humans onboard. But it is still, in reality, basically about a human being pulling a trigger. In other words, it is nothing all *that* new. And furthermore, if such technologies lead to fewer rather than more casualties, how is it possible to view them as representing a frightening, revolutionary turn of events?

I personally find the latter understanding of the emerging military technologies discussed within this volume, unmanned aerial vehicles (UAVs or drones) among them, rather implausible. Considering the range, capabilities, and ease of production that characterize many of these technologies, saying they represent little that is entirely new morally speaking is akin to claiming that the printing press was not as revolutionary as we think – it merely did what copying by pen could do, only faster. Admittedly, the newness may turn out to represent real improvement, morally and otherwise. But that conclusion can only be drawn through careful analysis, probing questioning, and serious reflection upon emerging experiences.

As already indicated, we find ourselves *in medias res* – right in the middle of complex change and ever new developments. It is not so evident what is truly new, what will remain for the future, or exactly where the major moral problems lie. It is a sober reminder that the development of everything from anesthetics to the automobile triggered a host of dire warnings as to moral and practical problems ahead. Some warnings were prudent enough; others seem simply laughable or wrongheaded in hindsight.

Yet, this is no excuse not to initiate these debates as soon as possible, and to let them be informed by what we already know. We have rich and nuanced resources at our disposal for discussing military force and military technologies, and we need to utilize those resources in order to start raising the right questions and express the right worries. The alternative is to move into the future with a blindfold firmly across our eyes.

Raising the right questions is the exact aim of this volume, built on two special issues of the *Journal of Military Ethics* (*JME*), from late 2010 and early 2013 respectively. The work and thought of some of the finest minds in military ethics *and* emerging technologies can be found within these pages. Both in those cases where these thinkers agree between themselves and in those cases where they strongly disagree, they help us see more clearly what are the basic questions we have to ask, as unmanned, autonomous, and cyber technologies are rapidly introduced into the world of warmaking.

Martin L. Cook, my co-editor of *JME*, and I have taken great pride in having our journal be a main venue for this debate. Indeed, several of the weightiest contributions to the debate about emerging military technologies have arguably first appeared in *JME*, including these special issues, over the past four to five years. Hence, those seeking to find a compilation of the most important questions that this overall debate raises should need look no further than this volume.

The debate is, of course, still ongoing and will be for many years. Different conclusions may come to be drawn from the ones found herein. But the fundamental questions will remain the ones asked in this volume, summed up well by George Lucas in his original introduction to the 2010 special issue: Is it possible that the development of new, faster, and seemingly less costly technologies will lead to war becoming not an option of last resort, but the option of least resistance? Which brings us back to my initial question, closely linked to how these technologies are to be practically used and morally evaluated: How truly new are these technologies, and what does their newness mainly consist in?

We will not know the full answers to those questions for quite a while. But that makes the questions even more important.

As editor of *JME*, I wish to thank our two guest editors for the original *JME* special issues, George R. Lucas, Jr., and B.J. Strawser, as well as one of *JME*'s two managing editors, Timothy (Tim) Demy, for so expertly guiding these articles first to journal publication and subsequently to appearance in this book. Special thanks to Tim for taking the initiative to have the articles gathered into a book. Thanks are also due to the fine, anonymous reviewers who contributed so much of their time and energy to evaluating these articles, as well as to my co-editor Martin L. Cook and our other managing editor, Nicole M. Hartwell, for helping out in so many ways in the original publication processes. And finally and most importantly, great thanks are due to the authors whose work is represented within these pages.

Introduction

Timothy J. Demy, George R. Lucas, Jr. and Bradley J. Strawser

We live in an age of enormous and rapid change. Technology advances so quickly that we are no longer surprised to discover that a system or device that was cutting edge only a few short years ago, or even months ago, is now obsolete. So too have the means of waging war changed at a similar pace – newly emerging military technologies have transformed the landscape of contemporary warfare and the structure of today's armed forces. What are not as readily apparent are the ethical considerations that should accompany the research, development, and utilization of the new technologies.

The present volume is an attempt to answer in part the question of how emerging military technologies bisect military ethics. The format of the volume is, itself, a reflection of the speed of technological development. The core of volume is a compilation of two special issues of the international *Journal of Military Ethics* (Volumes 9:4 from 2010 and 12:1 from 2013). Additionally, other articles from the journal have been included in order to give a fuller presentation of emerging technologies. Yet more articles might well have been included but the fast-moving nature of the subject matter required a *terminus ad quem*. Regrettably for the volume but fortunately for the field of military ethics there is always 'one more' new and valuable article that one would like to see included but which publication deadlines prohibit.

However, each of the essays in the journal volumes address core ethical considerations with respect to emerging military technologies and it is because of those analyses that this book is presented in hopes of reaching a broader audience. In so doing, perhaps some of the time gap can be bridged between technological development and ethical reflection on emerging weapons of war.

The promises and perils of technology with respect to warfare are not a new challenge. The dreams and nightmares have been present in every age and every conflict. What *is* new in contemporary warfare is the rapidity of technological development, the emergence of new technologies as opposed to the improvement of existing technologies of war, and the convergence of science and technology to create new domains of warfare such as warfare in the cyber-realm. What was once incremental development of new weapons and weapons systems is now exponential development not dissimilar to results argued for in 'Moore's Law' with respect to computer hardware. This concept has also been extended to other realms of science, technology, and the human experience by Ray Kurzweil's (and others advocating technological singularity) 'law of accelerating returns.'

In his popular and influential landmark work *Wired for War: The Robotics Revolution and Conflict in the 21st Century* (2009), P.W. Singer summarized the increasing use of robotics and 'digital warriors,' noting that the human "monopoly of warfare is being broken. We are entering the era of robots at war. . . . War just won't be the same." (42) He is correct. One of the major challenges with respect to emerging technologies and war (and other areas of life) has always been the lag time between scientific and technological development and application and legal and ethical standards of implementation and use.

Technology, like any instrument of power, has a neutral quality about it and can be used for either good or ill. Even at the earliest stages, there are always prior decisions regarding funding, design, and development before a particular technology is functional. Once beyond the R&D phases and acquisition phase, questions of how, when, and if a weapons system will be used become part of the tactical, operational, and strategic decisions political and military leaders must make for their military forces and nation. From a legal and ethical perspective technologies of war do not remove the possibility of virtue, but they do increase the possibilities of vice.

In the essays that follow, readers are encouraged to consider the importance of ethical reflection regarding present and future weapons of war. While some today argue that the just war tradition is no longer a viable framework for the realities and experiences of twenty-first century warfare, the editors believe that the core principles of it remain valid. But the tradition is not static and emerging military technologies require careful reflection with respect to present realities and future possibilities of the tragedy of war. The following articles are presented as a partial response to the need for such reflection.

Part I

The Ethics of Killer Applications: Why Is It So Hard To Talk About Morality When It Comes to New Military Technology?

P.W. SINGER
The Brookings Institution, Washington, DC, USA

ABSTRACT *We live in a world of rapidly advancing, revolutionary technologies that are not just reshaping our world and wars, but also creating a host of ethical questions that must be dealt with. But in trying to answer them, we must also explore why exactly is it so hard to have effective discussions about ethics, technology, and war in the first place? This article delves into the all-too-rarely discussed underlying issues that challenge the field of ethics when it comes to talking about war, weapons, and moral conduct. These issues include the difficulty of communicating across fields; the complexity of real world dilemmas versus the seminar room and laboratory; the magnified role that money and funding sources play in shaping not just who gets to talk, but what they research; cross-cultural differences; the growing role of geographic and temporal distance issues; suspicion of the actual value of law and ethics in a harsh realm like war; and a growing suspicion of science itself. If we hope better to address our growing ethical concerns, we must face up to these underlying issues as well.*

Introduction

We live in a world of 'killer applications.'

It used to be in history that every so often a new technology would come along and change the rules of the game. These were technologies like the printing press, gunpowder, the steam engine, or the atomic bomb. Indeed, such technologies were so rare in history that many other ones were oversold as being 'revolutionary,' when they actually were far from it, such as the Rumsfeld Pentagon's infatuation with 'network-centric warfare.'

What is different today, though, is the incredible pace of technologies' emergence. It used to be that an entire generation would go by without one technologic breakthrough that altered the way people fought, worked, communicated, or played. By the so-called "age of invention" in the late

1800s, they were coming one every decade or so. Today, with the ever-accelerating pace of technologic development (best illustrated by Moore's Law, the finding that microchips – and related developments in computers – have doubled in their power and capability every 18 months or so), wave after wave of new inventions and technologies that are literally rewriting the rules of the game are bursting onto the scene with an ever increasing pace. From robotic planes that strike targets 7,000 miles away from their human operators to 'synthetic life,' man-made cells assembled from DNA born out of laboratory chemicals, these astounding technologies grab today's headlines with such regularity that we have become almost numb to their historic importance.

Looking forward, the range of technologies that are already at the point of prototyping are dazzling in their potential impact, both in war and beyond. Directed energy weapons (aka Lasers), the proliferation of precision guided weapons ('smart' IEDs), nanotech and microbotics (*The Diamond Age*), bio-agents and genetic weaponry (DNA bombs), chemical and hardware enhancements to the human body (*IronMan* meets *Captain America*), autonomous armed robots (*Terminators*), electromagnetic pulse weaponry (*The Day After*, *Ocean's 11*), and space weaponry (*Star Wars*) all may seem straight from the realm of science fiction, but are on track to be deployable well before most of us have paid off our mortgages.

What makes such technologies true 'Killer Applications' was that they rock an existing understanding or arrangement back to its fundamentals. A prototypical example from the field of business is how the iPod device changed the way people bought music. That is, what makes a new development a 'killer app' is not merely the incredible, science fiction-like capabilities it might offer, but the hard questions it forces us to ask.

The most difficult of these questions are not about what is possible that was unimaginable before. Rather, the more challenging and perhaps more important questions are about what is proper. True killer apps raise issues of right and wrong which we did not have to think about before.

These issues of ethics are not just fascinating; the disputes they raise can often have immense consequences to foreign affairs and international security. The science fiction of a submarine being used to attack shipping (Jules Verne), for example, not only became reality, but dispute over the right and wrong of it was what drew the United States into World War I, ultimately leading to its superpower rise. Similarly, H. G. Wells' concept of an 'atomic bomb,' which inspired the real world scientists of the Manhattan Project, helped keep the Cold War cold, but continues to haunt the world today.

Why are these questions of ethics and technology so difficult, especially in the realm of war? Why is it that we are both 'giants' as General Omar Bradley once put it, when it comes to the technology of battle, but at the same time 'ethical infants?' As he remarked in a speech on the eve of Armistice Day in November 1948: 'The world has achieved brilliance without wisdom, power without conscience. Ours is a world of nuclear giants and ethical infants. We know more about war than we know about peace, more about killing than we know about living' (Bradley 1948). To my mind, this comes down to seven key factors, or 'sins' so to speak, that make it so difficult to talk about the ethical ramifications of emerging technology in war.

The Disconnect of the Fields

My background is in the social sciences, rather than in engineering or natural science, which means many in those fields would joke I have no background in the Sciences. Going from my own academic field into another can, on occasion, be like crossing into a foreign land. What is simple in one field can be viewed as incredibly dense and expert in another. Whenever we cross fields or disciplines, the paths and norms of behavior that are familiar become confusing; even the very language and words we have to use become exclusionary.

Take the field of cyber-security, which literally has a language of zeros and ones, but also encompasses very complex issues of law and ethics. Recently, one of the virtual world companies discovered that an individual had come up with a very sophisticated program for stealing virtual currency from the accounts of users in the virtual world and then transferring it into real money abroad. The company documented the hacking effort and then reached out to the Federal Bureau of Investigation (FBI) to explore the legal means for responding. The FBI agent first met with then asked, 'So, um, this is on the Internet right?'[1]

This same sort of issue is a particular one for discussions of ethics and the sciences especially in war. As scary as someone trained in the fields of law, politics, philosophy or ethics might find a discussion of the design parameters of MQ-X unmanned combat air system, the scientists or engineers behind that design might find entering the discussions of the ethical dilemmas with people who are trained to do so equally daunting. As one robotics professor I interviewed put it, 'Having discussions about ethics is very difficult because it requires me to put on a philosopher's hat, which I don't have.'[2]

The result is that we often steer away from such discussions and stay within our own worlds. A good example of this comes from a survey conducted of the top 25 stakeholders in the robotics field, as identified by the Association of Unmanned Vehicles and Systems International (the field's professional trade group). When asked whether they foresaw 'any social, ethical, or moral problems' that the continued development of unmanned systems might bring, 60 percent answered with a simple 'No.'[3]

Frankly, if we continue with this ostrich attitude, both the policy and ethical discussions will be uninformed, or even worse, guided by those who do not have scientific fact on their side. And, in turn, scientists and engineers with real world ethical issues will be abandoned.

Consider the following series of questions:

- From whom is it ethical to take research and development money? From whom should one refuse to accept funding?
- What attributes should one design into a new technology, such as its weaponization, autonomy or intelligence? What attributes should be limited or avoided?

- What organizations and individuals should be allowed to buy and use the technology? Who should not?
- What type of training or licensing should the users have?
- When someone is harmed as a result of the technology's actions, who is responsible? How is this determined?
- Who should own the wealth of information the technology gathers about the world around them? Who should not?

These are all examples of real questions emerging in the robotics field that are not so much about the technology as the ethics. Most importantly, ethics without accountability is empty, so each of these questions also has potential legal ramifications. But, if a young robotics graduate student wanted to do the ethical thing in answering these questions, he or she would have no code to turn to for guidance, the way, for example, a young doctor or medical researcher would in their field.

This is perhaps to be expected, as robotics is a much newer field than medicine. The key is not whether there is a gap, which is to be expected in any cutting-edge field, but what is being done to fill this gap. For example, the scientists working in the Human Genome Project set aside 5 percent of their annual budget to push discussions on the 'ethical, legal, and social implications' of their work (Moore 2002). The result is not that we have solved these tough issues related to the field of genetics. But we are certainly far better equipped to handle these debates, with the tenor and content of the discussion far past some of the inanity that dominated the project's early years.

Yet very few other fields are doing something similar. Indeed, more common is the sort of troubling attitude perhaps best illustrated by an email I once received after a talk at a leading engineering school. A professor wrote to chastise me for 'troubling' his students 'by asking them to think about the ethics of their work.' Both Socrates and Asimov are likely chuckling.

Applied Ethics: The Revenge of the Real World

Technology is often described as a way to reduce war's costs, passions, and thus its crimes. The poet John Donne, for example, claimed that in 1621 how the invention of better cannons would mean that wars would 'come to a quicker ends than heretofore, and the great expence of bloud is avoyed' (Dunlap 1999: 5).

Almost 400 years later, we still hear similar discussions when it comes to a variety of technologic efforts. *The Economist* and *Discovery Magazine* both recently covered attempts to create an 'ethical governor,' essentially software that weapons might be programmed with to make them act ethically in war, even more morally than their human masters. As Ron Arkin, a professor of computer science at Georgia Tech, who is working on such a project observes: 'Ultimately these systems could have more information to make wiser decisions than a human could make. Some robots are already stronger, faster and smarter than humans. We want to do better than people, to ultimately save more lives' (Bland 2009).

It is a noble sentiment, but also one that ignores the seamy underside of war, which may be becoming even darker in the twenty-first century. My own past books, for example, looked at the emergence of a global private military industry and the growing role of money and greed in war, and then at the sad reality of child soldiers; contrary to our idealized visions of war and who fights it, one of every ten combatants today is a child.

When we own up to the reality of war, rather than how we wish it, we see the double-edged sword of technology. We see that, for example, while human enhancements research is taking us beyond the prior limitations of the human body, those resulting biological enhancements do not take us past our all too human limitations and the inherent flaws – or original sins – that have also characterized us, such as our capacity for arrogance, greed, and hate. Similarly, just as a fork can be a tool for eating as well as plucking out eyeballs, we recognize that a non-lethal weapon can chase away Somali pirates, but also can be used by Japanese fisherman to chase away environmentalists protesting their illegal slaughter of endangered whales.

Too frequently discussions of new technologies assume that the 'fog of war has been lifted' (as the technophile thinkers who once surrounded Donald Rumsfeld argued) by either the perfection of our technology or the perfection of our souls. Instead, war in the twenty-first century shares the same qualities with past centuries: it is a mess, and maybe even more of a complicated mess.

So, for example, when someone asserts confidently that new war technologies will lead to less bloodshed or greater compliance with established moral principles, we should check such sentiments with a look through a different, dirtier lens. While scientists might note the fact that such promises often prove empty (the ethical generator for example, remains a 'black box' all design concept, but no reality), we should also recognize that even if such fantasies were to come true, we would still have problems.

For example, an enduring aspect of war is that regardless of how novel and advanced the technology, the enemy still has a vote. Making all this even more difficult today is that contemporary terrorist and insurgent groups are doing all they can to take advantage of the very laws they are supposed to follow. Charles J. Dunlap, Jr. has described this tactic of deliberately violating the Geneva Conventions that divide soldier from civilian as fighting a form of 'lawfare'. In Somalia, for example, fighters would fire at US soldiers from behind non-combatants. One Ranger recalled how a gunman shot at him with an AK-47 that was propped between the legs of two kneeling women, while four children sat on the shooter's back. The Somali warrior had literally created for himself a living suit of non-combatant armor (Edwards 2005: 288). Another example in the Kosovo War was a tank that was parked in a school yard, while another was a tank that drove through a town on an ethnic cleansing mission with women and children riding on top. In the 2006 Lebanon War civilians were blackmailed by Hizballah into launching remote-controlled rockets on their farms to rain down on Israeli cities.

Ethicists and lawyers could fill pages arguing back and forth when one should use force in response to such scenarios. To argue that the problem is going to be easily solved with some imaginary, yet un-invented artificial intelligence is a bit of a stretch. That is, even if we could invent a system that always followed the Geneva Conventions, in the mess of modern war, the problem is that applying them is just not a simple zero versus one proposition in terms of programming language. The laws themselves are often unclear about what to do, and even more so, under siege by the people supposed to follow them. In short, in the harsh reality of war, there are no silver bullet-technologic solutions for ethics.

The Dirty Role of Money

Whether it is a research laboratory, think-tank seminar room, or congressional hearing room, there is one topic that is generally considered impolite to talk too much about: Money. Yet there is perhaps no more important factor in determining who gets to talk, who does not, what gets worked on, and what does not.

My own experience perhaps can illustrate. Last year I was invited by the US Naval Academy's Stockdale Center for Ethical Leadership to give the annual ethics lecture to the student body about some of the questions that robotics was raising for warfighters. A day before the event, the center received an angry letter, expressing shock that it would invite such an 'evil,' 'unethical' person to corrupt the young minds of future leaders and arguing that it should dis-invite me.

The writer of the letter was the chief executive officer (CEO) of a private military firm, whose employees had been identified in several US military reports as being involved in the abuses at Abu Ghraib prison, one of the worst scandals in American history. But I had apparently done something far worse than billing the taxpayer for acts of abuse for which soldiers would later be court-martialed, but contractors would escape accountability because of a gap in the law. I had argued in an editorial piece that the contractors of this CEO's multi-billion dollar firm should be held to the same standards as those in uniform, and that the government should 'investigate the issue, bring people to justice, and ensure that lessons are learned so that the same mistakes are not repeated' (Singer 2004).

Normally, such a letter would be ignored, or the rich irony of it would merely prompt laughter. But the administrators of the center had to take it seriously for the sole reason that he was a wealthy individual that the school had previously been cultivating as a donor. Money, as we know, has a right to talk louder in our world than perhaps it should. Or, as the American writer Napoleon Hill might have commented about my CEO friend, 'Money without brains is always dangerous.'

I have a deep respect for how the Stockdale Center responded to this threat. The center did not let money shut down conversation, but instead investigated the situation and decided there was no merit to silencing my talk on ethics simply because a millionaire was upset. We all know that many other places would not have responded the same.

My point here in reflecting on this incident is not to rehash old academic war stories, but that I wonder what lessons the very first American conservatives might have for us. The authors of the *Federalist Papers,* who helped craft and defend the US Constitution, warned about the role of any private interests not responsive to the general interests of a broadly defined citizenry.[4] Among the Founding Fathers' worries for the vitality of democracy was that, when private interests move into the public realm and the airing of public views on public policy is stifled, governments tend to make policies that do not match the public interest.

I think this is something to keep in mind when we reflect on the issues of ethics and technology, which of course also touch on vast amounts of money sloshing about. How do we handle discussions that call into question multi-million dollar interests versus those that do not? Who has a louder bully pulpit?

This issue of money does not just shape the public debate, but goes all the way down to the individual decisions that a scientist or researcher working on such emerging technologies has to wrestle with. Benjamin Kuipers, a computer scientist at the University of Texas perhaps best described this real world dilemma that we often do not reflect upon:

> DARPA [Defense Advanced Projects Agency] and ONR [Office of Naval Research] and other DOD [Department of Defense] agencies support quite a lot of research that I think is valuable and virtuous. However, there is a slippery slope that I have seen in the careers of a number of colleagues. You start work on a project that is completely fine. Then, when renewal time comes, and you have students depending on you for support, your program officer says that they can continue to fund the same work, but now you need to phrase the proposal using an example in a military setting. Same research, but just use different language to talk about it. OK. Then, when the time comes for the next renewal, the pure research money is running a bit low, but they can still support your lab, if you can work on some applications that are really needed by the military. OK. Then, for the next round, you need to make regular visits to the military commanders, convincing them that your innovation will really help them in the field. And so on. By the end of a decade or two, you have become a different person from the one you were previously. You look back on your younger self, shake your head, and think, 'How naïve'. (as quoted in Singer 2009: 172–173)

The same sort of financial pressures, either positive or negative, also happen outside the laboratory. Those who study in such areas as ethics or policy often similarly depend on some sort of financial support from foundations or donors. The more money such a donor has, the more likely they are able to get people to research the questions they want, in the way they want them answered, and the more likely that their agenda will be advanced.

Indeed, this issue of money and bias can prove to be a problem even when a donor is guided by the loftiest ideals of charity. A good illustration of this comes from a meeting with a representative of one of the world's leading foundations for academic research about a potential initiative that would look at ways to establish global norms in cyber-security. They explained how their foundation thought the topic was interesting, but that their board 'had not yet decided whether cyber issues were important or not'. They felt that they would be in a position to decide whether cyber-security was an important issue in 'about three or four years.'[5]

While science works on the cutting-edge, donors to the social sciences and ethics tend to want to sponsor what is already established (with the irony that they are least often there at the point in time when their support might have the most impact). The sad truth is that if you are an ethics or policy researcher seeking funding, which, in turn, can be crucial in determining such things as whether you get tenure or not, you are better off starting new projects on old problems, rather than new projects on new problems.

Your Ethics Are Not My Ethics

One of the most important ripple effects of a technology in terms of its impact on foreign policy takes place when culture encounters ethics. This issue is playing out right now in the vastly different perceptions of unmanned aerial systems inside America and 7,000 miles away, where they are actually being used daily in war. While we use such adjectives as 'efficient' and 'costless' and 'cutting edge' to describe the Predator in our media, a vastly different story is being told in places like Lebanon, where the leading newspaper editor there called them 'cruel and cowardly' or in Pakistan, where 'drone' has become a colloquial word in Urdu and rock songs have lyrics that talk about America not fighting with honor. This issue of narrative, of differing realities when it comes to talking about the exact same technology, is hugely important, to the overall 'war of ideas' that we are fighting against radical groups and their own propaganda and recruiting efforts. It helps explain how painstaking efforts to act with precision emerge on the other side of the world through a cloud of anger and misperceptions.

But this issue of perception is something that goes beyond just drone strikes. We live in a diverse world and as *Star Trek* creator Gene Roddenberry put it, 'If man is to survive, he will have learned to take a delight in the essential differences between men and between cultures. He will learn that differences in ideas and attitudes are a delight, part of life's exciting variety, not something to fear.'[6]

Yet we must also acknowledge that that these differing cultural understandings can have a huge impact, creating differing perceptions of what is ethical or not.

We see this greatly illustrated with the differing perceptions of robotics in East and West. In Western culture, going back to its very first mention in the play *RUR* in 1921, the robot is consistently portrayed as the mechanical servant that wises up and then rises up. The technology is portrayed as its something heartless to be scared of. Yet, the very same technology is consistently viewed exactly opposite in Asian culture, going back to first mention in post-World War II anime comics as *Astro Boy* in which the robot is consistently the ethical hero, rather than the would-be Terminator to be feared. 'The machine is a friend of humans in Japan. A robot is a friend, basically,' tells Shuji Hasimoto, a robotics professor at Waseda University in Tokyo (Jacob 2006: 7).

This difference is not just something that comes out of popular culture. As opposed to the strict Western division between living and dead, the traditional Japanese religion of Shintoism holds that both animate and inanimate objects, from rocks to trees to robots, have a spirit or soul just like a person. Thus, to endow a robot with a soul is not a logical leap in either fiction or reality. Indeed, in many Japanese factories, robots are given Shinto rites and treated like members of the staff (Hornyak 2006).

The result is that popular attitudes over what is acceptable or not when it comes to this technology and how to use it diverge widely. In Asia, 'companion' robots for the elderly or babysitter robots for children are marketed with little controversy. By contrast, Rodney Brooks, an Massachusetts Institute of Technology professor and Chief Technical Officer of iRobot, explains that the concepts would not work in the US, for the simple reason that Americans find them 'too artificial and icky' (Singer 2009: 168).

One can see similar differences in perceptions, influenced by culture in a wide variety of emerging areas, from work in genetics, enhancements, and even what were once cutting edge but are now quite normal human medical treatments – normal, that is, only in some societies. The Japanese, for example, may love their robotics but are often quite horrified by organ transplants, to the extent that the very first heart transplant doctor in Japan was prosecuted for murder.[7]

The same cultural attitudes flow out and influence what different cultures think is acceptable in war or not. The issue of arming an autonomous weapons system is hugely controversial; controversial, that is, to Western minds. By contrast, in South Korea, it is not. The country's military forces sent two robotic sniper rifles to Iraq in April 2004, with little public debate. Indeed, Samsung not only manufactures the Autonomous Sentry Gun, a 5.5 millimeter machine gun with two cameras (infrared and zooming) and pattern-recognition software processors that can identify, classify, and destroy moving human targets from 1.25 miles away, but even made a promotional commercial extolling it, set to jazzy music.

This issue of differing senses of right and wrong behavior can even spark conflict. Much of the recent dispute in cyber-security between the US and China, especially as it relates to the Google incident, is woven within different cultural attitudes towards privacy and individual rights. Similarly, US State Department officials like Hillary Clinton often make a point to extol the power of 'social networking.' They fail to realize that others describe the West's very same push for it as its own form of cyber-warfare.

The Distance Problem

Our codes of law and our sense of ethics have long focused on our intent and actions. Take this quote from Aristotle: 'We do not act rightly because we have virtue or excellence, but rather we have those because we have acted rightly.'[8]

The challenge of this new wave of technologies is that the concepts of virtuous intent and action are being modulated in ways that Aristotle, and even the Geneva Conventions, crafted several thousand years later, did not imagine. Advancements used to be distinguished by the increase they created in a weapon's power, range, or capacity. The World War II era B-24 bomber, for example, was an improvement over the B-17 because it could carry a greater number of bombs over a greater distance. Similarly today, the MQ-9 unmanned Reaper plane is an improvement over the MQ-1 Predator because it can carry a greater number of bombs over a greater distance. But there is one more fundamental difference: The Reaper drone is smarter. That is, the newer technology can do much more on its own; take off and land autonomously, fly out mission way-points without human intervention, and carry targeting software that even allows it not only to pick out small details (such as footprints in a field), but also to start to make sense of them (such as backtracking the footprints to their point of origin).[9]

Such improvements present dilemmas for which our prevailing laws of war, best exemplified by the 1949 Geneva Conventions, may not be fully prepared. In essence, our new technology does not remove human intent and human action, but it can move that human role geographically and chronologically. Decisions now made thousands of miles away, or even years ago, may have great relevance to the actions of a machine (or to its inactions) in the here and now. This does not mean that there is no ability to talk about ethics or apply the law, but rather that this aspect of distancing makes it far more complex and difficult, especially in fields that very much focus on the role of the 'commander on the scene' and concepts of individual intent and responsibility for one's actions.

The Conventions date from the middle of the twentieth century; the year they came out, the average American home cost $7,400, and the most notable invention was the 45rpm record player. But while there is little chance of the global cooperation emerging for them to be updated and ratified anytime soon, technology is guaranteed to advance. Under what is known as Moore's Law, for instance, the computing power and capacity of microchips (and thus the many devices like computers that use them) has essentially doubled every 18 months of so, while continually going down by similar measures in price. If Moore's Law holds true over the next 25 years, the way it has over the last 40, our technologies will be a billion times more powerful in their computing. Indeed, even if we see just 1 percent of the improvement we have experienced historically, our technology will be a mere 10,000,000 times more powerful than today.

This may be the most challenging part of the distance dilemma, that while our understanding of law and ethics moves at a glacial pace, technology moves at an exponential pace. Thus, the distance between them grows further and further.

The Suspicion of Ethics

In the spring of 2010 I had the honor of sharing a panel with two US military Judge Advocate General officers at a session hosted by the Institute for Ethics and Public Affairs at Old Dominion University, Norfolk, Virginia. The session's focus was to be on the ethical questions of using new robotic technologies in war. However when a person in the audience stood up and asked a more fundamental question of the three of us, 'What would you tell a mother who has lost her son to one of these terrorists in Iraq as to why we should even care about something like ethics or the laws of war? Why should we even bother?'

The question took the discussion to a whole deeper level. For all the discussion in academic journals and seminar rooms, many people wonder about something far more fundamental: why do we even care about trying to figure out right from wrong in a realm like war where so much wrong happens?

It is a difficult issue and cuts to the heart of why we even care about ethics and morality to begin with in any human endeavor. Linked to the specific realm of war, two factors carry weight in my mind. First, the son was a 'serviceman' and that is a powerful term. It meant he was not fighting out of anger or hate, but serving something beyond. When he joined the US military, he took an oath to serve the Constitution and respect the laws, including those of war. It is this notion of service, this sense of ideals, that is why we honor servicemen and women, and even more so it is the essence of what distinguishes them from their foes. I think they serve on the side of right, not just because they are my countrymen, but because the other side does not respect the laws of war, and thus is the equivalent of barbarians, which is how historically we have viewed those who willfully violate the law.[10] This sense of service is not just why we fight, but is why our way of fighting can be made just.

The second factor is something we do not like to talk about in issues of ethics, but very much matters in war: raw self-interest (something we may try to sugarcoat by terming it as 'pragmatism'). Rather than it being an advantage to break the laws of war, the facts show something different. In the history of war, the side that has fought with a sense of ethics, respected the laws, and fought as 'professionals,' has tended to win more often than those that do not. That is, in the historic record of over 5,000 years of war, professionals, almost always triumph over those who are willing to behave like barbarians. Indeed, this distinction was found to be a key event in Iraq, where that son was killed. Even though the US forces in Al-Anbar province were far more alien to the local society than the members of Al-Qaeda, and many in the media complained that our forces were stymied by a webwork of laws and lawyers, we won in the end for this very reason. These extremists, these twenty-first century barbarians, carried the seeds of their own downfall, by the very fact that they were wanton with violence and our forces were not. Eventually this was what persuaded the local tribes to turn on them, which was the actual key to the victory in the 'surge.'

But while this may be the case, the narrative that ethics is only a hindrance remains powerful and popular. Those who care about ethics must acknowledge they face a basic problem of convincing folks both inside and outside the military that ethics is worthwhile, not just because to be selfless is morally praiseworthy, but also because we are self-interested and want to win.

Beware of Magic

When the warriors of the Hehe tribe in Tanzania surrounded a single German colonist in 1891, they seemingly had little to worry about. But he had strong magic, a box that spat out bullets (what we call the 'machine gun'). Armed with such seemingly mystical power, he killed almost 1,000 spear-armed warriors (Ellis 1986: 89).

As English physicist and science fiction author Arthur C. Clarke famously put it, 'Any sufficiently advanced technology is indistinguishable from magic.'[11] And in no realm is this truer than in war, where we do not just merely view advanced technology sometimes as magical, we also fear it for that very reason. We fear what we do not understand.

While we might think such times have past, the problem of technology's magical side continues to bedevil us even in the twenty-first century, perhaps even more so as technology truly performs magical feats, even while large parts of the world live their lives no differently than they did centuries past. For instance, an official with the US military's Joint Special Operations Command recounted a meeting with elders in the tribal region of Pakistan, the area where Al-Qaeda leaders were reputed to be hiding out, and the site of more than 175 drone strikes in the last few years. One of the elders was enamored of the sweet-tasting bread that was served to them at the meeting. He, however, went on to tell how the Americans had to be working with forces of 'evil,' because of the way that their enemies were being killed from afar, in a way that was almost inexplicable. 'They must have the power of the devil behind them.'

As the official recounted with a wry chuckle, 'You have a guy who's never eaten a cookie before. Of course, he's going to see a drone as like the devil, like black magic.'[12] Again, the elder felt we were doing something evil not because of civilian casualties or the like, but for the very reason that he did not understand. As Marian Anderson once put it, 'Fear is a disease that eats away at logic.'[13]

But this suspicion of advancement is not just limited to distant, tribal regions. It is something that is increasingly playing out here in the US, and is harming our ability to have effective discussions on policy and ethics in the twenty-first century. We are seeing what CNN (Cable News Network) has characterized as a growing American 'fear of science' or what writer Michael Specter explored in his book, *Denialism*. As Specter puts it, the problem now when it comes to discussions that involve the intersection of science and public policy is that, 'when people don't like facts, they ignore them.'[14]

We see this in all sorts of areas, from the widespread fear of vaccines, to the useless trust placed in the multibillion industry of dietary supplements, to the climate change debate. Indeed, a major political party nominated for Vice President of the United States a person who described the scientific method as 'snake oil.'[15] But we should not be too hard on them, they are simply reflections of a populace in which 11 percent of Americans cling to the belief that Elvis really is still alive, and 18 percent that the Inquisition did have it right and the Sun revolves around the Earth.

Our challenge is thus often not only to make sure the policymaker and public understand the ethical issues of emerging technologies, but that they also basically accept the scientific principles underlying those technologies as well.

Conclusions

These challenges are certainly daunting, but by no means do they imply that any discussion of ethics, morality, and technology is hopeless, and that there is no way to think, speak, and act ethically when it comes to emerging technologies in war. Rather it is the very opposite. The difficulty makes the project all the more important and the efforts to solve them all the more worthy.

Rather our success or failure in navigating the moral dilemmas of a world of killer apps will depend on recognizing that these very problems are part and parcel of the discussion. We must own up to these challenges, face them, and overcome them. Otherwise we will continue to spin in circles. And we had better act soon. For the thread that runs through all of this is how the fast-moving pace of technology and change is making it harder for our all too human institutions, including those of ethics and law, to keep pace.

Notes

[1] Author interview with Linden Laboratory executive, 27 March 2010.

[2] Author interview with Prof. Illah Nourbakhsh, 31 October 2006.

[3] Survey released at AUVSI-San Diego, 27 October 2009, conducted by Kendall Haven.

[4] See especially Federalist Papers #10 and #51, available for download at http://avalon.law.yale.edu/subject_menus/fed.asp; Internet.

[5] Author meeting with Foundation grant officer, 13 April 2010.

[6] Quotation of *Star Trek* creator and producer, Eugene W. Roddenberry recorded in Santa Barbara, CA (1971): http://memory-alpha.org/wiki/Template:Gene_Roddenberry_quotes. [Accessed 26 October 2010].

[7] For more information see Lock 2002. The next transplant in Japan was not performed until 30 years later; see also http://news.bbc.co.uk/2/hi/health/287880.stm.

[8] From *Nicomachean Ethics,* Book II, 4.

[9] For more on this, see Singer 2009: ch. 5.

[10] For more on this, see Peters 1994.

[11] This is known widely as 'Clarke's Third Law,' after Arthur C. Clarke, 'Profiles of the Future' (1961): see, e.g., http://www.quotationspage.com/quotes/Arthur_C._Clarke/; Internet [accessed 26 October 2010].

[12] Author interview with JSOC official, 10 March 2009.

[13] See, e.g., http://quotationsbook.com/quote/14686/; Internet [accessed 26 October 2010].

[14] See http://articles.cnn.com/2010-04-13/opinion/specter.denying.science_1_organic-food-genetically-supplements?_s =PM:OPINION; Internet [accessed 26 October 2010]; see also Specter 2009.

[15] See, e.g., http://www.cbsnews.com/stories/2010/02/09/politics/main6189211.shtml; Internet [accessed 26 October 2010].

References

Bland, E. (2009) Robot Warriors will get a Guide to Ethics, Discovery News, 18 May 2009, accessed 26 October 2010, available at: http://www.msnbc.msn.com/id/30810070/; Internet.

Bradley, O. (1948) An Armistice Day Address: 10 November 1948, accessed 26 October 2010, available at: http://www.opinionbug.com/2109/armistice-day-1948-address-general-omar-n-bradley/; Internet.

Dunlap, C. J. (1999) *Technology and the 21st Century Battlefield: Re-complicating Moral Life for the Statesman and the Soldier* (Carlisle, PA: Strategic Studies Institute, US Army War College, 1999), accessed 26 October 2010, available at: http://www.strategicstudiesinstitute.army.mil/pdffiles/pub 229.pdf; Internet.

Edwards, S. J. A. (2005) *Swarming and the Future of Warfare*. Originally presented as the author' thesis Doctoral – Pardee Rand Graduate School), RAND.

Ellis, J. (1986) *The Social History of the Machine Gun* (Baltimore: Johns Hopkins University Press). Federalist Papers, accessed 26 October 2010, available at: http://avalon.law.yale.edu/subject_menus/fed.asp; Internet.

Hornyak, T. N. (2006) *Loving the Machine: The Art and Science of Japanese Robots* 1st edn. (Tokyo; New York: Kodansha International).

Jacob, M. (2006) Japan's Robots Stride into Future, *Chicago Tribune*, 15 July 2006, p. 7.

Lock, M. (2002) *Twice Dead* (Berkeley, CA: University of California Press)).

Moore, J. A. (2002). The Future Dances on a Pin's Head: Nanotechnology: Will it be a Boon – or Kill Us All? *Los Angeles Times*, 26 November 2002, accessed 26 October 2010, available at: http://www.wilsoncenter.org/index.cfm?fuseaction=news.item&news_id=14638; Internet.

Peters, R. (1994) The New Warrior Class, *Parameters*, 24(2), pp. 16–26.

Singer, P. W. (2004) The Contract America Needs to Break, *Washington Post*, 12 September 2004, accessed 26 October 2010, available at: http://www.brookings.edu/opinions/2004/0912defenseindustry_singer.aspx; Internet.

Singer, P. W. (2009) *Wired for War: The Robotics Revolution and Conflict in the 21st Century* (New York: Penguin).

Specter, M. (2009) *Denialism: How Irrational Thinking Hinders Scientific Progress, Harms the Planet and Threatens our Lives* (New York: Penguin).

Biography

Peter Warren Singer is Director of the 21st Century Defense Initiative and Senior Fellow in Foreign Policy at The Brookings Institution (Washington, DC). He is author of numerous articles and three books, including *Corporate Warriors: the Rise of the Privatized Military Industry* (Cornell UP, 2003), *Children at War* (Pantheon, 2005), and most recently the *New York Times* bestseller, *Wired for War: The Robotics Revolution and Conflict in the 21st Century* (Penguin, 2009). For more information visit www.pwsinger.com

With Fear and Trembling: An Ethical Framework for Non-Lethal Weapons

PAULINE KAURIN
Pacific Lutheran University, Tacoma, WA, USA

ABSTRACT *This article augments and enlarges the process of framing a more systematic and holistic ethical approach to non-lethal weapons that also provides caveats and restrictions – along the lines of* jus in bello *principles in the just war tradition – on their use. Rejected as morally impermissible is the use of non-lethal weapons as: (1) a way to circumvent or make irrelevant classical moral distinctions; (2) an 'easy' technological fix to complex moral and strategic problems; and (3) a method to make war more palatable and easier to use as both a military and political option. Non-lethal weapons can be ethical, and in fact may be ethically preferable to conventional weapons, only if they are used consistently with the following criteria (in strict order of priority): (1) to provide the military with more flexible response time and options, allowing them more time and space to carefully make the strategic and ethical judgments necessary in war and to respond with appropriate and proportional force; (2) to reduce unnecessary suffering on the part of non-combatants; (3) to facilitate the eventual restoration of peace; and (4) to minimize combatant casualties.*

> 'At the heart of the promise of non-lethal weapons is the possibility of minimizing pain and death by providing flexible response tools for policing and military purposes.'
> (Rappert 2002: 53)

> 'The main purpose and outcome of war is injuring' (Scarry 1985: 63)

Introduction

What if one could wage war largely without killing people? What if one could wage war in such a way as to dramatically reduce both combatant and non-combatant casualties? To many it would seem intuitive that weapons which promise generally not to kill, but rather to stun, stop, or otherwise disable persons and materiel without irreversible effects would be ethically preferable;

they would reduce combat casualties, reduce collateral damage, spare non-combatants from unnecessary suffering, and facilitate the restoration of the peace while still allowing the military to achieve their objectives. Is it not obvious that *not* killing people is ethically preferable to killing them?

These questions have taken on increased urgency with the current conflicts in Iraq and Afghanistan, where operations occur amidst large civilian populations and it is essential to military success, morality, and public perception that the amount and extent of both combatant, but especially non-combatant, casualties be reduced. Increasingly commanders are asking for and making greater use of non-lethal weapons (NLW) in these types of operations; a recent Department of Defense Non-Lethal Weapons Program document observes, 'NLW provide tools for controlling force escalation – pre-empting conflict where possible, de-escalating tension, applying the right force at the right time and facilitating the return to peace' (Department of Defense Non-Lethal Weapons Program 2007: 20). Lasers are being used at check points, non-lethal rounds are used by soldiers guarding the US embassy in Afghanistan, and grenade sting balls may be used against rioting detainees at Guantanamo Bay, Cuba. Military and civilian leaders alike are calling for more money for research on NLW and greater integration of NLW into the current force structure.[1]

Given these developments, it is essential that the ethical discourse and frameworks keep pace with and inform these developments. While there has been substantial discussion of both the ethical problems with, and the potential practical and ethical advantages of NLW, there is much less discussion of systematic ethical frameworks and rules for thinking about the use of NLW, particularly in asymmetrical conflicts. In 2007, Eric Partin furthered this project by arguing that systematic thinking about NLW in terms of rules of engagement (ROE) was missing and necessary, arguing for a seamless approach to ROE regardless of the type of weapon or situation involved, 'The ROE must cover the entire spectrum of events that could cause a soldier to use no force, non-lethal force or deadly force' (Partin 2007: 8–9).

My intention is to augment and enlarge this process of framing a more systematic and holistic ethical approach to NLW that takes account of both the possible benefits and advantages of these weapons, but also addresses and provides restrictions – along the lines of *jus in bello* principles in the just war tradition (JWT) – on their use. Typically the two most central *jus in bello* principles, which also find expression to varying degrees in international law, are (1) the principle of discrimination (POD) and (2) the principle of proportionality (of means.) The principle of discrimination stipulates that all reasonable efforts must be made to discriminate between combatants and non-combatants in the actions of war. The principle of non-combatant immunity (NCI) is related to this principle and requires that only combatants be intentionally targeted in warfare. The principle of proportionality (of means) requires that the harm to be inflicted by a certain weapon or method of war be outweighed by the good or benefit to be achieved by the action. Despite the potential that such weapons promise in terms of making warfare more humane, I reject as morally impermissible any use that sees these weapons as: (1) a way to circumvent or make irrelevant moral distinctions like the combatant/non-combatant distinction; (2) an 'easy' technological fix to complex moral and strategic problems; and/or (3) a method to make war more palatable and easier to use as both a military and political option.

Non-lethal weapons can be ethical, and in fact *may* be ethically preferable to conventional weapons, *only* if they are used consistent with the following criteria (in strict order of priority): (1) to provide the military with more flexible response times and options, allowing them more time and space to carefully make the strategic and ethical judgments necessary in war and to respond with appropriate and proportional force; (2) to reduce unnecessary suffering on the part of non-combatants; (3) to facilitate the eventual restoration of peace; and (4) to minimize combatant casualties. War is an enterprise that should be entered into with fear and trembling, with serious debate, discussion and consideration both of the aims to be achieved and the likely (and unlikely) consequences. Non-lethal weapons should be examined for their potential to ameliorate the *effects* of war, *never* to make the *resort* to it more ethically, politically or militarily attractive.

Nature of Non-Lethal Weapons

To set the stage for the discussion, it is necessary to take up the issue of what constitutes a non-lethal weapon, which is by no means an uncontroversial matter. Non-lethal weapons go by a variety of terms: non-lethal weapons, soft-kill, less-than-lethal, sub-lethal; terms all intended to denote a difference between these weapons and conventional weapons, which while not always lethal, are designed to be used in a lethal manner. As critics point out, NLW *can* be lethal, but they are not *intended* to be so. One definitional approach centers on the physical properties of the weapon and the operational characteristics as having a potentially radical break with traditional warfare (Lewer & Schofield 1997: 6–7). These weapons are not intended to kill, permanently injure or maim; any effects are intended to be temporary, minor, and reversible. Another definitional approach to NLW is that they are intended to have one or both of the following characteristics: (1) they have relatively reversible effects on personnel or material and (2) they affect objects differently within their area of influence (Alexander 1999: 5).

While some kinds of non-lethal weapons (batons used by police forces, herbicides, tasers, sounds and lights as weapons of war, sanctions, and psychological or information forms of warfare) have been in existence for quite some time, the current generation of NLW were first tactically employed by US Marines in Somalia in the 1990s. The US Directorate on Non-Lethal Weapons was established in 1996 to head up the Department of Defense efforts to develop, evaluate and employ NLW in US military operations. NLW have been effectively used in a crowd situation in Kosovo in 2000, as well is in Operations Iraqi Freedom, and Enduring Freedom. Current NLW include everything from low-impact bullets, foams, nets, lights, noise, to gas grenades. In current development and testing are a variety of other kinds of weapons: directed energy systems, lasers (used as a deterrent or to temporarily blind), electromagnetic power to degrade (especially electronic) equipment, light weapons (which can send conflicting messages to the brain, causing the individual to feel very confused and unsteady), sticky foams (which can quickly spread over an area and immobilize persons and materiel), pheromones (which could attract bugs or other animals to overrun an area), and vinyl nets deployed by mines capable of stopping vehicles. Of particular interest for the future are systems that can disable or neutralize vehicles (especially at checkpoints), optical distracters (lasers with reversible optical effects), focused acoustics, Active Denial Technology (ADT), and laser induced plasmas (JNLWP).

According to their proponents, the aim of all these weapons is to provide increased flexibility and response time on the part of troops and to find ways to neutralize a battlefield threat (or a person who could be potentially threatening) without having to resort to lethal force. The benefits of their use are highlighted in environments like the current action in Iraq: military units are operating in urban or mixed areas where it is not clear which individuals are combatants and which are non-combatants, at vehicle checkpoints where it is advisable to be able to assess the intent of individuals and vehicles from a safe distance (given the prevalence of improvised explosive deceives and suicide bombing tactics) and in areas where one wants to reduce the impact of war on non-combatants to help 'win hearts and minds' to help facilitate the restoration of the peace and foster long term stability in the region.

Caveats and Critiques – Fear and Trembling

If NLW are to be developed in a way that will not run foul of the ethical and legal considerations essential to conducting just and legal warfare there are three major caveats that must be observed. These caveats are rooted in realistic concerns that develop in a significant way or are clearly articulated in the current discourse around NLW. Moreover, these caveats reflect and are designed to address systemic problems in the discourse on NLW, not merely isolated rhetorical overstatements by impassioned proponents of these weapons.

The first and the most crucial caveat is that under *no* circumstances can the development and use of NLW go forward in any way that attempts to use technology to circumvent or weaken the principle of discrimination. Despite the seductive view of technology as a panacea that will allow us to avoid the hard moral and practical choices of war, there can be absolutely no substitute for rendering of moral judgments and distinctions made by human agents in war. To try to avoid this is to enter dangerous moral ground, essentially abdicating moral responsibility for the harm, injury, and death that are an essential part of war. The marine manning a checkpoint in Iraq has to make moral and practical determinations about the intent and potential hostility (or not) of the persons approaching his checkpoint. While certain kinds of weapons might give him more time or options other than lethal force, he still has to make a judgment about whether to treat the approaching individuals as objects of war or objects of protection. In short, he has to make a moral judgment; the weapons involved do not change that basic fact.

Col. George Fenton, Director of the US Joint Non Lethal Weapon Directorate observes that he '. . . would like a magic dust that would put everyone in a building to sleep, combatants and non-combatants' (Fidler 2002: 36). Regardless of whether such a scenario is possible, this quote (and others like it from proponents of NLW) reveals something important about the underlying thinking and assumptions operating in the NLW discourse. The question is not really whether it is possible, but whether it is moral. Once we look at the issue in this light, a desire to ignore or circumvent the crucial moral questions that are an essential part of any discussion of justification of warfare seems to be emerging. In reference to Vietnam and Somalia, John Alexander (1999: 24), arguably one of the most vocal proponents of NLW, argues that 'non-lethal weapons offer a chance for soldiers to avoid facing such dilemmas in the future. The psychological burden of killing the innocent or even tacitly supportive civilians should not be thrust upon young troops when viable alternatives are available'. Another writer observes that, '. . . such combat-at-close-quarters with civilians and hostages part of the mix – could call for non-lethal chemical weapons to sort out the real "bad guys"; from non-combatants, human shields and those forced to take up arms' (Knickerbocker 2003: 1).

These quotes, representative of the arguments frequently made by proponents of NLW, demonstrate clear attempts to avoid, or at least substantially mitigate, facing certain crucial moral questions. First, such views hold that these moral questions (especially the question of discrimination between combatant and non-combatant) lose their ethical import since NLW tend not to kill – one could incapacitate people and then sort out who is who later. Second, they also reveal an important misperception: that somehow *weapons* could have the ability to sort out or help sort out the combatants from the non-combatants. Clearly that is a task for moral judgment, which requires humans – well trained, thoughtful and ethically minded humans. Third, there is frequently a claim that NLW are *by their nature* discriminate, or more discriminate than conventional weapons, which ostensibly frees up the humans using them from having to engage in certain kinds of moral and legal judgments.

Is there something about NLW that do render them *inherently* discriminant? Although some weapons are clearly indiscriminate, what makes a weapon discriminate is the moral judgment of the person using it. In order to be ethically justifiable, NLW must be able to be used in ways that make it possible to discriminate between combatants and non-combatants, thus upholding the principle of discrimination. While one can argue that NLW *can* cause less severe damage, that argument addresses the scale and scope of destruction (which is important relative to *proportionality*), not *discrimination*. In the case of gases, foams or other agents that could spread over a large area and whose effects are difficult to target and control where they go, it is perfectly possible that NLW could do less damage and still be indiscriminate.

NLW can also be intentionally used against non-combatant populations and targets in violation of both JWT and international law. This raises the crucial question of what justifies targeting persons who are not combatants, regardless of the effects. To answer this question we must separate two issues around the question of discrimination, two issues which are frequently conflated in the NLW discourse. First, there is the question of discrimination in *targeting* required by the principle of discrimination. As soon as we entertain the permissibility of targeting some non-combatants, we have stepped over a crucial moral line. The principle of discrimination relates not just to killing non-combatants, but to *intentionally targeting* non-combatants and/or engaging in indiscriminate attacks (Lewer and Schofield 1997: 84–85). The principle of non-combatant immunity, which is often discussed in connection with the principle of discrimination, does not just protect non-combatants from death, but it directs soldiers to treat non-combatants differently from soldiers (Mayer 2007: 229). While the doctrine of double effect does allow harm to non-combatants, it can only be permissible when the harm is *unintentional* and outweighed by the benefit of the military action in question; and, furthermore, combatants must be expected to incur risk to minimize non-combatant harm.[2]

In addition, it is generally allowed that non-combatants have a right to self-defense, are by definition not a threat (at least not in the same way that another combatant is), do not consent to taking part in war and, therefore, have not abdicated their right to life or autonomy.[3] Not allowing non-combatants to make choices or exercise their own autonomy is a serious matter and is *harm* in itself; attacking non-combatants – even if to save them from a greater harm – is a violation of NCI. Soldiers are prohibited from treating non-combatants as part of the battleground just because they happen to be in the way. In such cases, Chris Mayer (2007) aptly observes that targeting non-combatants is really for the convenience (as opposed to necessity) of the military, and not justified for any moral reason.

> Even if using NLW against the non-combatants saves more lives in the long run, something, specifically a military action, is being done to the non-combatants for military purposes. It makes the idea of NCI meaningless because the restrictions can be set aside anytime a military force can show that less harm will occur if it is permitted to use NLW against non-combatants.

The prohibition here is not simply on the intentional killing on non-combatants, but on the intentional *targeting* of them. Therefore, it is not clear that NLW offer any ethical advantage because they, just like conventional weapons, can be targeted just as easily against non-combatants as they can against combatants. This concern can be seen quite clearly in the case of the Moscow theatre hostage situation in 2002 where gas was used to ostensibly incapacitate the hostage takers, but ended up killing both hostage takers and hostages alike. Even had no one been killed, this would have still been problematic as the hostages (non-combatants) were targeted in the same way as the hostage takers (combatants); they were treated as objects of war in violation of the principle of discrimination. Mitigating the *effects* of war, which is essential to upholding the principle of proportionality, does not address the moral violation of the principle of discrimination.

Second, there is the question of discrimination in *effect*, which could include concerns around collateral damage and the doctrine of double effect. Arguably this part of the discussion is really more properly considered as connected to the principle of proportionality. In any case, I would want to make clear that these considerations must be considered *apart* from discrimination in terms of targeting. It is in this sense of discrimination that NLW could have an ethical advantage, because on the whole, they tend to do less lethal and serious damage in general, which should mean less harm to non-combatants. When used properly, that is with good intelligence, moral judgment and restraint, they would likely produce less collateral damage.

However, as soon as they are used in conjunction with traditional lethal weaponry, this argument may go out the window. This is a particularly telling objection since many proponents *do* see NLW being used in conjunction with conventional weaponry as a 'force multiplier'.[4] If one were to use NLW to soften up targets prior to the application of lethal force, this could and probably would cause more damage in the end, not less. In reference to the use of NLW by police and troops in Northern Ireland, Victor Wallace argues that in low-intensity conflicts, the evidence suggests that the number of people killed may be greater (than wounded) when firearms are in use against individuals who are immobilized, incapable of defending themselves, or in a confined space (where NLW are used first to immobilize or confine.) This means that the availability of NLW does not necessarily reduce the kill rates by police forces and may in fact make them worse (Wallace 2002: 142). These points illustrate some of the difficulties with the argument that NLW are more humane and better able to discriminate or are better able to be used discriminately. In fact, we see that use of NLW, especially as a force multiplier, could lead to the inflicting of unnecessary suffering, even if these weapons do not kill.

The second caveat is motivated by a more subtle concern underlying these arguments: that NLW are increasingly been seen as a technological way to erode or avoid dealing with moral principles and distinctions in war. Since soldiers are not killing people, one might claim, making these discriminations do not carry the same moral weight and urgency. However, the development and use of NLW should not be seen as an easy technological fix to increasingly complex combat and strategic situations – humanitarian interventions, peacekeeping, nation-building, anti-terrorism, domestic policing, or assistance-rendering missions. In other words, some are looking for a technological solution to what is essentially a moral problem: war as the infliction of injury and harm to exercise one's will over another.[5]

Presumably many of the situations in which proponents claim NLW would be especially helpful are *more* dangerous in some ways than conventional combat. If one thinks about the military providing security for and assisting in humanitarian food distribution, the presence of large crowds that may contain both combatants and non-combatants in a tense and anxious state makes for a dangerous situation. The military will want to be attentive to force protection, there will be non-combatants that need protection, there may be combatants (hidden or disguised) who want to provoke an incident, and the potential for escalation in violence is high. In this case and others like it NLW may be helpful, but are not the technological panacea for which many are hoping.

The third caveat is that the development and use of NLW should *not* be seen (either directly or indirectly) as a way to make war and its effects more palatable or easier to use ethically, militarily, or politically. Should these weapons make the resort to force easier, more publicly palatable, less risky, less costly (in any way), then we have crossed a crucial moral line. The resort to force should always be morally difficult and entered into seriously, with clear and realistic expectations about its ramifications. Take, for example, a case where a NLW is used that causes no direct harm, but gives the enemy leadership in effect a private warning without being exposed to public ridicule from their internal supporters or like-minded bystanders (Alexander 1999: 141). What about war being a public act? What about the requirement in JWT that there be a public declaration of war? In cases like this, low-risk weapons could make interventions more tempting and increase conflict rather than reduce it. This could possibly legitimize what might otherwise (if lethal force were used) be an unwarranted use of force and make the resort to force more routine. David Fidler (2002) argues that NLW may expand rather than limit 'just causes' for using force, reversing the trend in international law to limit the legitimate use of force; anticipatory self-defense might be viewed more favorably if NLW are used, and they could be used to enforce sanctions. Christian Enemark (2008: 204) raises the question about whether such a mode of warfare '. . . would be superior ethically as to bestow upon the holder of NLW technology a greater right to use force in the first place'.

At this point, one might object that having more wars might not be such a bad thing if those wars were, on the whole, less damaging than the ones we have now; one could argue that more wars with less harm (especially to non-combatants) would mean less overall harm.[6] This point is well taken if we think of overall harm purely in terms of the physical harm done to persons, the environment, or the societies involved. However, warfare and the easy resort to force could have the kind of morally corrosive effect that easy resort to other forms of violence (domestic violence, torture, corporeal punishment) have on people and the fabric of a society or culture. While the initial levels of violence – especially with the moral use of NLW – might be less, the easier the resort to force, the more tempting the danger of escalation and with escalation, the more resultant harm. In addition, the kind of relatively easy resort to warfare does not take into account the immeasurable and serious harm done to the autonomy of the individuals involved by virtue of their being viewed and treated as potential objects of war, without first crossing the usual moral thresholds set up to protect the autonomy of persons and nations.

In spite of these concerns, it is possible that there is a 'perfect storm' emerging: the fact that war currently involves killing and dying, combined with the CNN and Somalia effects, does currently make politicians think hard about using force, and there is already a battle against the public perception of the possibility and desirability of a 'costless' or 'bloodless' war. It may be desirable (politically, militarily, and diplomatically) to use force to exert our will and look strong, without looking 'bad' (mean, immoral, uncaring) by killing or seriously harming others, especially non-combatants. Furthermore, the public is presumed to have a low tolerance for combatant casualties. The 1991 Gulf War with its short duration and low casualties created unrealistic expectations among the public and the military about what could be done and how little it would cost (Mandel 2004: 33–34). Therefore, in contemporary conflicts the foremost consideration has become to minimize casualties, to the point where success or failure has hinged on this rather than the question of whether the action was justified or not. This obscures the crucial fact that there are moral limitations on the use of force that are not just concerned with the effects of the use of force. NLW could make this situation significantly worse by providing what seems to be an easy answer to a difficult problem.

Another concern here revolves around risk aversion and the nature of war. General John Shalikashvili cautions, 'I do not want to see us evolve to a point where we have expectations in this country of a war where nobody gets killed on either side and where we don't have any collateral damage on the other side' (Shalikashvili quoted in Mandel 2004: 165).

There is substantial reason to think that we are already fairly far along this road. While many acknowledge that the 'bloodless' war is not currently possible, very few are seriously addressing the question of whether the concept is a good one, even though the concept seems to be driving more and more discourse around war in general and NLW in particular. To return to our opening questions: what if you *could* remove permanent suffering and death from war – is that a good idea? Is it still war? Elaine Scarry (1985: 63) argues that the basic nature and structure of war involves injuring; reciprocal injuring is part of the analytical definition of what makes it different from other pursuits, like games, sports or criminal activity. This desire, even if unrealistic, for an increasingly 'bloodless' war may increasingly turn war into a spectator sport, especially for the public who has minimal contact and participation with war, and increasingly numb citizens to the moral and other costs of war. War does and *should* cost. NLW should make war more humane, but never costless. Even if it were possible to make it costless, it would be morally problematic to do so.

Limited Justification for NLW

Even given the above caveats, the fact that NLW are designed to be able to affect certain military aims and objectives in ways that minimize harm to persons, materiel and the environment, it would seem that the case for them would be ethically straightforward to make. Lethal force is currently justified in warfare, so it appears that using something short of lethal force to achieve the same ends would be ethically and strategically preferable. However, lethal force is not morally justified, *per se*, but only under certain circumstances (*jus ad bellum*) and to be used in certain ways (*jus in bello*). In both the just war tradition and international law, the resort to lethal force in warfare is legitimate only in certain circumstances, for example, where justice cannot be obtained in any other way, or to protect those that cannot protect themselves. Warfare is not condemned as *never* morally and legally appropriate; it can be appropriate provided that certain moral and legal distinctions are kept in view – war versus murder, soldier versus murderer, combatant versus non-combatant, legitimate military target versus prohibited targets.

If, as even proponents will admit, NLW can and do cause harm to people, the environment and materiel, then it makes sense to claim that a similar argument should also apply to NLW. In the qualified and restricted defense of NLW below, my aim is to give a justification for NLW that preserves, as closely as possible, similar moral distinctions that undergird the current thinking in JWT and law about lethal warfare. NLW can be morally justifiable and possibly *even ethically superior* to lethal weapons provided that the following four criteria are followed. Note that these criteria are listed in strict order of import and priority. My argument is that these criteria should restrict and inform the use and further development of NLW in a similar way as the *jus ad bellum* and *jus in bello* inform the resort to and conduct of war.

First, and most importantly, NLW are justified to provide more flexible response times to make strategic, moral and practical judgments about the facts of the situation, in particular to distinguish between combatants and non-combatants, as well as to have time to consider and respond with appropriate and proportional force. These types of weapons should augment (never replace) the moral education and training of soldiers to maintain and support international law, conceptions of military honor and the just war tradition. To return to our example of the checkpoint at Iraq, using a laser as a warning device seems a good candidate since it provides a warning system and used correctly can buy the soldier more time to make a discernment about the combatant status and/or the intent of the individuals involved while maintaining reasonable distance for force protection. In other words, it buys him more time to use force in the most moral manner possible.

One important question is why this criterion takes priority, even over the minimization of non-combatant harm, which seems to be a primary consideration in contemporary warfare, and arguably important to the moral justification of both the principle of discrimination and of proportionality. NLW could give more options in terms of anticipating violence, as well as ways to slow down the escalation to lethal force, if not interrupt it at a crucial decision phase. In terms of humanitarian intervention, peacekeeping, or asymmetrical conflicts, where one is operating in areas where combatants and non-combatants are mixed, or where one is operating in urban or suburban contexts or in other kinds of fluid and fast changing environments, having time to make clear and accurate moral and strategic judgments and being able to control the level of violence in a given situation will be immeasurably helpful in maintaining the moral distinctions in war as a practical, workable reality.

In all of these cases, NLW provide the possibility of slowing down or altering the trajectory of violence such that the parties can make the necessary moral discriminations and judgments essential to the ethical waging of war. If one fails to make the correct moral judgments, there are serious and even catastrophic consequences which will impact the other three criteria; this criterion has priority because establishing it will provide the moral foundation to fulfill the other criteria effectively. In the case of the Haditha incident in Iraq in 2005, 15 non-combatants and 8 alleged insurgents were killed, ostensibly in retaliation for a road side bomb that went off killing at least one US Marine. The marines reported taking fire from a nearby house and consequently took the house, resulting in at least some of the non-combatant deaths. To this day, there is controversy about exactly what happened and why, but given the difficult and confusing context, it is possible that if there had been options other than lethal force considered to deal with a perceived, but unclear threat, the trajectory of violence might have been slowed down, allowing for greater discrimination and a less lethal outcome.[7]

Second, to reduce unnecessary suffering on the part of non-combatants, NLW may be particularly effective. Even the most precision-targeted lethal weaponry causes things to blow up and explode, which means that, especially in contemporary mixed environments, non-combatants will be subjected to unnecessary suffering – on their own person, to their physical environment, and to their communities with ensuing adverse political and economic ramifications. If weapons exist that can achieve military objectives in a manner that reduces the kind of physical and other damage to non-combatants associated with war, clearly those have an ethical advantage. The just war tradition, the Geneva and Hague Conventions, and other international and domestic law all make clear that restrictions on war are designed to avoid any unnecessary suffering being inflicted on non-combatants. Non-lethal weapons, in so far as they do less damage, do damage that is more reversible, and insofar as they open up response time for other less lethal or even non-violent (diplomacy, negotiation) options, and can reduce unnecessary suffering of populations that are not legitimate military targets. In addition, non-lethal weapons could be conceptualized as being helpful on a macro scale to deter opponents, to support diplomatic efforts, and to contain and limit acts of aggression that could lead to a larger war. In combat they may be used to minimize collateral damage, destroy the opponents' lethal capabilities with minimal non-combatant damage and deaths, minimize environmental impact, minimize reconstruction, and to maintain moral credibility (Lewer & Schofield 1997: 56).

Third, another core tenet of JWT and international law is that restrictions on warfare are designed to facilitate the restoration of the peace after the conflict. The use of unrestricted lethal force can create incredible physical, environmental, political, and economic hardships for combatants and non-combatants alike, but total warfare can also create much deeper problems on both the individual and societal level. In wars where particularly brutal tactics have been used, these 'atrocity' stories are told and kept alive within a family and often provide a context and motivation for private vengeance by other family members or friends. They also contribute to the societal narrative of the future relationship with the adversary, providing contexts and motivations for group hatreds, policies of estrangement, resistance to reconciliation, and a desire for revenge that may feed into more conflict or another war. All of these things can, and historically speaking have, interfered with the restoration of the peace after the conflict.[8] Any weapons that can reduce the brutality of tactics and minimize all impacts of war, but especially the kinds that leave lasting psychological and social scars, can more easily facilitate the restoration of the peace afterwards.

Lastly, NLW could be ethical if they minimize combatant casualties and suffering. However, it is important to stress that this is the last criterion, such that if NLW do not fulfill the first three it cannot be argued that NLW should be pursued and are ethical on this last ground alone. It is not the case that as long as you could keep death, pain, and suffering to a minimum, there would be no other important moral questions in war. Even with a minimization of casualties, there are other kinds of harm (violation of autonomy for one) that must be considered, both as an integral element and an effect of war. Much was made in the wake of the first (1991) Gulf War about the relatively low numbers of casualties incurred by the American and coalition forces, but clearly there was a more complicated picture about the war (damage to the environment, effects on the political dynamic in Iraq and the larger region) and the harm created by the war that the level of harm (or relative lack thereof) and suffering incurred by the combatants.

An important litmus test for the just war tradition and the question of justified resort to war is the following: are you willing to suffer casualties in the pursuit of the objectives? If not, that objective and/or the motivation beyond the objective is probably of questionable ethical justification, at least to the extent it requires lethal force to bring it about. Insofar as NLW can be justified using the other three criteria, it would make sense that such weapons would also have (largely as a beneficial by-product) the effect of reducing combatant casualties. If NLW have less impact on non-combatants and their environment, it should be the case that they will have less drastic impacts on the materiel and personnel of the opposing army, as well as one's own. International law and the just war tradition have prohibited weaponry that inflicts unnecessary suffering (relative to military necessity) on combatants, and so NLW, since they reduce overall impact on combatants, will reduce unnecessary suffering as well. In particular, if NLW reduce the amount of killing that combatants have to engage in, this will have beneficial effects – especially in terms of the psychological and moral impacts of war on the individual soldiers.

Given that it is the combatants who are asked to kill and be killed, maimed, injured, and in other ways irreparably changed and impacted by war, it seems that those who ask this of them *do* have a moral obligation to take seriously the job of assessing when and how we deploy these soldiers and the tools with which we equip them. To the extent that NLW can help soldiers wage war in a moral fashion, can buy them more time to make critical moral judgments, and can minimize harm to all parties involved, they should be pursued and deployed within the current ethical and legal frameworks of just warfare.

At this point it is important to emphasize that my arguments in favor, under fairly limited conditions, of NLW all rely on the assumption that NLW limit the extent and scope of the damage incurred in warfare. This argument *in no way* changes any of the core moral and legal distinctions that have traditionally been a part of ethical warfare. The above line of argument provides a limited structure under which NLW development and use might proceed in a way that is consistent with the just war tradition, international law on war, military law, conceptions of military honor and professionalism, as well as general social expectations about waging a just war. In one important sense the use of NLW would not and should not represent a new way of war: this is no revolution in war, at least not from a moral and legal perspective. We still have to deal with the old moral questions, albeit in a new context. There are moral problems with war (even with extensive use of NLW and low casualties) as a means to resolve or address political and social problems. Further, there are questions about the moral and existential status and nature of the warrior culture, the place of war in the identity of social groups and states, and the place of war in the general human condition that would have to be asked and addressed regardless of what the effects of particular weapons of war might be.

Conclusion

Where do we go from here? I have argued for a qualified yes to the question of whether non-lethal weapons are morally preferable to conventional, lethal weapons by trying to augment and develop one possible ethical framework for thinking about the use of NLW in contemporary warfare. Having a conceptual structure that takes into account the prohibitions and possible moral problems as well as the moral permissions and justifications brings this discussion into line with calls, like Eric Partin's on NLW and ROE, for a more holistic and integrated approach to NLW. It is also the beginning of what needs to be a much more in-depth and substantive discussion about how to bring training and use of NLW in line with just war tradition and international law, since it seems that we have moved past the question of *whether* they will be used; the question is now *which ones* and *how* will they be used. NLW can and should be pursued with fear and trembling, and only if they can meet certain criteria in line with just war tradition and international law and provided that certain tendencies currently prevalent in the discourse in NLW are rejected and avoided as a basis for either argument or policy making. What is needed is a much more careful and nuanced discourse about NLW, not simply a discussion about their effects, but about how NLW effect the fundamental moral questions, principles, and distinctions in warfare.

Acknowledgements

An earlier version of this paper was completed thanks to funding from the Kelmer Roe Humanities Grant for student and faculty research at Pacific

Lutheran University. I owe great thanks to Calvin Moore for his comments, conversation and support as a part of that project. This paper was also presented at the International Society for Military Ethics and I am indebted to George Lucas and David Perry for critical insights.

Notes

[1] For one example see Kennedy (2002).
[2] For an excellent discussion of this issue and its intersection with NLW and the principle of proportionality of means, see Gross (2008).
[3] See Walzer (1977: ch. 9).
[4] See www.jdnlw.com, accessed 17 August 2009.
[5] See Scarry (1985: ch. 3).
[6] Thanks to both one of the anonymous reviewers and Dr Henrik Syse for this point.
[7] For a fuller discussion of the incident, see McGirk (2006).
[8] Two obvious examples would be the Balkan conflicts and the US Civil War and the reconstruction that followed.

References

Alexander, J. (1999) *Future War: Non-Lethal Weapons in the Twenty-First Century Warfare* (New York: St Martin's Press).

Department of Defense Non-Lethal Weapons Program (2007) *Escalation of Force Options*, accessed 17 August 2009, available at: http://www.au.af.mil/au/aul/bibs/nonlethal08.htm; Internet.

Enemark, C. (2008) 'Nonlethal' Weapons and the Occupation of Iraq: Technology, Ethics and Law, *Cambridge Review of International Affairs*, 21(2), pp. 199–215.

Fidler, D. (2002) Non Lethal Weapons and International Law: Three Perspectives on the Future, in: N. Lewer (Ed.), *The Future of Non-Lethal Weapons: Technologies, Operations, Ethics and Law*, pp. 26–38 (London: Frank Cass).

Gross, M. L. (2008) The Second Lebanon War: The Question of Proportionality and the Prospect of Non-Lethal Warfare, *Journal of Military Ethics*, 7(1), pp. 1–22.

Kennedy, H. (2002) US Troops Find New Uses for Non-Lethal Weaponry, *National Defense*, 86(580), pp. 26–28.

Knickerbocker, B. (2003) The Fuzzy Ethics of Non-Lethal Weapons, *Christian Science Monitor*, 14 February.

Lewer, N. & Schofield, S. (1997) *Non Lethal Weapons: A Fatal Attraction?* (London: Zed Books).

Mandel, R. (2004) *Security, Strategy: The Quest for a Bloodless War* (Boulder, CO: Lynne Rienner).

Mayer, C. (2007) Non-Lethal Weapons and Non-Combatant Immunity: Is it Permissible to Target Non Combatants?, *Journal of Military Ethics*, 6(3), pp. 221–231.

McGirk, T. (2006) Collateral Damage or Civilian Massacre in Haditha?, *Time*, 19 March.

Partin, E. (2007) *It's Called a Weapon for a Reason: The Need for Non-Lethal Weapons Specific Rules of Engagement* (Newport, RI: Naval War College).

Scarry, E. (1985) *The Body in Pain* (New York: Oxford University Press).

Walzer, M. (1977) *Just and Unjust Wars* (New York: Basic Books).

Biography

Dr Pauline Kaurin holds a PhD in Philosophy from Temple University, PA, and is a specialist in military ethics, just war theory, philosophy of law and applied ethics. She is Assistant Professor of Philosophy at Pacific Lutheran University in Tacoma, WA, and teaches courses in military ethics, warfare, business ethics, and philosophy of law. Recent publications include: Nothing New in Guantanamo Bay: Precedent and Prisoners of War, in: T. Shanahan (Ed.), *Philosophy 9/11: Philosophy and Terrorism* (Open Court Press, 2005); Five Steps to Non-Combatant Protection, in: J. Lango, H. VanderLinden and M. Brough (Eds), *Rethinking the Just War Tradition* (SUNY Press, 2007); and To Debate or Not to Debate: A Question of Torture, in: M. Morgan (Ed.), *A Day that Changed Everything: The Impact of 9/11 on Religion and Philosophy* (Palgrave MacMillan, forthcoming 2009).

Do Non-Lethal Capabilities License to 'Silence'?

SJEF ORBONS
Faculty of Military Sciences, Netherlands Defence Academy, Breda, The Netherlands

ABSTRACT *Most contemporary conflicts can be characterized as 'wars or conflicts amongst the people'. International military forces deployed in such conflicts are confronted with complex operational environments where the distinction between combatants and non-combatants is often impossible to make. At the same time, there is a moral requirement imposed on Western coalition forces to perform in a humane manner and to keep casualties to a minimum. Non-lethal weapons are expected to enable military forces to accomplish their mission without having to kill or destroy. The extent to which these weapons meet that promise in real world conditions is the subject of debate. A defence technology assessment of non-lethal weapons is proposed that validates their utility and moral implications by considering the user, the weapon and the target in the context of the deployment situation. It will be illustrated that the technical parameters of the non-lethal weapons are no guarantee that the weapons will perform as intended as the user and target introduce many factors of uncertainty in real world situations. Although such uncertainties undermine the moral acceptability of non-lethal weapons, it is argued that the principle of non-lethality is compliant with the Just War Tradition principle of proportionality. The principles of non-combatant immunity and discrimination need to be re-calibrated given the human-centricity of many military intervention operations.*

Introduction

In the post-Cold War security environment, military intervention forces have increasingly been facing situations where the use of lethal force in the presence of the civil population is prohibitive.[1] The complex circumstances under which peacekeeping intervention forces have to accomplish their tasks are further exacerbated by the irregular methods and tactics warring factions employ to deny traditional military practices and weapons used by peacekeeping forces. The blurred distinction between combatants and non-combatants, including warriors' chameleon behaviour, has become especially troublesome with the operational scene often being the urban environment. The nature of this new operational context can be described as a 'war amongst the people' (Smith 2006: 267). It is the people that have become the critical ground in most contemporary conflicts.

Military options during such UN-led interventions have mostly been constrained by political mandates to which the missions are submitted. Forces are facing the dilemma between either doing nothing or talking and using lethal fire as the last resort. They perform their tasks and operations under scrutiny of the media and any mistake, including the – unintended – killing of innocent victims, is magnified and may politically jeopardize the accomplishment of the mission.

Soon after the end of the Cold War non-lethal weapons (NLWs), some of which were already in use with police forces for several decades for domestic purposes, have been claimed as promising to close the conceptual gap in the spectrum of military capabilities to meet the challenges in contemporary conflict patterns. Since then, a broad range of technological concepts has been explored, developed and tested for employment by military forces in current operations. However, despite the promises ascribed to NLWs and the efforts invested in such concepts, only a very few new devices have been fielded so far.

Today, more than 15 years after the emergence of the military requirement for non-lethality, considerable scepticism and even opposition still exists concerning the military employment of NLWs, including ethical controversy on their applicability and acceptability as military instruments. It is still the subject of intensive debate whether non-lethal weapons are as benign as to provide the military with a license to 'silence' when the use of traditional military force is prohibitive or undesirable. The ethical dimension of the debate is informed by the Just War Tradition, to explore to what extent non-lethality as a military capability can be justified under Rupert Smith's new paradigm of 'war amongst the people'. The moral dimension resides in the military organization and its personnel that field and operate NLWs, as their practical application depend on the norms, values and virtues incorporated in military education and training.

Some analysts discuss issues of ethics and morality with NLWs based on assumptions that these weapons function precisely as expected (Mayer 2007). This article takes as the departure point that NLW performance is not perfect and effects can be different from those intended.

This article therefore endeavours to discuss and explain factors that underlie the discrepancy between the design expectations of NLWs and their military utility and how the associated uncertainties affect the ethical acceptability of NLWs and the moral implications of their use. After a brief discussion of the military rationale of non-lethality and the official policies that define and guide the development and use of NLWs, this article addresses two main questions:

(1) *Do NLWs always perform as they are designed for?* One of the problems with NLWs is that while they are designed and intended to attain a well defined non-lethal effect, this is not always accomplished under real operational circumstances. This article introduces an assessment approach that takes into account factors other than merely technological to enable a

more comprehensive validation of NLWs. Three NLW-devices have been selected, to which this assessment approach will be applied.

(2) *Given the outcome of the above assessment, can NLWs use be in compliance with the jus in bello principles of discrimination and proportionality from the Just War Tradition?* The technology assessment approach also informs the moral problematic which is associated with the application of NLWs. The author contends that while NLW design, employment and effects may raise moral concerns on the one hand, they can also alleviate existing moral dilemmas facing politicians and military operators on the other.

The Military Rationale of Non-Lethality

In Western societies, drawing from their democratic culture and tradition, ethical and moral issues increasingly have come into play when dealing with conflict. War has begun to have a human face, or from an antagonist's point of view: respect for human life was becoming a critical vulnerability for Western nations. The desire for a society without war, which emerged after the deadly wars in the first half of the twentieth century, was seemingly accomplished with the military introduction of 'clean' weapons that were far less destructive than the previous generations. The ethical imperative emerged that armed conflict should be far less lethal than before.

In war amongst the people the utility of lethal precision weaponry that had excelled during the 1991 Gulf War would appear to be limited soon thereafter. The expectation of many analysts, such as forecasting that precision strike technology could mitigate war's terror and even prevent hitting innocent civilians inadvertently, would soon appear overoptimistic (Freedman 1998). Others question the utility of precision-guided munitions if certain conditions are not met. For instance, when fighting non-state terrorists or insurgent groups in urban settings, the effectiveness of precision weaponry may end up being at its lowest because some targeting discriminations are not simply a function of advanced technology (Mandel 2004). Although sensors and intelligence systems constantly improve, it is highly unlikely that the level of detail necessary to obtain perfect knowledge of battlefield conditions needed for a long range precision strike will be achieved. Furthermore, the role of the media continuously increases in shaping world opinion. Therefore it would be incumbent on the military to make an extra effort to limit collateral casualties, which should include rapid development of area effect NLWs (Alexander 2007a).

The notion of non-lethality has been expressed in NLW definitions contained in official national and international policy documents on NLWs.[2] For the purpose of this article, a definition of NLWs has been taken from a formal NLW policy document published by the North Atlantic Council:

> Non-Lethal Weapons are weapons that are explicitly designed and developed so as to incapacitate or repel personnel, with a low probability of fatality or permanent injury, or to disable equipment, with minimal undesired damage or impact on the environment. (NATO 1999: 1)

The definition stresses the *intended* effect of an NLW. However, earlier experiences with NLWs, in particular of police organizations that have extensively used a growing set of NLWs, indicate that while non-lethals are an accepted core capability in policing, casualties and other unwanted effects still occur. These adverse effects have triggered a continuous debate between human rights organizations and law enforcement organizations on the moral and ethical acceptability of NLWs. While this debate generally focuses on the domestic use of NLWs, the sensitivity of the issue has the potential to be further compounded when peace-supporting military forces employ NLWs in a civil environment in foreign states.

The NATO policy document provides guidance for the development and use of NLWs. It includes aims and requirements that express expectations on non-lethality. NLWs should enhance the capability to accomplish military missions and tasks in situations and conditions where the use of lethal force, although not prohibited, may not be necessary or required. Furthermore, they should enhance the capability to limit or control escalation, to discourage, delay or prevent hostile action and to improve force protection. The policy also specifies that NLWs may be used in conjunction with lethal weapons to enhance the latter's effectiveness and efficiency across the full spectrum of military operations.

The latter requirement seems to be at odds with the original rationale of non-lethality and this policy clause immediately triggers the debate on the ethical acceptability of NLWs: NLWs would appear to be nothing else than just another means to raise the effectiveness of one's own operations, irrespective of casualty reduction considerations. Hence this widened applicability of NLWs disregards the original benign motive to introduce NLWs, which precisely is to minimize fatalities, permanent injury and undesired damage. The explicitness in the NATO policy of a 'dual use' role of NLW potentially fuels political and public concern about the way NLWs will be used by the military.

Although the policy does not explicitly state that NLWs are also to be used against non-combatants, it appears from scenario analysis and technical studies performed by NATO bodies that such options are fully taken into consideration. Whether non-lethal engagement of civilians by military forces is compatible with current international law and ethical standards is the subject of ongoing debate. It is questionable whether the intentional use of NLWs against non-combatants, ignoring the principle of discrimination and the associated non-combatant immunity (NCI), is ethically acceptable. It can be argued that military forces using NLWs should not be exempted from those principles (Mayer 2007; Kaurin 2008). Others contend that NLWs ought to be taken into consideration to diminish the disproportionate harm to civilians caused in contemporary conflicts (Alexander 2007a; Gross 2008).

With the rationale of non-lethality in the military domain being translated into formal policies to guide its military implementation, the question arises whether the intent of NLWs as expressed in those policies is feasible when put to the test in real world circumstances. It is conceivable that NLWs cause unwanted human suffering or even death due to the dynamics, particularities and responses of the actors in relation to the political, operational and physical environment in which the military needs to act. The ideal of achieving political objectives relying on non-lethality is degraded by the friction resulting, among other things, from non-cooperativeness and from hostile counteractions.

Between Expectation, Intent ('The Ideal') and Operational Reality ('Friction')

The emergence and institutionalization of the military rationale of the concept of non-lethality and the associated questions on its real world performance and moral desirability can be put in a broader perspective based on the notions of ideal and friction applied to warfare. The idealist approach has been expressed in a maxim from Sun Tzu, the Chinese strategist who lived over 2000 years ago. He was the first person who considered restraint from killing and destruction, and thus limiting casualties and damage to the environment, of strategic importance. In his book *The Art of War* the maxim reads as follows:

> The general rule for use of the military is that it is better to keep a nation intact than to destroy it. It is better to keep an army intact than to destroy it. It is better to keep a division intact than to destroy it, it is better to keep a battalion intact than to destroy it, it is better to keep a unit intact than to destroy it. (Sun Tzu [ca. 500BC] 1963: 77–79)

The concept of non-lethality is quite compatible with Sun Tzu's thesis. From his perspective, non-lethality in warfare can be interpreted as the ultimate form of controlled use of military force: it strives to achieve a strategic or operational objective without destroying or killing. It is the idealistic approach to warfare.

In 1831, in his famous opus *On War*, Von Clausewitz focuses on the natural features of warfare (Von Clausewitz [1831] 1984). He introduces the phenomenon of friction in warfare and argues that friction is the only concept that enables one to distinguish real war from war on paper. Friction, he contends, is the accumulation of many difficulties and is inconceivable unless one has experienced war. Friction is a force that can never be defined by theory. To Von Clausewitz the key challenge in the conduct of warfare is to remain faithful in action to the principles that have been laid down by ourselves.

When related to the contemporary situation and the introduction of non-lethality, the concept of friction can help understand that while NLWs are designed to achieve a well-defined effect, it is still submitted to many factors of influence and chance that erode the effect in practice. Only from real world experience can one learn how far the principles of non-lethality and intended effects of NLWs are affected by specific situational conditions or even denied by friction due to resistance and countermeasures.

During the early days of the contemporary debate on the promises of non-lethality, the idealist approach was adopted by analysts who believed that war could now be waged and opponents subdued without, or with minimal, killing (Morris & Morris 1990; Toffler & Toffler 1994; Lewer & Schofield 1997). Their vision was inspired by a climate of euphoria that accompanied the new world order which followed after the Cold War. The idea of bloodless or humane warfare persisted throughout the 1990s, describing NLWs as instruments to eliminate the incivility of modern warfare (Coker 2001).

The belief in benevolence in warfare through NLWs also relies on the implicit premise in the idealist's views that non-lethal technologies and concepts in themselves are equivalent to non-lethality in military practice. Other analysts who adhere to a more realistic approach question such expectations of NLWs as being premature. They stress the lack of objective data to test the effectiveness of NLWs. These data should be generated by combat testing, exercising and military experience, but are in short supply. Bloodlessness of warfare by the use of technology to defeat an enemy without casualties appeals to our sense of morality but is, unfortunately, unrealistic (Siniscalchi 1998). The concept of 'weapons of mass protection' is considered one-sided, based on predominantly technologically determined expectations, emphasizing technological capabilities while operational limitations are de-emphasized (Grin 2000).

Obviously, the tension between the ideal (expectation and intent) and friction (uncertainty and reality) can be projected on to the assessing of NLW designs on their performance on the ground. Three NLW concepts have been selected to consider the feasibility of their intended effect.

Realities on the Ground: Assessing NLWs

NLWs, like any other technology, could be assumed and presented as neutral tools. In contrast, one school of thought, as expressed by Rappert, suggests that there is a subtle and complicated inter-relation of technology and action. Rappert argues that technology enters into, and helps reconstitute, existing patterns of social relations. The meanings ascribed to technologies in particular circumstances bear on the manner in which these technologies are used (Rappert 2002: 63). In an earlier study, Grin argued that an assessment of military technologies has to take into account the operational circumstances in which they are applied and judge their implications against the background of the security–political situation (Grin 1990).

Technology assessment (TA) is a research analysis approach that seeks to validate innovative technological concepts within a contextual framework to provide a balanced judgement of its utility and political desirability. More specifically, defence technology assessment (DTA) considers new military concepts within the operational context, including human and procedural factors, in which the system is to be applied (Smit et al. 1992).

The effect of NLW use, therefore, not only depends on the technical specification of the NLW, but also on the skills of the operator to employ the weapon in the prescribed manner. Still, optimal employment of NLWs against individuals or crowds is no guarantee that determined opponents are always incapacitated. Their attitude and response may even be based on an entirely different code.

These considerations highlight that the question of how lethal or excessively harmful non-lethality can become under real circumstances largely depends on the appropriateness of its application. Ongoing efforts in the technological domain to ensure reliability of the NLWs themselves have steadily improved the safety of the effect when properly employed. However, the risk of serious harm increases when the operator is forced to use the NLW beyond the design conditions. Technological system innovation may be supportive to prevent improper use by overruling the human operator. The probability of unintentional misuse much depends on the professionalism, preparedness and skills of the operator. Intentional abuse or disproportionate use is always possible with NLWs in the hands of those who disregard the human rights of the target individual.

A DTA of non-lethal technology concepts should therefore take into account all factors that interplay and together determine the operational outcome of an NLW used. A scrutinized analysis, also considering the operational realities, should, complementary to the weapon-technological characteristics of the NLW, address issues specifically related to the user/operator as well as to the target, as has been visualized below. All three components are presented as complexes, because of the numerous and partly unforeseeable factors affecting the responses and performance of each of them.

The *weapon & technology* complex represents the physical and technical parameters and performance characteristics of the NLW, the immediate and longer term impact the weapon is designed and expected to have on the human target or target group, the manageability of target effects given the type of NLW and technology, and the sensitivity of the NLW effect for operator performance and intent. The technical properties and intended effects of the spectrum of NLWs vary widely and may, therefore, be validated from an ethical perspective.

The *user complex* includes the factors that influence the attitude and performance of the military organization and personnel that deploy and employ the NLW. Factors such as skills, experience, training and education, concept of operation, guidance, rules of engagement, mental and physical state, judgement and perception of the potential target population or target individual are determinants of the user complex. These factors also shape, implicitly and explicitly, the moral attitude and approach of the military organization and its personnel.

The *target complex* incorporates factors of influence that depend on the individual or group exposed to the NLW engagement. These include the attitude, mental and physical state of the individual(s), perception of the user and its organization and background, motivation, experience, preparedness for armed engagement and against NLWs, specific countermeasures and target group dynamics.

The three components are interlinked into a *system* that in itself is embedded in a specific *context* coloured by the situation and circumstances in which the NLW is deployed. As the analysis will point out, the incompleteness and uncertainty of the system components, as well as the difficulty to capture and anticipate the operational (and strategic) context of NLW deployment make performance prediction and control of NLW use under field conditions a fuzzy process. In turn, this unpredictability will complicate the judgement about the moral implications of the employment of NLWs.

The analysis provided here is not exhaustive, but rather intended to be indicative with an aim to assess whether NLWs do what they are claimed to.

Three NLW Devices

The spectrum of NLW effects is very diverse. NLW taxonomies and overviews of NLWs, including in some cases hundreds of NLWs, have been widely published (Rademaker 2003; Davison 2007). Categories of effectors specifically designed to be used against humans include kinetic and mechanical weapons, chemical agents, electrical devices, light emitters, directed energy and sound generators. What these effects have in common is that they inflict interaction with the physiology of the human being, with the intent to manipulate or influence the behaviour of an individual or even deprive a person of the control over his body. Each of the devices has a more or less critical window of design effect. Within these margins there is a high probability that the energy transmitted to the target individual produces the desired effect, i.e. a partial or full incapacitation or compliance of the individual or of a group of individuals. Outside the margins the effect may either be too weak or too strong. In the first case the target individual(s)'s behaviour is insufficiently affected by the NLW, in the latter there is a considerable risk that the target will be seriously or permanently injured, or even killed. Hence a residual risk remains that the effect will be excessive.

Three examples from the spectrum of anti-personnel NLWs have been selected for this paper for a more scrutinized review. Two of them are first generation NLWs already in use for several decades, while the third is a technologically highly innovative second generation NLW design:

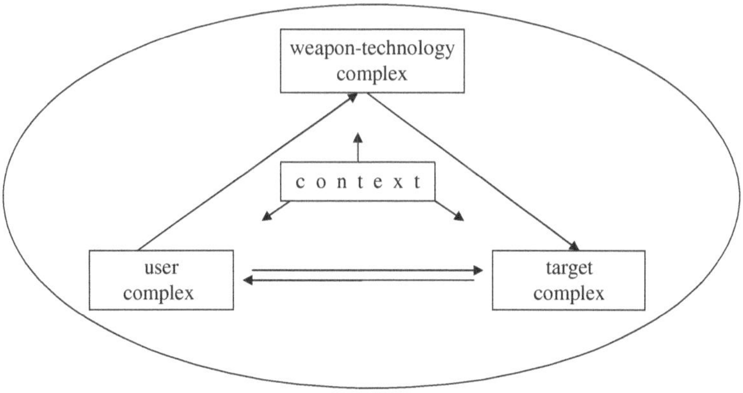

Figure 1. A simplified Defence Technology Assessment approach of NLWs.

- the *baton round*, a widely employed kinetic impact weapon against individual target persons;
- *CS gas*, better known as 'tear gas', a chemical agent in use as an area denial weapon against crowds of individuals; and
- the *Active Denial System* (ADS), a directed energy weapon emitting millimetre wave radiation against individuals or small groups of individuals.

Baton Round

The baton round is a kinetic energy projectile that for several decades has been deployed by law enforcement organizations worldwide. The baton round is designed to cause pain, irritation and minimal injury to the subject. The physiological effect is directly related to the location of the body where the projectile strikes the subject to induce blunt force trauma to the individual (Vilke & Chan 2007). The operator employs the weapon within a certain distance margin to the target. Firing the round from close distances has the potential to cause serious injury, while extended firing ranges may not result in insufficient stopping power to the target. In addition, the aiming accuracy, depending on technical performance of the weapon as well as the operator skills, declines with increasing range. This raises the risk that the round hits critical parts of the body, in particular the head. Adverse weather conditions and low visibility contribute to the probability of aiming errors. The desired target response is to dissuade or prevent a violent or potentially violent person from the intended course of action, thereby neutralising the threat.

CS Gas

CS gas is an irritant agent which can be used to disperse large gatherings or to temporarily incapacitate individuals. These agents are commonly dispersed as gases, liquid, smoke or aerosols. Dissemination can be achieved through explosive devices, spraying or by pyrotechnics.

The effect of CS is irritation and inflammation of the skin, airways and mucous membrane tissues, starting within minutes of exposure and continuing as long as the individual is exposed to the substance. The sensation is typically one of skin burning. These effects worsen with increased concentration and duration of exposure. The symptoms improve after the exposure is removed, and gradually disappear within an hour. Some types of CS are insoluble and can remain active for several weeks and therefore require decontamination of buildings or other items. This is critical due to the high-flammability of CS (Vilke & Chan 2007). CS can also be applied as a hand-held liquid irritant projector that fires a stream of irritant agent in solution, thus making it a better directive against individual persons (Davison 2006). High temperature and humidity make the effect of exposure to CS more severe and the eyes are strongly affected by CS and may not open for several hours. Some people can develop a tolerance to CS, especially after being exposed regularly (Sutherland 2008).

The Active Denial System (ADS)

The Active Denial System (ADS) is a directed electromagnetic (EM) energy weapon. It is a millimetre wave (MMW) emitter at a frequency of 95Ghz. The system uses an antenna to direct a focused beam toward a selected target individual.

The MMW energy strikes the subject and reaches a skin penetration depth of less than a millimetre. It produces a heat sensation that within seconds becomes intolerable and forces the targeted individual to instinctively move. The sensation immediately ceases when the individual moves out of the beam or when the operator turns off the beam. When the system is operated properly there is minimal risk of injury because of the shallow penetration depth of energy into the skin at the short wavelength, the safety features designed into the system, and normal human instinctive reactions (US Department of Defense Joint Non-Lethal Weapons Program 2007). The heating effect experienced by target individuals is caused by the stimulation of pain nerves in the upper skin (Murphy et al. 2003). Only if the subject remained in a sustained beam while ignoring or resisting the pain sensation could permanent damage occur. From experimental research the risk of eye injury is found to be very low, as a subject will close or shield their eyes well before the heat induced pain threshold is attained. Nevertheless, it is to be evaluated whether long duration low-level exposure may affect vision (Annati 2007).

The vehicle-based ADS incorporates computerized control systems to modulate the beam transmission to achieve a safe and effective non-lethal repel effect. The operational range can be chosen between 15 and 500 metres. The MMW beam can be slightly widened to target three or four individuals grouped closely together (Magnuson 2006). The ADS is to be positioned in such a way that an uninterrupted line-of-sight over a considerable sector is ensured. The system's accuracy is advertised to enable discrimination between single target persons who are at least 2 metres apart from each other. Such accuracy levels can generally be attained at moderate engagement ranges and degrades with extending distances. Range-to-target measurement accuracy is a critical factor for selecting the correct emission power level (Alexander 2007b). ADS fielding is foreseen for 2010.

The three NLW devices will be submitted to the assessment framework in order to consider their performance in a conceivable real world environment.

The Weapon & Technology Complex

Many efforts have been and are still being undertaken to enhance the technical performance and reliability of non-lethal devices. What experience teaches us is that NLWs cannot be claimed to be technically perfect. As has been pointed out earlier, the margins between effective performance of the devices and unacceptable weapon impact are small. Even a slight malfunction can have severe consequences.

Baton rounds should strike the human body only at specified parts. Yet, the ballistic accuracy of the projectile to ensure safe impact appears to be limited.

A considerable percentage of the projectiles fired during tests demonstrated unacceptable deviations from the aiming point (Kendig 2001). While this assessment might appear somewhat dated, the necessarily blunt front end of the baton round, required to prevent penetration of the skin, compromises accuracy. Whatever the design of a baton round or any other kinetic projectile, the need to optimize between the engagement range, the impact energy of the round, and its ballistic accuracy, results in a residual risk that can hardly be reduced with current technology.

The experiences with baton round use in Northern Ireland since the early 1970s show a steady decline in casualties. This is, to a large extent, due to improved weapon design resulting in a more reliable performance. Between 1994 and 2001, a total of 13,500 baton rounds were fired with no fatalities (Burrows 2002). However, more recent evaluations of the use of such rounds conclude that the risk of unintended but serious harm is still real (Northern Ireland Office 2004).

CS gas is designed and deployed as an area effect weapon. The technical limitations associated with the use of CS gas are related to the inherent impossibility to manage the amount and concentration of CS gas an individual in the deployment area inhales. As the CS gas doses an individual receives depends on a range of factors, including his proximity to the location of the releasing CS grenade or canister, the user's ability to control the individual doses of the agent is very constrained. Usually considerable variations of concentrations in a given area occur, in particular during the first seconds after the agent has been released. Apart from the enhanced risk for individuals suffering from respiratory illnesses, such as asthma, who may be affected critically by the agent even at low doses, there may also be unknown long term effects to individuals who have inhaled considerable amounts of CS gas.

Other factors, including the environmental circumstances in the area of deployment, can vary considerably, which potentially leads to variations of effect upon the target group. In Northern Ireland, where CS gas has commonly been used in urban areas, the combination of wind and specific built-up configurations has complicated the predictability and control of agent dispersal (Thornton 2007). Due to its indiscriminate effect against generally heterogeneously composed crowds, CS impact on individual crowd members may vary widely, ranging from very limited or no effect on some individuals to excessive or even toxic incapacitation of others.[3]

The ADS system is still in a testing phase. As an anti-personnel directed energy device, its performance and reliability under practical circumstances is still unknown. An extensive testing programme including the exposure of thousands of volunteers has optimized the design. Tests are usually run under controlled conditions. Despite the highly successful performance of the system during extensive testing series, the technical malfunction of system components or improper tuning of energy levels transmitted may result in adverse effects against the target individuals. Given the extended engagement ranges of up to hundreds of metres, combined with the invisibility of the skin heating effect of the ADS, feedback of a possible harmful engagement caused by technical unreliability may be difficult to detect in the absence of close observers. One of the factors that influences actual radiation energy levels received by a target person is his physical environment. If the operational setting is a confined space, as is often the case in urban missions, radiation reflections may occur which raise the energy level locally to unacceptable intensities. Furthermore, the response performance of the ADS control system to adjust the amount of energy transmitted and other tuning parameters of the radiation beam may be insufficient to follow the behaviour of individuals and group dynamics in the target zone. Such problems can be further exacerbated by poor visibility conditions.

This brief account of the potential technical uncertainties and risks of NLW devices to perform beyond the non-lethal mode illustrates that even when operators of NLWs act as trained and expected, and target persons behave or respond as anticipated, the demarcation between effectiveness in terms of temporary incapacitation and undesirable harmful effects is fragile. It also shows that the effect of area denial weapons such as CS gas are fundamentally less controllable than those NLWs which are aimed against selected individuals, like baton rounds and ADS. The implication of the former category is that innocent civilians nearby may be harmed. Hence, from an ethical perspective, area denial weapons like CS gas should not be used if innocent people will be affected who are not part of the crowd or riot but nevertheless cannot leave the risk zone. This comparison of the two categories of NLWs also tells us that NLWs, from their technical characteristics alone, have to be judged on ethical acceptability on a one by one basis.

As will be argued hereafter, in real world circumstances technical imperfection may be a lesser risk than the human factor, i.e. the attitude and performance of the holder and operator of the NLW.

The User Complex

The baton round is intended to provide the policeman or soldier with an option to keep an aggressive or dangerous individual at safe distance. In Northern Ireland the weapon was used to prevent protesters being able to hit police and military forces when hurling stones and Molotov cocktails. Without the availability of the baton round, on many occasions they would have been authorized to use lethal weapons for self defence.

The reliability and precision of CS gas when used in crowd and riot control situations depends, amongst others, on factors related to appropriate use by trained personnel. Such factors include the availability of a suitable range of capabilities to dispense CS gas and the selection of the most effective method of delivery for the situation at hand.

General agreement exists among experts in the field of NLWs that these weapons have been, and potentially will be, misused by regular police and security forces as well as irregular actors in various parts of the world. International organizations as Amnesty International and Human Rights Watch have frequently reported on human rights violations in a range of countries. Most of the cases documented refer to domestic situations, where police or security forces have applied NLWs in an excessive and unlawful manner. Such abuses included CS gas in many cases, and also baton rounds, as well as electro-shock weapons. In particular, the use of CS gas in confined and even closed spaces, which is prohibitive given the unacceptably extended periods of exposure to high concentrations of the gas, has been reported.

Human Rights Watch made public the abuse of CS gas in Venezuela. It reported that tear gas powder was being sprinkled on the faces and bodies of individuals in custody, as well as about the use of tear gas and pepper spray against prisoners kept in closed vehicles. Gas used in confined spaces is more concentrated and lasting in its effects than when used in the open and could cause individuals who are incapacitated or unable to move to suffocate. It is also potentially fatal to those with lung or heart ailments (Human Rights Watch 2004).

There are situations, however, in which the use of CS gas in confined spaces is regarded as an effective and acceptable method to resolve a crisis situation in the absence of alternatives that would not put the security forces in considerable danger themselves. In Northern Ireland, in prison barracks where IRA detainees were held, the prisoners took control over their own prison building in 1972. As they had blocked their prison physically and were using improvised arms, security forces decided to chase the occupants out of their huts by inserting CS grenades before overwhelming the prisoners. The tactic worked, with no prisoners suffering from enduring after-effects of the exposure to CS.[4] A similar use of CS gas in a confined space has occurred as an approved practice applied during inmate uprisings in US prisons. Here too, CS gas has been used to prevent security personnel from being exposed to high risk and also to avoid the use of lethal arms in these situations.

Baton rounds have also been abused as a weapon of punishment in, amongst others, Northern Ireland and in the conflict between Israel and the Palestinians. This has led to fatalities in both conflict areas. Wright speaks of a systematic abuse in both situations where baton rounds have been used on a regular basis rather than in urgent cases. He lists a number of ways the abuse can take form:

> These abuses include doctoring projectiles to enhance lethality, breaching guidelines on use only as a last resort, firing below the minimum distance, firing at areas of the body that should not be targeted (such as head, face, neck, or chest), shooting out of moving vehicles, use as street punishment during zone clearing operations. (Wright 2002: 80–81)

Besides the numerous written accounts, the internet provides an abundance of footage of abuse of a variety of NLWs. To mention but one example, in March 2007, a video was posted showing Israeli soldiers firing rubber bullets indiscriminately from a range above 50 metres at a crowd of demonstrating Palestinian women. Various women were reported to have been severely injured, in part because some bullets had a steel kernel inserted in a rubber casing to give the round a false non-lethal appearance (Gaelli 2007).

Abuse is a risk that may also arise in the context of suppression of people fighting for a just cause. Whether or not abuse may occur depends on a range of factors and circumstances, of which the attitude and intention of the user of the weapon is the most important. Such behaviour is also determined by the enforcement 'culture' of the forces that deal with the conflict situation. Furthermore, it also depends on the way they view the non-cooperative 'other side', and on their professional skills, self control and self confidence. It is in the operator's hand, or else in the unit commander's directions, whether the effect is tuned such that the NLW performs within the design safety margins, or turns into an instrument of excessive harm, torture or lethal force. In physical terms this basically comes down to selecting the intensity of the energy or concentration of agent delivered to the target individual or group, the duration and repetition of the effect and, for a number weapon categories, the part of the body at which the effect is aimed.

The technological status of the NLW is usually of limited relevance for the operator's options to misuse it, as long as there is no automated self-control mechanism built into the device, similar for instance to controlling the speed of a car in cruise control mode. First generation NLWs, like the baton round or CS gas, allow the operator to freely choose the mode of employment. The ADS, as a new generation NLW, is integrated on an advanced platform and supported by a computerized weapon control system, facilitating the system to be set to disable operation beyond the predefined safety margins of engagement distance, radiation energy level and pulse emission duration. The operator can only select the targets to be engaged and decide when to activate the beam.

The likelihood of abuse of NLWs to occur is correlated with the types of effect at the user's disposal: electro-shock weapons, and to a certain extent, sound directed energy weapons such as the ADS and lasers, bright light sources and chemical agents, leave no traces on the victim, which makes after-action investigation into alleged abuse or excessive use of NLWs harder. In particular the electro-shock weapons have been reported in numerous cases to have facilitated human rights violations, including torture. In a 1997 report Amnesty International documented electric shock torture in 50 countries worldwide since 1990 (Wright 2002).

What is easily overlooked or even ignored by security forces and police is the disproportional vulnerability of children to NLWs. The design parameters of most NLWs are based on what is required to temporary incapacitate an adult of average size. Using the same weapon indiscriminately against children mixed in crowds has caused fatal injuries or lasting damage to their body. The ADS, as a next generation NLW, may be one of the very few non-lethal system concepts exempted from this concern, as the radiation only interacts with the upper skin, hence it is indifferent for the size of the target individual.

Whereas the possibility of intentional abuse of NLWs appears obvious, the unintentional improper employment of NLWs is of significance as well. This may occur in situations where the military or the police are pressed or feel forced to act. When, for instance, a crowd turns into a mob and threatens the security forces, personnel may use their weapons for self defence, including NLWs. Inaccurate aiming of the weapon or being pressed to use the weapon too close to the target enhances the risk of non-lethals becoming lethal or causing serious harm. Apart from kinetic energy weapons, such as the baton round, this also may occur with riot control agents such as CS gas. The initial intended effect of deploying NLWs, i.e. supporting the control of disorder using appropriate levels of violence, may not be achieved and occasionally rather the opposite may happen. This illustrates that the friction which Clausewitz refers to in conventional war is also present in military operations other than war. A specific example is the excessive use of CS gas by the British Army in Northern Ireland in a particular confrontation between Unionists (Catholics) and Loyalists (Protestants) that has been claimed to have decisively destroyed the confidence of the Unionists in the Army as their safeguard (Thornton 2007).

There is a growing concern that the increasing availability of NLWs to security forces as a complementary capability to lethal options may give rise to a tendency that operators become more 'trigger happy'. It is easier to physically incapacitate an individual or group to control behaviour than accomplishing compliance through negotiation or warnings. This translates into a trade-off between the security forces preparedness to accept risk and the exposure to forced action of the opposing or non-cooperative party. The moral consequence of this trade-off can be linked to Mayer's analysis of the justification of NCI in relation to the question of targeting non-combatants non-lethally:

> Combatants must make a concerted effort and accept risk to minimize noncombatant harm. (Mayer 2007: 223)

While Mayer considers non-combatants as complete 'outsiders' in the context of armed exchanges between warring parties, it becomes clear that minimizing harm to civilians in a 'war amongst the people' context is a moral imperative as well, even when intentional – non-lethal – targeting of them is an option for the military intervention forces. What tends to be overlooked in the ethical debate is that military intervention forces operating in today's conflict zones predominantly act as stabilization forces with a focus on de-escalation of force. They should not be defined as combatants in the traditional sense, bound as they are by strict rules of engagement (RoEs), prohibiting the deployment of a major part of their military power. Recognizing this shift in role and status of the military force would redefine the relative notions of combatants and non-combatants.

The Target Complex

The reaction of a person struck by a baton round will vary depending on the area of the body struck and the degree of motivation of the individual. Responses may vary more when individuals are under the influence of alcohol, drugs or suffer from a mental disorder. Furthermore, very motivated individuals may be capable of continuing their initial resisting behaviour. There is evidence that it may require repeated firing to stop and incapacitate a very aggressive person.[5] Crowds may anticipate the use of baton rounds by security forces with improvised body protection. Countermeasures against CS gas are possible by covering the face with wet tissue or even gas masks. Protection against the ADS energy beam is possible by wearing a radiation reflective suit on the body.

The level of preparedness of target groups significantly depends on the history of the conflict or crisis in which the NLW is being deployed. This is an illustration of the importance of context for the effectiveness of NLW use. Enduring conflicts are often characterized by fixed patterns of exchange between security forces and resistance groups. This may go as far as to trigger an arms race between both sides. This was the case in Northern Ireland where, after a short while, CS gas was considered largely ineffective against the intended targets and was even counterproductive as extended release of CS gas in populated areas did harm to local residents who were not involved in the confrontation.[6] Ironically, whereas CS gas is designed to be used in open air, CS gas effects turn out to be more controllable in confined spaces as it is easier to tune doses.

The development of counter tactics is also an outcome of prolonged and repetitive engagement between security forces and non-cooperative target individuals and groups. The use of cover in dense built-up areas denies NLWs like the baton round and the ADS the ability to exploit the necessary stand-off distance due to lack of free line-of-sight. Militants in crowds may use innocent civilians as a fence and perform their aggressive act from behind.

In general, mixed crowds composed of passive bystanders, peaceful demonstrators, militant persons and ringleaders pose a considerable challenge to security forces in selecting the right approach. Selective engagement with NLWs is often complicated by chaos and by the impossibility to discriminate between aggressors and non-violent group members.

The mental state of an individual or crowd, the motives for their resistance or fight and their perception of the forces and their political masters in charge of the security mission potentially have a deep impact on the susceptibility for and response to the employment of violence, including NLWs. Such contextual factors of NLW deployment may profoundly affect the impact of NLW use. If the use of (non-lethal) violence by security forces is seen as fundamentally unjust by the target population and wider public, it is possible that such use may escalate rather than de-escalate violent resistance. Hence, the target complex embodies frictional factors that potentially deny the NLWs' intended effect.

Assessment Synthesis

The discussion and assessment of the three 'system components' above has demonstrated that situational factors to a considerable degree determine the tactical and operational effect of the three NLWs considered. It appears that the general expectations of NLWs as expressed from an idealist point of view will often be thwarted by the Clausewitzian frictional factors encountered in the contextual reality. The fact that NLWs are designed to be effective in a narrow window of technical opportunity is by no means a guarantee that this window will always present itself in a real world scenario. It proves to be fairly easy for a creative opponent to escape that window of permissible non-lethal exposure. Upon that, it can not be excluded that military security forces would not, intentionally or unintentionally, operate the NLW beyond the safety margins of the designed non-lethal window. The planners of non-lethal force employment are facing fundamental uncertainties about the impact of its actual use. Any wrong outcome of NLW use causing excessive harm, in particular against civilians, will be perceived as disproportionate use of force and judged as morally reprehensible.

The result of a DTA applied to NLWs may vary widely, as it depends on the particular type of non-lethal effect considered. It is difficult to generalize about NLWs as there is so much diversity in their characteristics and breeds that they are often not really closely related. Some analysts even suggest that it might be intellectually cleaner not to speak of a category of NLWs at all (Koplow 2006). Such an approach supports the preference for a conceptual separation of the notions of 'non-lethality' and of 'non-lethal weapon', as has been adopted in this paper.

Whereas from a functional and technological point of view, non-lethality would seem instrumental to cope with the tactical, operational and even strategic problems of the impossibility of discrimination between benign and hostile targets in irregular scenarios, the question remains whether indiscriminate use of non-lethal force would be legitimate. As we have seen, in contemporary conflict scenarios the civilian population can often no longer be considered an outsider in relation to the type of mission and tasks the military intervention force is performing. To give another example, when a crowd in an urban scenario does not behave aggressively, but at the same time there is a high probability that hostile and armed elements hide in that crowd, would it be acceptable to neutralize or disperse the crowd by means of 'mass-incapacitation'? In this situation, the NLW is used pre-discriminately, because after the mass-incapacitation has been achieved, forces could approach and single out the opposing militants (Enemark 2008). A particular group of NLWs that would facilitate such an operation can be labelled as area effect weapons. Their intent is to engage a group of individuals that are gathered within a certain space, either in the open or indoors. Advocates of the operational need for this capability argue as follows:

> What are needed with absolute certainty, are NLWs that can reliably incapacitate, or otherwise initiate control of targeted persons dispersed over a fairly large area. Applications of NLWs in counterterrorism, counterinsurgency, peace support and other law enforcement-like operations, make these systems imperative. (Alexander 2007a: 4)

Resolving uncertainty through indiscriminate targeting is also a relevant issue in asymmetric conflict scenarios involving singular actors. In one case, international security forces in Afghanistan have been confronted with situations in which military convoys passing through villages or towns were attacked by vehicle borne suicide Improvised Explosive Device (IED) attacks.[7] In response to the threat, force protection procedures resulted in various civilian casualties when car drivers or motor bikers did not respond to warning signs and warning shots from the passing military vehicles to respect the minimal safe distance from the convoy. Non-lethality is being considered by operational commanders as a method to be able to stop or neutralize individuals indiscriminately. In an effort to accomplish this, a stepwise approach is considered, with increasing levels of non-lethal physical effect when the target would continue to approach the convoy.

Whether or not the capabilities called for are ultimately feasible, they inform the debate on the ethical implications of engaging a population indiscriminately to reduce the uncertainty problem beleaguering current military missions.

NLW and Just War Tradition

The paradigm shift in the nature of conflict defines new roles and challenges for the military. The new roles are to a large extent other than war-fighting. These changed circumstances have implications for the interpretation of the Just War Tradition. In the absence of overt war, the *jus in bello* principles of proportionality and discrimination, as well as NCI, need to be re-calibrated. The ongoing NLW debate, in particular the ethical and moral ramifications of the new military options, is a clear indication of this need.

NLWs have already been deployed and used by the military forces of a variety of countries in a range of international intervention missions. This includes the use of NLWs against civilians in particular circumstances, and in situations where the discrimination between combatants and non-combatants is not possible. Some analysts criticize the use of NLWs as a technical fix to reduce this uncertainty problem. Kaurin (2008) argues against the use of NLWs as a way to circumvent or make irrelevant moral distinctions like combatants/non-combatants. However, the DTA approach in this article tells us that the military forces are morally still committed to accept a considerable level of risk and show restraint before using force, even with a non-lethal option at their disposal. While NCI obliges the military to treat non-combatants as different from militants, the blurring distinction between combatants and non-combatants requires the operational commander to take decisions on the use of force in situations where the universal immunity of non-combatants in the classic sense is no longer a workable concept.

Those in favour of the permissible use of NLWs against non-combatants argue that the intentional harm to civilians requires that the principle of NCI be subjected to a 'lesser evils' test that compares the small intentional harm with a greater level of non-intentional harm that comes from lethal force. If the former is significantly less than the latter, then there are moral grounds to targeting civilian non-combatants with NLWs (Gross 2008). NCI needs a new interpretation in order to effectively serve as an effective moral compass for the military commander and operator in complex situations when the use of NLWs is contemplated. The DTA method can be used to match NCI 'new style' with the application of non-lethal force against non-combatants, in part by checking the moral standards educated and trained to the military forces equipped with NLWs.

Abuse and proliferation are by-products of non-lethal technologies and weapons that deserve due attention and response. Immoral uses of NLWs will occur, and have the potential of becoming political showstoppers for non-lethality, even when regular forces would generally use NLWs in an operationally and morally rightful manner. Any technology, including dedicated non-lethal ones, has the potential of human rights violations, although particular technologies which leave no marks pose a particular challenge to finding evidence of abuse. Besides abuse of NLWs, it is important to assess the likelihood of an NLW being unreliable regarding its non-lethal performance. It would appear ethically inappropriate to field any NLW that poses a significant risk of excessively harming people, albeit unintentional.

One of the issues interfering with the rationale of non-lethality is the permissibility and acceptability of intentionally inflicting pain, harm and suffering against civilians. The principle of proportionality of a certain military action is usually tested by assessing whether the significance of the military objective of the action outweighs the unavoidable civilian casualties. Asymmetrical warfare, by its very nature, carries a greater risk of disproportionate, unnecessary and intentional harm than conventional fighting between two relatively well-matched forces. NLWs may not solve these problems entirely, but do offer an avenue for limiting disproportionate harm to civilians in contemporary armed conflict (Gross 2008). In analogy to the DTA approach to consider the military utility of NLWs contextually, a judgement of NLWs on moral and ethical acceptability and on proportionality could be performed contextually as well.

Some analysts consider the use of NLWs by military forces directed against civilian populations as a radical change in military strategy (Enemark 2008). For the appreciation of the moral acceptability of such a strategic shift it is relevant to refer to Fidler's proposal suggesting an amendment in international law that reflects the changing nature of conflict and its implications for the position of NLWs:

> The Selective Change Perspective ... accepts that the nature of conflict [has] changed and will continue to change. Consequently, rules for NLW can be different from those for lethal weapons, but NLW needs to be clearly defined to enable change of particular parts of international law regimes on a case by case basis. (Fidler 2001: 198)

Fidler requires a clear definition of (an) NLW, in order to enable that a change in international law can be formulated. DTA tells us, however, that it is not only the NLW itself that has to be clearly defined, but also the attitude and approach of the user who employs it. A change in international law, therefore, seems premature for NLWs with effects that are 'user sensitive'.

Conclusions

In Western societies, drawing from their democratic culture and tradition, moral and ethical issues increasingly have come into play when dealing with conflict. In the Western way of war, in particular in wars of choice rather than necessity, respect for human life has become a key issue. A need has emerged for more fine-tuned effects of weapons as to make them perform non-lethally. The Just War Tradition, in particular the *jus in bello* component expressed in the requirements of proportionality and discrimination, needs to be adjusted to the human-centric operational environment.

From an idealist perspective, NLWs are technologically designed and expected to make military operations more benign and even bloodless. In reality the performance of NLWs is submitted to operational friction, comprising many factors of influence and chance that erode the expected effect. Contemporary conflict is therefore characterized by uncertainty.

A Defence Technology Assessment of NLWs considers the user, the weapon and the target as complex components of a system that is exposed to the influences of the specific context of the scenario of deployment. The DTA application demonstrates that apart from technical constraints, NLW effectiveness is strongly affected by independent factors within the user and target complexes. The target population may respond in widely diverging ways, depending on situational circumstances. The risk of serious harm or fatalities predominantly depends on the training, skills and attitude of the military user. Hence, a home-tested functioning technological concept in itself is no guarantee for successful operational utility in real scenarios. Instead, the DTA approach would help to unravel the operational complexity and thus enable an uncertainty reduction. This approach also informs the ethical and moral issues that guide NLW use.

The occurrence of abuse of NLWs is impossible to rule out, as any technology can be employed in a harmful manner, especially with types of NLWs that leave no traces on the victim.

In this respect it would be ethically appropriate to consider the infliction of temporary harm as a lesser evil. The concept of non-lethality appears to be implicitly compliant with the imperative of proportionality. Ultimately, a partial review of International Humanitarian Law would facilitate the mitigation of the risk of inflicting innocent casualties by allowing a wider deployment of non-lethality. Such a review would provide the military with a license to non-lethally and temporarily silence a violent or risk situation in international intervention missions. It is therefore recommendable that a thorough assessment of NLW performance in a contextual framework precedes amendment of law.

Acknowledgements

I want to thank Professor John Grin for his suggestions and constructive remarks on a draft of this article.

Notes

1 The article is based on a paper that I presented at the international conference Emerging Technologies, Ethics and Leadership, organized by the Netherlands Defence Academy, 5–7 November 2008, Amsterdam.

[2] A distinction is made between the notions of non-lethality and non-lethal (weapons or capabilities). Non-lethality refers to wider strategic and operational concepts that guide the use of force that is supposed to deliver non-lethal effects. Non-lethal refers to the properties and effects of weapons. Compare Rappert (2002: 55, 73).

[3] In an interview with the author, J. Bolle stated that in US prisons the spraying of CS gas to control turmoil in gatherings of inmates was less effective and led to replacement of the agent by the much stronger pepper spray, also dispersed as an area effect aerosol (Ossendrecht, Netherlands Police Academy, October 2007).

[4] Interview by the author with Col (rtd) J. Wilson, who gave an account of the prison incident that took place in the 1970s (Upavon, HQ UK Army, 16 December 2008).

[5] Interview by the author with a commander of an arrest team in the Dutch armed forces. The name of the interviewee is not disclosed for security reasons (March 2008).

[6] Interview by the author with Brig (rtd) T. Longland (Shrivenham, DCDC, 18 December 2008).

[7] Anecdotal accounts of such situations have been taken from interviews by the author in 2007 and the first half of 2008 with Dutch officers who have served with the Dutch Battle Group of Task Force Uruzgan in Afghanistan. Respondents are not quoted in this paper for security reasons.

References

Alexander, J. (2007a) When Precision is Not Enough, in: *Proceedings of the 4th European Symposium on Non-Lethal Weapons*, 21–23 May, EWG (Ettlingen, European Working Group on Non-Lethal Weapons).

Alexander, J. (2007b) Private conversation with the author (Ettlingen, 23 May).

Annati, M. (2007) Non-Lethal Weapons Revisited, *Military Technology*, 31(3), pp. 82–87.

Burrows, C. (2002) Operationalizing Non-Lethality: A Northern Ireland Perspective, in: N. Lewer (Ed.), *The Future of Non-Lethal Weapons – Technologies, Operations, Ethics and Law*, pp. 99–111 (London: Frank Cass).

Coker, C. (2001) *Humane Warfare* (London: Routledge).

Davison, N. (2006) *The Early History of 'Non-Lethal' Weapons*, Occasional Paper No. 1 (University of Bradford, Bradford Non-Lethal Weapons Research Project).

Davison, N. (2007) *The Contemporary Development of 'Non-Lethal' Weapons*, Occasional Paper No. 3 (University of Bradford, Bradford Non-Lethal Weapons Research Project).

Enemark, C. (2008) Non-Lethal Weapons and the Occupation of Iraq, *Cambridge Review of International Affairs*, 21(2), pp. 199–215.

Fidler, D. (2001) Non-Lethal Weapons and International Law, *Medicine Conflict, and Survival*, 17(3), pp. 194–206.

Freedman, L. (1998) *The Revolution in Military Affairs* (International Institute for Strategic Studies, London: Oxford University Press).

Gaelli, J. (2007) *Rubber Bullets 101*, accessed 12 September 2008, available at: http://www.youtube.com/watch?v=rTx4vbTYr-w&feature=PlayList&p=AA07D1C64EF21E6D&playnext=1&index=12; Internet.

Grin, J. (1990) *Military–Technological Choices and Political Implications* (Amsterdam: VU University Press).

Grin, J. (2000) Bloodless War or Bloody Non-Sense? Technology Assessment's Role in Uncovering Old Propositions behind New Airpower Concepts, in: J. Grin & A. Grunwald (Eds), *Vision Assessment: Shaping Technology in 21st Century Society towards a Repertoire for Technology Assessment*, pp. 71–96 (New York: Springer).

Gross, M. (2008) The Second Lebanon War: The Question of Proportionality and the Prospect of Non-Lethal Warfare, *Journal of Military Ethics*, 79(1), pp. 1–22.

Human Rights Watch (2004) *Letter to President Hugo Chávez Frías*, accessed 12 September 2008, available at: http://hrw.org/english/docs/2004/04/12/venezu8423_txt.htm; Internet.

Kaurin, P. (2008) With Fear and Trembling: A Qualified Defense of 'Non-Lethal' Weapons (Tacoma, Pacific Lutheran University), accessed 5 June 2008, available at: http://www.usafa.edu/isme/ISME08/Kaurin08.html; Internet.

Kendig, T. (2001) *Non-Lethal Weapons Testing Proves Many Inaccurate* (Pennsylvania, Penn State Intercom), accessed 13 October 2008, available at: http://www.psu.edu/ur/archives/intercom_2001/May24/weapons.html; Internet.
Koplow, D. (2006) *Non-Lethal Weapons – The Law and Policy of Revolutionary Technologies for the Military and Law Enforcement* (New York: Cambridge University Press).
Lewer, N. & Schofield, S. (1997) *Non-Lethal Weapons: A Fatal Attraction? Military Strategies and Technologies for 21st Century Conflict* (London: Zed).
Magnuson, S. (2006) Non-Lethal Weapon Readied for Battlefield, *National Defense*, 90(626), pp. 18–19.
Mandel, R. (2004) *Security, Strategy and the Quest for Bloodless War* (London: Lynne Rienner).
Mayer, C. (2007) Non-Lethal Weapons and Non-Combatant Immunity: Is it Permissible to Target Noncombatants?, *Journal of Military Ethics*, 6(3), pp. 221–231.
Morris, J. & Morris, C. (1990) *Toward a Nonlethal Strategy* (West Hyannis Port, MA: Morris & Morris).
Murphy, M. R., Merritt, J. H., Mason, P. A., D'Andrea, J. A., Blick, D. W. & Scholl, D. M. (2003) Bioeffects Research in Support of the Active Denial System (ADS): A Novel Directed Energy Non-Lethal Weapon. Paper presented at the European Working Group NLW 2nd Symposium on Non-Lethal Weapons, Ettlingen, 13–14 May.
NATO (1999) *NATO Policy on Non-Lethal Weapons* (Brussels: North-Atlantic Council).
Northern Ireland Office (2004) Patten Report Recommendations 69 and 70 relating to Public Order Equipment – A Research Programme into Alternative Policing Approaches Towards the Management of Conflict (Fourth Report), available at: http://www.nio.gov.uk/index/nio-publication.htm; Internet.
Rademaker, M. (2003) *Technology Overview of Non-Lethal Weapons* (Den Haag: TNO-FEL).
Rappert, B. (2002) Towards an Understanding of Non-Lethality, in: N. Lewer (Ed.), *The Future of Non-Lethal Weapons – Technologies, Operations, Ethics and Law*, pp. 53–74 (London: Frank Cass).
Siniscalchi, J. (1998) *Non-Lethal Technologies: Implications for Military Strategy*, Occasional Paper No. 3 (Center for Strategy and Technology, Air War College, Air University, Alabama).
Smit, W. A., Grin, J. & Voronkov, L. S. (1992) Assessing Military–Technological Development Paths: Examples and Methodologies, in: W. Smit, J. Grin & L. Voronkov (Eds), *Military Technological Innovation and Stability in a Changing World*, pp. 161–166 (Amsterdam: VU University Press).
Smith, R. (2006) *The Utility of Force – The Art of War in the Modern World* (London: Penguin).
Sun Tzu [ca. 500BC] (1963) *The Art of War*, transl. S. Griffith (New York: Oxford University Press).
Sutherland, R. G. (2008) *Chemical and Biological Weapons – Political and Technical Aspects* (Stockholm, SIPRI Policy Paper No. 23).
Thornton, R. (2007) Getting it Wrong: The Crucial Mistakes Made in the Early Stages of the British Army's Deployment in Northern Ireland (August 1969 to March 1972), *The Journal of Strategic Studies*, 30(1), pp. 73–107.
Toffler, A. & Toffler, H. (1994) *War and Anti-War – Survival at the Dawn of the 21st Century* (London: Warner).
US Department of Defense Joint Non-Lethal Weapons Program (2007), accessed 20 October 2007, available at: https://www.jnlwp.com/ads.asp; Internet.
Vilke, G. & Chan, T. (2007) Less Lethal Technology: Medical Issues, *Policing: An International Journal of Police Strategies & Management*, 30(3), pp. 341–357.
Von Clausewitz, C. [1831] (1984) *On War*, M. Howard & P. Paret (Eds) (Princeton, NJ: Princeton University Press).
Wright, S. (2002) The Role of Sublethal Weapons in Human Rights Abuse, in: N. Lewer (Ed.), *The Future of Non-Lethal Weapons – Technologies, Operations, Ethics and Law*, pp. 75–86 (London: Frank Cass).

Biography

Sjef Orbons joined the Ministry of Defence of the Netherlands in 1989. He is currently a researcher at the Netherlands Defence Academy and preparing a PhD thesis on the topic of non-lethality. Previous staff positions include the Department of Defence Concepts of the MOD and the NATO Defense College in Rome. He received his Master of Science from Eindhoven University of Technology.

Ethical Blowback from Emerging Technologies

PATRICK LIN
Ethics + Emerging Sciences Group, Philosophy Department, California Polytechnic State University, San Luis Obispo, USA

ABSTRACT *The military is a major driver of technological, world-changing innovations which, like the Internet, often have unpredictable dual uses and widespread civilian impact ('blowback'). Ethical and policy concerns arising from such technologies, therefore, are not limited to military affairs, but can have great implications for society at large as well. This paper will focus on two technology areas making headlines at present: human enhancement technologies and robotics, representing both biological and technological upgrades to the military. The concerns we will raise are about (1) the unintended effects that these technologies could have on society and (2) the possibly self-defeating nature of these military programs in making war itself more, rather than less likely to occur. Given that society also can influence military programs (e.g., through civil protests) it is important for the military to attend to such concerns.*

Introduction

The first ethics concern with any new military weapon or system is usually whether its development or use would violate existing international law. However, emerging technologies can also raise much larger worries, given their revolutionary nature. For instance, using robots for dangerous jobs promises to help reduce the number of casualties on the deploying side. But might this effect prove counterproductive to the larger goal of peace, to the extent that these machines also lower the political costs associated with declaring war, and so make engaging in war a preferred or convenient method of conflict resolution easier rather than more difficult? Toward the same goal of soldier survivability and reducing abuses and accidents, at some point we may be able to eliminate human emotions such as fear or anger or hatred – but must these enhancements (and others, such as super-strength) be temporary or reversible, considering that warfighters usually return to civilian life?

These and other questions not only determine how such technologies might be used by the military, but how they might also impact society at large, should we end up, for instance, entering into ever more armed conflicts because they are 'risk free' or, in quite another sense, if armies of biologically enhanced and advantaged veterans suddenly flooded back into the civilian workforce. *'Blowback'* is the term frequently invoked to draw attention to these unintended and largely unanticipated effects that radical or innovative new technologies originally designed for military purposes might come to exert on civilian life. Given that the feedback loop also routes back in the other direction – from social criticism to military changes – the broader impacts of war technologies should be a concern for the military. To that end, this paper will examine the intertwined relationship between the military and society, particularly how the use of human enhancement (also called human performance optimization) technologies and advanced robotics in war may cause a blowback of significant and unintended effects that humanity will need to confront.

Enhancement technologies and robotics are particularly significant, because they represent opposite directions to the same goal of creating a super-warfighter: one is an engineering approach that works with mechanized systems, and the other is a biomedical approach that works with organic systems. Our discussion also applies broadly to other emerging military technologies, such as to both non-lethal and cyberweapons. However, we will focus on robotics and enhancement technologies here, since they more directly impact the human dimension – by either replacing human combatants from the theater of war or upgrading them.

The Role of the Military in Society

To appreciate the ethical threads that run through both society and the military, let us first briefly acknowledge the relationship between the two. Since the beginning of human history, we have been engaged in violent conflicts of varying scale, from individual fights to tribal battles to multinational wars. We would like to think that rational individuals as well as societies strongly prefer to live in peace rather than in war, which has terrible costs. Even a purely self-interested person, who may naturally want to gain power and property, can understand that the benefits of peace outweigh the disvalue or 'incommodities' of war (Hobbes 1651). But we also recognize that sometimes it may be necessary to go to war, for instance, in self-defense, and that not everyone is rational, at least to our standards. In that regard, a standing military is useful to defend ourselves, even if its historic use includes ignoble purposes such as conquest and empire expansion.

Around 350 BC, Aristotle pronounced, '[We] make war that we may live in peace' (Aristotle, *Nicomachean Ethics*: X.7), variations of which are attributed also to great military leaders from Julius Caesar to Carl von Clausewitz. If that is true, then the fates of the military and society are inextricably interwoven. Of course, this relationship also has been disputed: Nobel laureate Albert Einstein insisted, 'You cannot simultaneously prevent and prepare for war' (Nathan & Norden 1960). We need not resolve the debate here, since the important point for this discussion is merely that the military plays an influential role in society as a matter of fact. Further, neither Einstein nor others seem to dispute the basic point that the purpose of war is – or ought to be – to achieve peace, as implied by Aristotle, Clausewitz, and others. As Hobbes recognized, society cannot endure in a constant state of war.

In the United States, for example, the relationship between the military and society is substantial. The military and defense-related programs in the US – which ranks first in national military spending, with China a distant second place with about one-seventh of the US spending level – command nearly a third of the entire national budget (Office of Management and Budget 2010; GlobalSecurity.org 2010). The 2010 fiscal year budget for those activities is approximately $1 trillion and accounts for about 40 percent of US taxpayer dollars (Internal Revenue Service 2010).[1] A significant portion of this funding is directed at research and development (R&D). The Fiscal Year 2010 US budget for R&D is approximately $143 billion, of which about 60 percent goes toward national defense, not even including related areas such as space technologies, basic science, and energy. In comparison, only 20 per cent of the total R&D budget is earmarked for health-related projects (Office of Management and Budget 2010).

These staggering sums are perhaps understandable: Insomuch as life is a continuous struggle, one of our most basic and greatest needs is security – from the cold and rain, from hunger, and from enemies, both real and perceived. At least for the moment, we have solved many of these insecurities, such as with housing, heating, and food science, and we are making steady progress in medicine. But the intentions and attitudes of our enemies continue to vex us; the consequences of losing a conflict are usually dramatic; and the temptations of power are most difficult to resist. For many, these and other influences make heavy investments in military R&D seem not only sensible but also urgent and necessary.

Ethical and policy issues arising from the use of military force, or from the development of new military technologies, can also spill over to the civilian realm. This is the part of the feedback loop between military and society on which we will primarily focus. A second part of the feedback loop, however, is that society can also effect changes in the military: Public opinion is a significant force and can manifest itself in several forms, whether these be anti-war protests (e.g., protesting the US-led war in Iraq, or the detention and interrogation of suspected terrorists at Guantanamo Bay Naval Base), or significant public policy movements (such as the grassroots movement to effect an international ban against the production and use of landmines by the United Nations in 1997), or, at present, to regulate the use of unmanned aerial vehicles (UAVs) or military robotic aircraft (Bowcott 2010). Though important, we will not be concerned with the dynamics of this second part of the feedback loop here.

Military Technologies

It would be difficult to imagine our modern life apart from the numerous inventions originally inspired, for better or worse, primarily for military purposes. For instance, it is well known that the Internet has deep roots in the military's ARPANET (Advanced Research Projects Agency Network), a packet-switching network developed in the 1960s by the US to allow communications in the event that a nuclear attack disabled the telephone and telegraph circuit-switching networks that were then the principal means of communication. The global positioning system (GPS) was also created by the US Department of Defense in the 1970s as a necessary improvement on navigation systems, especially for ballistic missiles, and is now integrated into automobiles, mobile phones, and other computing devices worldwide. In fact, the ubiquitous computer itself has origins in military necessity: Germany's Z3 (1941), Britain's Colossus (1943), the Harvard Mark 1 (1944), ENIAC (1946), and others performed number-crunching operations that humans could not easily or ever do, such as aerodynamics analysis, codebreaking, thermonuclear calculations for the design of the hydrogen bomb, and more.

The list of other technologies invented or greatly improved upon by the military is long, but to mention a few more here: communications technologies, including walkie-talkies and radio telephony (to contact ships and airplanes from land), as some of the first attempts at mobile or wireless communication; nuclear power, as introduced by nuclear weapons, both fission and fusion; microwave power, which was accidentally discovered by a Raytheon engineer working on an active radar set; radar, which comes from the US Navy's acronym RADAR, or radio detection and ranging, but developed in the 1930s by several nations independently and in great secrecy; medicine, including the practice of triage, surgical advances, and prosthetics; and gunpowder, invented in ancient China. And, of course, the military extensively uses and enhances vehicles for land, air, sea, and space, including rockets and missiles. For instance, the first automobile is believed to have been a French military tractor created in 1769, and several auto companies today can trace their lineage to military operations, such as BMW, Saab, and Jeep.

Whether or not these technologies would have been invented or advanced anyway, without military research, is of course a matter of considerable historical debate. Notwithstanding, for the purposes of this paper, it is enough that we acknowledge that society today is, in fact, very much shaped by technologies from the proving grounds of the military. So to that extent, we can expect new and emerging military technologies to make their way into society. And where technology gives rise to challenges that society must confront – from mutually assured destruction (MAD) with nuclear weapons to intellectual property, privacy, 'cyberbullying', and many other issues with the Internet – by looking at military research today, we can begin to identify and work through the ethical and policy challenges they will bring tomorrow.

Today, we find a broad range of inventive and strategic projects funded by the US military, such as: cyborg insects; robots that 'eat' (that is, derive energy from ingesting local waste products, or even human corpses constituting battlefield casualties); energy weapons; telepathic communication; quantum computing; submersible aircraft; exoskeletons; enhanced warfighters; dynamic armor; invisible shields and cloaks; accelerated therapeutics; real-time language translations; and programmable matter (DARPA 2010a,b). For the sake of brevity, as mentioned earlier, we will focus only on human enhancement and robotics technologies in the analysis that follows.

Human Enhancement

Broadly understood, human enhancement or optimization is about endowing an individual with an ability that goes beyond the typical level or statistically normal range of functioning for *Homo sapiens* (Daniels 2000). For instance, anabolic steroids may give an individual greater strength and speed than most others (at least of the same size, build, etc.), thereby enabling the competitor to break athletic records. We may also understand human enhancement to be distinct from therapy, which is about treatments aimed at pathologies that compromise health or reduce one's level of functioning below this species-typical or statistically normal level (Juengst 1997). Thus, anabolic steroids given to a muscular dystrophy patient would be a form of therapy and not human enhancement, since the goal is to help return the patient to a level of strength that he otherwise would have had without the affliction.

And while many instances of human enhancement and therapy may be biomedical in nature, for example, Ritalin or modafinil which is used by patients with attention-deficit hyperactivity disorder (ADHD) as well as by otherwise healthy students who want to focus more intently on their studies, some may be technological, for example, the same carbon-fiber legs given to South African runner Oscar Pistorius could also be surgically attached to a healthy runner looking for a competitive edge. This suggests another distinction. Human enhancement technologies are not just mere external tools, such as a desktop computer or binoculars. Instead, they are either integrated into the body or so closely worn or connected that they confer a similar 24/7, 'always-on' advantage as an internal or organic enhancement that transforms the person.

As critics have noted, however, these and other conceptions of human enhancement may have definitional exceptions. For instance, would vaccinations count as (preventive?) therapy or an enhancement to the immune system? There are good reasons to support either answer, suggesting that the line between enhancement and therapy is blurry (Bostrom & Roache 2009). We will not address this continuing controversy here, but let us proceed on the assumption that 'human enhancement' and 'human optimization' are intelligible concepts and delineate at least some technological or biomedical applications from their therapeutic counterparts, including the ones that the military categorizes as such.

With that understanding, the US military supports a host of research projects designed to optimize or enhance the warfighter, including (DARPA 2010b, unless otherwise noted):

- Accelerated Learning: Best practices in learning through neuroscience and statistical modeling and analysis.

- Crystalline Cellulose Conversion to Glucose (C3G): To enable humans to eat grass and other currently non-digestible plants.
- Enabling Stress Resistance: Molecular neurobiology, neuro-imaging, and molecular pathway modeling to reduce stress and its negative effects.
- Exoskeleton for Human Performance Augmentation (EHPA): Mechanized suit to give humans super-strength and endurance, for example, in lifting (Heary 2010).
- Human-aided Optical Recognition/Notification of Elusive Threats (HORNET): Neuro-optical binoculars that can detect threats, even if the user does not consciously recognize them as such (Northrop Grumman 2010).
- Peak Soldier Performance (Metabolic Dominance): A program to enable humans to maintain high performance levels with, for example, a bodysuit to reduce core body temperature and a nutraceutical or dietary supplement to boost health and immunity (DARPA 2009).
- PowerSwim: Underwater propulsion devices to enable humans to swim like fish and penguins.
- RealNose: Sensors that can detect scents and chemicals as accurately as a dog can.
- Synthetic Telepathy: Brain-computer interface that records, transmits, and translates EEG (electroencephalogram) brain scans to enable communication by thought alone, that is, without visual means or audible speech (UC Irvine 2009).
- Z-Man: To enable humans to climb up walls like lizards.

While these and other projects may currently be in a research phase or involve devices too large to integrate with the human body, technologies such as nanoelectronics are shrinking the size of computers, making it foreseeable that these devices eventually can be embedded or wearable, thus falling in the domain of human enhancement, as opposed to mere tools. Innovations in the civilian sector may have dual-use potential, serving also military purposes, such as an electronics-packed contact lens that can acts as a digital display for augmented reality (Parviz 2009), in addition to advanced prosthetics and pharmaceuticals already mentioned, such as carbon-fiber legs and Ritalin.

Robotics

Robotics[2] too poses a definitional challenge, as obvious as a definition might seem to the layperson. There is still a lack of consensus among roboticists on how they define the object of their craft. For instance, an intuitive definition could be that a robot is merely a computer with sensors and actuators that allow it to interact with the external world; however, any computer that is connected to a printer or can eject a CD might qualify as a robot under that definition, yet few roboticists would defend that implication. A plausible definition, therefore, needs to be more precise and distinguish robots from mere computers and other devices.

As with human enhancement, we do not presume we can resolve this great debate here, but it is important that we offer a working definition. In its most basic sense, we understand 'robot' to be an engineered machine that senses, thinks, and acts:

> Thus a robot must have sensors, processing ability that emulates some aspects of cognition, and actuators. Sensors are needed to obtain information from the environment. Reactive behaviors (like the stretch reflex in humans) do not require any deep cognitive ability, but on-board intelligence is necessary if the robot is to perform significant tasks autonomously, and actuation is needed to enable the robot to exert forces upon the environment. Generally, these forces will result in motion of the entire robot or one of its elements, such as an arm, a leg, or awheel. (Bekey 2005).

This definition does not imply that a robot must be electromechanical; it leaves open the possibility of biological robots, as well as virtual or software ones. But it does rule out as robots any *fully* remote-controlled machines, since those devices do not 'think', for example, many animatronics and children's toys. That is, most of these toys do not make decisions for themselves; they depend on human input or an outside actor. Rather, the generally accepted idea of a robot depends critically on the notion that it exhibits some degree of autonomy or can 'think' for itself, making its own decisions to act upon the environment.

Thus, the US Air Force's Predator unmanned aerial vehicle (UAV), though mostly teleoperated by humans, makes some navigational decisions on its own and therefore would count as a robot. By the same definition, the following things are not robots: conventional landmines, toasters, adding machines, coffee makers, and other ordinary devices. Weaponized machines such as TALON SWORDS and Modular Advanced Armed Robotic System (MAARS) also do not count as robots, properly understood, since they do not seem to have any degree of autonomy, as their manufacturer admits (Qinetiq 2010).

War robots with fierce names, for example, Harpy and Global Hawk, currently perform a range of duties in the military theater, such as spying or surveillance (air, land, underwater, space), defusing bombs, assisting the wounded, inspecting hideouts, and attacking targets – in other words, they perform the 'three Ds': jobs that are dull, dirty, or dangerous. Those used or under development by the US military include the following (US Dept. of Defense 2009, unless otherwise noted):

- Crusher: A six-and-a-half ton unmanned ground vehicle (UGV) that is armored and armed, capable of navigating extreme terrain, including through walls.
- Battlefield Extraction-Assist Robot (BEAR): A UGV that can lift and retrieve combat casualties, carrying them out of harm's way.
- BigDog: Quadruped robot that can traverse difficult terrain, such as rubble. The much smaller version is the LittleDog (Boston Dynamics 2010a). The bipedal and anthropomorphized or humanoid version is the PETMAN (Boston Dynamics 2010b).

- PackBot: This family of robots includes models that famously have saved thousands of lives by detecting and defusing improvised explosive devices (IEDs).
- Dragon Runner: A surveillance ground robot designed for urban environments, such as operating in buildings.
- Reaper: A long-distance, high-altitude surveillance UAV that is armed with Hellfire anti-armor missiles. The smaller version is the Predator – both have gained the most attention, given their current and controversial use in Iraq, Afghanistan, and Pakistan.
- Hummingbird: A tiny robot or 'nano-UAV', at 10 grams or the weight of two nickels, that is highly maneuverable and can be used for surveillance in urban environments (AeroVironment 2009).
- Seahorse: An unmanned underwater vehicle (UUV) launched from torpedo tubes that can conduct oceanographic surveying and bottom mapping.
- Seafox: An unmanned surface vehicle (USV) or robotic boat that can be used to patrol waterways, detect mines, rescue overboard sailors or pilots, and other missions.
- Surface Mine Countermeasures (SMCM): Mine-detecting UUV, launched from torpedo tubes.
- Phalanx Close-In Weapons System (CIWS): An immobile or fixed robot, mounted on board a ship and with a full-auto mode, to identify, track, and shoot down incoming missiles. It is similar to the US Army's Counter Rocket, Artillery, and Mortar (C-RAM) robot.
- Chemical Robot (ChemBot): A soft, flexible, and amorphous robot, also called a 'blob-bot', that can identify and squeeze through small openings and then return to its previous size and shape.

Other nations are already using sentry robots to guard borders, for instance, in Israel and South Korea. These robots will challenge trespassers that they encounter and, if proper identification is not provided, can shoot them (Telegraph editors 2010). We can also expect space robots that serve military purposes in the future, if they are not already now (Singer 2009). And, intersecting with human enhancement, research with robot-human integrations or cyborgs can be seen today with advanced prosthetics, such as a bionic knee (Zamiska 2004), so we can expect that to increase in the future as well.

Again, human enhancement and robotics can be considered as two different approaches to the same goal of creating a better warfighter, one from a biological starting point and the other from a mechanized one. And this makes their ethical issues of special interest, to the extent that they are about an important human dimension of war, as discussed next.

Some Issues within Military Technology Ethics

Military technology ethics is not a new area of inquiry. For instance, in 1139, Pope Innocent II banned the use of crossbows, calling them 'hated by God', because they were so devastating, even by an untrained fighter, against the powerful, noble, and revered knight in mail armor (Headrick 2009). The ban, as we know, was ignored. In 1215, article 51 in the Magna Carta singled out crossbows as threats to society: 'As soon as peace is restored, we will banish from the kingdom all foreign-born knights, crossbowmen, sergeants, and mercenary soldiers who have come with horses and arms to the kingdom's hurt' (Constitution Society 2010). The crossbow, apparently, was a moral abomination that made warfare into a depersonalized and dishonorable 'point-and-click' affair, disrupting the existing code of ethics and balance of power.

With respect to human enhancement and robotics, some questions already have been raised in the areas of risk, policy, ethics, and philosophy. What, for example, are the health risks of military human enhancements such as steroids, Ritalin, and anthrax vaccines (Allhoff et al. 2009)? What is the safety risk of autonomous and armed robots, given reports of accidental deaths (Shachtman 2007) and wayward UAVs veering toward the US capital (Bumiller 2010)? How do military robots force us to reconsider policy, for example, the requirement to have 'eyes on target' – need they be human eyes (Lin et al. 2008)? Should consent be needed for military human enhancements, such as anthrax vaccines (Cummings 2002)? Do we become *less than human* with some forms of human enhancement, for example, drugs that can remove emotions such as fear (President's Council on Bioethics 2003)?

While there is much literature on the ethics of human enhancement (e.g., Savulescu & Bostrom 2009; Sandel 2007; Garreau 2005; Mehlman 2009), little discussion exists within a military context (e.g., Moreno 2006; Wolfendale 2008). The literature on robot ethics is growing, a good deal of which is about military applications (e.g., Sharkey 2007; Asaro 2008; Sparrow 2007; Borenstein 2008). As I mentioned above, the two 'big picture' issues that possibly arise from military human enhancement and robotics, which are also relevant to other technologies, are: (1) an unintended spillover of effects from military to society, given the close link between the two, and (2) the possibility of exacerbating a culture of war, which might prove contrary to peace as the goal of war in the first place.

Civilian Blowback

Society can be affected by military technologies in various ways. As it evolved from the military's original ARPANET, the Internet was adopted and built upon by civilian users, ushering in ubiquitous functions from email to marketing and shopping as we know them today. GPS did not need to evolve, but was simply incorporated into computing devices for civilian purposes, such as traffic navigation and location-based services. Military Jeeps and Humvees became trendy for the everyday driver. Nuclear weapons have fortunately not yet been adopted in any meaningful sense by non-military users, but they continue to cast a shadow of the threat of horrific harm across society, and this anxiety and fear have economic, political, and social costs.

Human enhancement technologies, however, may infiltrate society in an altogether different way. Depending on how they are implemented, enhancements that change the human warfighter's body or mind – whether some drug or, in the future, a neural implant or microcomputer embedded in one's head – ought to be a concern, because every warfighter is a potential veteran, returning to society from military service. Approximately 23 million veterans today live in the US, accounting for about one in 10 adults, plus three million active and reserve military personnel (US Dept. of Veteran Affairs 2010). Thus, the civilian blowback from military human enhancements may be significant and direct, given the size of the veteran population alone, even beyond the blowback that might be anticipated from adoption or development of dual-use capacities of such technologies by other civilians for non-military purposes.

For instance, suppose that some warfighters have been enhanced with prosthetics or gene therapy or pharmaceuticals that gave them super-strength. Assuming that many of these individuals will eventually return to the civilian workforce, we would suddenly be faced with a new class of individuals in society with that unique ability (again depending on specific details of the technology's implementation), representing a competitive advantage that may be disruptive to society (e.g., by displacing other civilian workers or skewing athletic contests). Other enhancements may raise similar issues: A job candidate with a neural implant that enables better data retention, faster information processing, or instant access to Google may consistently beat unenhanced candidates; a person with superhuman hearing or sight could circumvent existing privacy protections and expectations by easily and undetectably eavesdropping or spying on others; more students (and professors) using Ritalin may grab admission or tenure at all the best universities, reducing those opportunities for others; life-extension technologies may bankrupt current pension plans and increase pressure on energy, food, and jobs; and so on (Allhoff et al. 2009).

Thus, the precise manner in which military human enhancements are rolled out or implemented is important to consider, since some plans may impact society more than others. Some of the questions that need to be investigated include:[3]

1. As a firewall for society, should some enhancements be restricted to only special or elite teams, and which types of enhancements and why? Could and should these elite teams be prohibited from returning to society? Does being enhanced require a change either way in the length of a warfighter's service commitment? Is the ability to remain enhanced after leaving the service a legitimate recruiting incentive?

2. Must the enhancements be reversible? If so, should they be reversed routinely upon discharge? Should the warfighter be able to refuse to have them reversed upon discharge? If the enhancements are not reversed, what effects will they have on the ability of the enhanced warfighter to make a successful return to civilian life, and how can adverse effects be minimized?

3. What effect will this have on the US Department of Veterans Affairs (VA), insofar as enhancements are service-connected? Legally and ethically, how safe should these technologies be before they are deployed? What are the additional costs expected for VA hospitals to deal with adverse health effects from enhancements?

4. What kinds of social disruptions can be expected from enhanced warfighters? How do we evaluate issues of fairness, equity, and access (as alluded to in the above scenarios)? Would there be a divide between the enhanced and unenhanced, similar to that between the rich and the poor, and with what effects?

Military robots also pose spillover effects – working their way back home from deployment abroad onto a nation's own borders and eventually within the nation itself (Calo 2010a). As with Jeeps and Humvees, there seems to be a natural migration path of military robots to civil robots, especially those that would perform the same functions of security and surveillance. Police and security robots today are guarding borders and buildings, assisting in hostage situations, scanning for pedophiles and criminals, dispensing helpful information, reciting warnings, and more (Sharkey 2008). There is also a growing market for home-security robots, which can shoot pepper spray or paintball pellets and transmit pictures of suspicious activities to their owners' mobile phones (e.g., Wang 2009).

In the future, we can expect both military and civilian robots – again, driven in large part by military research – to be more autonomous, networked, and sensitive (Sharkey 2008). For instance, biometric tools can enable robots to identify individuals by recognizing faces, retina scans, or fingerprints. Molecular sensors can enable robots to sniff out drugs as well as suspects. Surveillance technologies can enable robots to detect weapons at a distance, see through walls, and eavesdrop on conversations. Integration with databases, for example, financial, medical, driving, and shopping, can enable robots to immediately pull up personal information on individuals they encounter. These scenarios, however, raise challenges to civil liberties and civil law, such as the following:

5. Would greater powers of surveillance by robots count as searches? Currently in the US, a search usually requires a judicial warrant and therefore has some oversight, and mere surveillance is not considered to be a search (e.g., *Katz v. US* 1967; *US v. Knotts* 1983). But a robot that can detect weapons or drugs hidden on a person's body – which cannot be detected by usual methods, namely, by the naked human eye – may force us to reconsider the distinction between surveillance and search.

6. How do such civil robots as well as home robots impact current notions of privacy, if they can be outfitted with surveillance and recording capabilities? Networked robots may allow the government to access homes and other spaces that have been historically protected, and

humans may be more likely to disclose information and grant access to an anthropomorphic robot than to other technologies (Calo 2010b).

7. On the battlefield, militaries are immunized from certain unintended killings of noncombatants, or collateral damage, and weapons manufacturers also are released from liability to a great extent; but in civil society, those immunities derived from international law, for example, Geneva Conventions, would no longer apply. So who would be liable for malfunctions, errors, and unpredictability with civil security robots, from accidentally running over children to firing upon misidentified individuals?

8. While replacing human soldiers with robotic ones seems to be desirable given potential (friendly) lives saved, replacing human workers in society with robotic ones may be more problematic, if those humans cannot find other work. So what is the economic impact of advanced robotics in society, whether in security roles or performing other labor? What human jobs might become obsolete?

9. What is the environmental impact of a growing robotics industry? In the computer industry, 'e-waste' is a growing and urgent problem (e.g., O'Donoghue 2010), given the disposal of heavy metals and toxic materials in the devices at the end of their product lifecycle. Robots as embodied computers will likely exacerbate the problem, as well as increase pressure on rare-earth elements needed today to build computing devices and energy resources needed to power them.

These represent only some of the civilian-blowback issues posed by the emerging technologies. By themselves, they are not arguments that we should not develop robotics and human enhancement technologies. For instance, the Internet created a host of issues related to intellectual property, privacy, information security, brick-and-mortar retailers, and so on, yet these are not persuasive reasons to think that we should never have developed the Internet. Rather, they are simply challenges that society needs to confront, and anticipatory thinking on these issues may help to mitigate negative effects.

Culture of War

A key issue that does suggest we should not pursue such technologies, however, has already been levied against military robotics. Some critics have argued that because the primary point of military robotics is to remove risk to ourselves in war, this lowers the barriers to starting or engaging in wars in the first place, including social, economic, and political costs from incurring casualties (e.g., Sharkey 2007; Asaro 2008; Sparrow 2007; Borenstein 2008). (A similar charge could be made against human enhancement technologies, though as far as we can tell, this has not yet occurred, perhaps because the advantage gained is incremental and not as dramatic as replacing human warfighters entirely with machines.)

Just war theory, and *jus ad bellum* specifically, is arguably premised on the fact that war is a moral disaster and should be engaged in only as a last resort (Walzer 1977). American Civil War general Robert E. Lee affirmed this sentiment: 'It is well that war is so terrible, or we would grow too fond of it' (Alexander 1907). If war were not so terrible – say, with robots and enhanced warfighters that can more quickly achieve victory with fewer friendly casualties – then we lose an important disincentive to engage in war, so the argument goes. This seems to be supported by human psychology, that we respond to positive and negative incentives, as at least a partial explanation of our motivations and behavior (Skinner 1953). Indeed, classical economics is built on such an assumption, for example, that demand for a product (or, in this case, for war) will decrease as its price goes up (Marshall 1890).

A problem with this argument, though, is that it may imply that war ought to be as terrible as possible, to more certainly ensure that we engage in it as a last resort (Lin et al. 2008). That is, we should not do anything that makes armed conflict more palatable, which includes reducing friendly casualties, improving battlefield medicine, and conducting any more robotics and human-enhancement research that would make victory more likely and quicker. Taken to the extreme, the argument seems to imply that we should *raise* barriers to war, to make fighting as brutal as possible, such as using primitive weapons without armor or even acid attacks, so that we would never engage in it unless it were truly the last resort. However, such a position appears counterintuitive at best and dangerously foolish at worst, particularly if we expect that other nations would not readily adopt a policy of relinquishment, which would put our side at a competitive disadvantage. Thus, it is unclear how literally or to what degree we should take the argument – are there good reasons not to make the cost of war as high as possible?

This problem aside, we can agree that weapons proliferation is real: As any given side develops some war technology, others will soon develop the same capability – negating whatever competitive advantage was initially created. For instance, while much of the international media focuses on the US military robots in the Middle East, more than 40 other nations have them as well, including China, Russia, Pakistan, and Iran (Singer 2009; Harris 2009). This proliferation too contributes to a culture of war, continuing a cycle of armed conflicts that can be won with the (temporarily) new technologies and keeping international tensions high.

This is a paradox. No one, we presume, really wants war; rather, we prefer to live in peace. Yet it nevertheless seems sensible in the interim for every nation to develop or acquire offensive weapons, such as tanks, and defensive tools like armor, with the aim of strengthening and protecting itself in an often hostile world. However, the relentless race for new and better weapons, by making war itself easier to resort to, could hold a lasting, true peace out of reach. This schizophrenic culture of war then raises questions, including the six following:

1. Ought we to develop military robotics and human enhancement technologies at all? Clearly, the technologies offer tremendous benefits on the battlefield, but that is beside the larger point here, which is about rational consistency in the enterprise of warfare. Are there moral limits to such research that we should observe? Even if we agree that, morally speaking, we should not pursue such military research, is it practical to stop – is it too late, having already opened Pandora's box?

2. How would the laws of war and rules of engagement need to evolve or be clarified to account for new technologies? For example, do current international laws pertaining to the treatment of prisoners of war also apply to enhanced warfighters who can tolerate greater physical or mental abuse? The Predator UAV pilots in Las Vegas, half a world away from the robots they operate, would seem to be classified as combatants and therefore lawful targets of attack even when they are at home with family, after their working shifts have ended and they have left the military base – does this align with our intuitions, namely, does the 'part-time' character of this job challenge existing rules on who may be targeted and when? This impacts society, in that international law may allow for enemies to attack the US in certain locations such as Las Vegas – bringing the 'War on Terrorism' back to American soil.

3. The use of military robots by the US is seen as cowardly and dishonorable by enemies in the Middle East, and this is expected to fuel resentment, anger, and hatred – and therefore recruitment efforts – by those enemies (Singer 2009; Shachtman 2009). This too appears to be counterproductive to the larger goal of lasting peace. Would enhanced warfighters be viewed the same way, and what effect would this have on mission success? Are there reasons to prohibit militaries from giving warfighters unnatural and inhuman enhancements, such as horns or an otherwise frightening appearance, to instill fear in the enemy, cf. Viking and Japanese samurai helmets?

4. If people respond to incentives, then it would seem that technologies that give one side a sizable competitive advantage would act as a negative incentive or deterrent for the other side against starting, provoking, or otherwise engaging in armed conflicts. For instance, nuclear weapons seemed to have prevented, at least, total wars if not some conflicts altogether. Other technologies, such as machine guns or Tomahawk cruise missiles, cannot be obviously credited with deterrence, since history shows a continuation of conflicts even with those weapons in play. So under what conditions are technologies effective deterrents – do robotics and human enhancement technologies satisfy those conditions? If they are not good deterrents, then they may further enable the cycle of war.

5. Relatedly, to the extent that new technologies are not effective deterrents, how might the enemy attempt to negate the effects of enhancements and

robotics? We cannot reasonably expect determined enemies to simply throw their hands up and surrender in the face of superior technology or even a superior military, and they are responding in new ways, such as attacking human UAV controllers or pilots, since it is so difficult to attack the UAVs in flight and potentially devastating to allow those UAVs even to leave the base (MSNBC editors 2010). Some enemies may resort to bringing the fight onto US soil, where the Predators are not, and this will have a chilling effect on society (Lin et al. 2008), for example, if terrorists work harder to obtain and detonate a 'dirty bomb' that could contaminate large cities, causing residents to flee.

6. What are the implications for space? We are on the verge of developing a commercial space industry, the first steps to extending society – giving us the means to leave the planet and perhaps re-colonize elsewhere. This is a long-term goal, and during which time we can expect military activity to increase in space, despite international restraint today. The existence of wars in space, including those in which robots are deployed, would be incredibly destructive to the progression of orderly space settlements and society. These problems may range from creating areas of conflict to more floating debris in space to land-grabs on other planets and moons, notwithstanding existing treaties.

Conclusion

The robotics industry is rapidly advancing and already giving rise to a great many additional questions in society of a non-military nature: for example, whether we are morally permitted to reassign our duty to care for elderly and children to robotic caretakers. The human enhancement industry, likewise, is quickly moving forward and creating new questions for society to confront: for example, whether cognition-boosting drugs such as modafinil should be allowed in schools or the workplace, echoing the steroids debate in sports. But as discussed above, many of the advances in these fields will come from the defense sector, as the military is often on the frontlines of technology innovation. How these technologies are actually developed and deployed by the military will, in turn, have ethical and policy implications for society, given a tight relationship between the military and society.

We need to attend to these issues now, rather than wait for the technologies to be developed first, as ethics is usually slow to catch up with science and technology, which can lead to a 'policy vacuum' (Moor 1985). Recall that the Human Genome Project was started in 1990, but it required the ensuing 18 years for the US Congress to finally pass a bill to protect citizens from discrimination based on their genetic information. Right now, society is still fumbling through privacy, copyright, and other intellectual property issues in the Digital Age, nearly 10 years since Napster and other peer-to-peer file-sharing sites were first shut down.

To be clear, this is not a call for a moratorium or even necessarily limiting the development of the technologies themselves. Military technologies promise great, perhaps irresistible, benefits for not only defense-related activities but also broader society as dual-use technologies. Yet we can already anticipate that they would disrupt current military and social systems. Disruption and change by themselves are not necessarily bad things. But failing to manage change and disruption may lead to negative effects that might have been mitigated with some forethought – including a public backlash against military programs as warranted, the second return path of the military-society feedback loop mentioned earlier. Alan Turing, one of the founding fathers of the technological age, foretold of our task: 'We can only see a short distance ahead, but we can see plenty there that needs to be done' (1950).

Acknowledgements

This work is supported by grants from The Greenwall Foundation and the US Office of Naval Research. Any opinions, findings, and conclusions or recommendations expressed in this material are those of the author and do not necessarily reflect the views of the funding organizations.

Notes

[1] To put this number in perspective, the wealthiest person in the US is currently Bill Gates, co-founder of Microsoft Corporation and the largest philanthropist in world history, whose net worth is estimated at $50 billion (Miller & Greenberg 2009) – which means that, every year, the US spends about 20 times Gates' entire fortune on defense programs.

[2] Some of the discussion in this section draws from my book in progress, *Robot Ethics: The Ethical and Social Implications of Robotics*, to be published by MIT Press in 2011 (Lin et al. forthcoming).

[3] Most of these questions arise from recent collaborative efforts and inquiry of the Consortium for Emerging Technologies, Military Operations, and National Security (CETMONS), headquartered at Arizona State University.

References

AeroVironment (2009) DARPA Awards AeroVironment Phase II Contract Extension for Nano Air Vehicle Development Program, accessed 19 September 2010, available at: http://www.businesswire.com/portal/site/google/?ndmViewId=news_view&newsId=20090701005345&newsLang=en; Internet.

Alexander, E.P. [1907] (1993) *Military Memoirs of a Confederate* (New York: De Capo Press).

Allhoff, F., Lin, P., Moor, J. & Weckert, J. (2009) *Ethics of Human Enhancement: 25 Questions & Answers*, a US National Science Foundation-funded report, accessed 19 September 2010, available at: http://www.humanenhance.com/NSF_report.pdf; Internet.

Aristotle [ca. 350 BC] (1998) *The Nicomachean Ethics*, Ross, W. D. (trans.), Akrill, J. L. & Urmson, J. O. (Eds), (Oxford: Oxford University Press).

Asaro, P. (2008) How Just Could a Robot War Be?, in: Briggle, A., Waelbers, K. & Brey, P. (Eds) *Current Issues in Computing and Philosophy* (Amsterdam: IOS Press).

Bekey, G. (2005) *Autonomous Robots: From Biological Inspiration to Implementation and Control* (Cambridge, MA: MIT Press).

Borenstein, J. (2008) The Ethics of Autonomous Military Robots, *Studies in Ethics, Law, and Technology*, 2(1): article 2, accessed 19 September 2010, available at http://www.bepress.com/selt/vol2/iss1/art2; Internet.

Boston Dynamics (2010a) LittleDog – The Legged Locomotion Learning Robot, accessed 19 September 2010, available at http://www.bostondynamics.com/robot_littledog.html; Internet.

Boston Dynamics (2010b) PETMAN – BigDog Gets a Big Brother, accessed 19 September 2010, available at http://www.bostondynamics.com/robot_petman.html; Internet.

Bostrom, N. & Roache, R. (2008) Ethical Issues in Human Enhancement, in: J. Ryberg, T. S. Petersen & C. Wolf (Eds), *New Waves in Applied Ethics* (New York: Palgrave Macmillan).

Bowcott, O. (2010) Robotic Warfare: Campaigners Call for Tighter Controls of Deadly Drones, *The Guardian*, accessed 19 September 2010, available at http://www.guardian.co.uk/technology/2010/sep/16/robot-warfare- conferences: Internet.

Bumiller, E. (2010) Navy Drone Violated Washington Airspace, *New York Times*, accessed 19 September 2010, available at http://www.nytimes.com/2010/08/26/us/26drone.html; Internet.

Calo, R. (2010a) Robots: The Law, podcast interview, accessed 19 September 2010, available at http://www.robotspodcast.com/podcast/2010/09/robots-the-law/; Internet.

Calo, R. (2010b) Robots and Privacy, in: Lin, P., Bekey, G. & Abney, K. (Eds) *Robot Ethics: The Ethical and Social Implications of Robotics* (Cambridge: MIT Press, forthcoming), accessed 19 September 2010, available at http://ssrn.com/abstract=1599189; Internet.

Constitution Society (2010) Magna Carta, accessed 19 September 2010, available at http://www.constitution.org/eng/magnacar.htm; Internet.

Cummings, M. L. (2002) Informed Consent and Investigational New Drug Abuses in the US Military, *Accountability in Research*, 9(2), pp. 93–103.

Daniels, N. (2000) Normal Functioning and the Treatment-Enhancement Distinction, *Cambridge Quarterly of Healthcare Ethics*, 9, pp. 309–322.

Defense Agency Research Projects Agency (2009) Interview with Jon Mogford, Program Manager, DSO, accessed 19 September 2010, available at http://www.darpa.mil/podcasttranscript_december2009.html; Internet.

Defense Agency Research Projects Agency (2010a) Ongoing Research Programs, accessed 19 September 2010, available at http://www.darpa.mil/off_programs.html; Internet.

Defense Agency Research Projects Agency (2010b), Programs (A–Z), accessed 19 September 2010, available at http://www.darpa.mil/dso/programs.htm; Internet.

Garreau, J. (2005) *Radical Evolution: The Promise and Peril of Enhancing Our Minds, Our Bodies – and What It Means to Be Human* (New York: Doubleday).

GlobalSecurity.org (2010) China's Defense Budget, accessed 19 September 2010, available at http://www.globalsecurity.org/military/world/china/budget.htm; Internet.

Harris, M. (2009) iRobot, Where the Terminator is Coming to Life, *Sunday Times*, accessed 19 September 2010, available at http://technology.timesonline.co.uk/tol/news/tech_and_web/article6389337.ece; Internet.

Headrick, D. (2009) *Technology: A World History* (New York: Oxford University Press).

Heary, J. (2010) Top 5 DARPA Projects of 2010, *Network World*, accessed 19 September 2010, available at http://www.networkworld.com/community/node/57992; Internet.

Hobbes, T. [1651] (2009) *Leviathan*, J. Gaskin (Ed), (New York: Oxford University Press).

Internal Revenue Service (2010) *Data Book, 2009* (Washington, DC: US Government Printing Office), accessed 19 September 2010, available at http://www.irs.gov/pub/irs-soi/09databk.pdf; Internet.

Juengst, E. (1997) Can Enhancement Be Distinguished from Prevention in Genetic Medicine? *Journal of Medicine and Philosophy*, 22, pp. 125–142.

Katz v. United States (1967) 389 US 347, accessed 19 September 2010, available at http://laws.findlaw.com/us/389/347.html; Internet.

Lin, P., Bekey, G. & Abney, K. (2008) *Autonomous Military Robotics: Risk, Ethics, and Design*, a US Department of Defense/Office of Naval Research-funded report, accessed 19 September 2010, available at http://ethics.calpoly.edu/ONR_report.pdf; Internet.

Lin, P., Bekey, G. & Abney, K. (forthcoming) *Robot Ethics: The Ethical and Social Implications of Robotics* (Cambridge, MA: MIT Press).

Marshall, A. [1890] (2006) *Principles of Economics* (New York: Cosimo).

Mehlman, M. (2009) *The Price of Perfection: Individualism and Society in the Era of Biomedical Enhancement* (Baltimore, MD: Johns Hopkins University Press).

Miller, M. & Greenberg, D. (Eds) (2009) The Richest People in America in Forbes.com, accessed 19 September 2010, available at http://www.forbes.com/lists/2009/54/rich-list-09_William-Gates-III_BH69.html; Internet.

Moor, J. (1985) What Is Computer Ethics?, *Metaphilosophy*, 16(4), pp. 266–275.

Moreno, J. (2006) *Mind Wars* (New York: Dana Press)).

MSNBC editors (2010) Al-Qaida: CIA Attack Avenges Drone Strikes, *MSNBC.com*, accessed 19 September 2010, available at http://www.msnbc.msn.com/id/34737634/; Internet.

Nathan, O. & Norden, H. (1960) *Einstein on Peace* (New York: Avenel Books).

Northrop Grumman (2010) Human-Aided Optical Recognition/Notification of Elusive Threats (HOR-NET), accessed 19 September 2010, available at http://www.es.northropgrumman.com/solutions/hornet/; Internet.

O'Donoghue, A.J. (2010) E-Waste is a Growing Issue for States, *Deseret News*, accessed 19 September 2010, available at http://www.deseretnews.com/article/700059360/E-waste-is-a-growing-issue-for-states.html?pg=1; Internet.

Office of Management and Budget (2010) *Budget of the US Government: Fiscal Year 2010* (Washington, DC: US Government Printing Office), accessed 19 September 2010, available at http://www.gpoaccess.gov/usbudget/fy10/pdf/fy10-newera.pdf; Internet.

Parviz, B. (2009) Augmented Reality in a Contact Lens, *IEEE Spectrum*, September 2009, accessed 19 September 2010, available at http://spectrum.ieee.org/biomedical/bionics/augmented-reality-in-a-contact-lens/0; Internet.

Plato [ca. 380 BC] (2004) *Republic*, trans. Reeve, C. D. C, (Indianapolis: Hackett Publishing Company).

President's Council on Bioethics (2003) *Beyond Therapy: Biotechnology and the Pursuit of Happiness* (Washington, DC: US Government Printing Office), accessed 19 September 2010, available at http://bioethics.georgetown.edu/pcbe/reports/beyondtherapy/beyond_therapy_final_webcorrected.pdf; Internet.

Qinetiq (2010) TALON Family of Military, Tactical, EOD, MAARS, CBRNE, Hazmat, SWAT and Dragon Runner Robots, accessed 19 September 2010, available at http://foster-miller.qinetiq-na.com/lemming.htm; Internet.

Sandel, M. (2007) *The Case Against Perfection: Ethics in the Age of Genetic Engineering* (Cambridge, MA: Belknap Press).

Savulescu, J. & Bostrom, N. (2009) *Human Enhancement* (New York: Oxford University Press).

Shachtman, N. (2007) Robot Cannon Kills 9, Wounds 14, *Wired.com*, accessed 19 September 2010, available at http://blog.wired.com/defense/2007/10/robot-cannon-ki.html; Internet.

Shachtman, N. (2009) Call Off Drone War, Influential US Advisor Says, *Wired.com*, accessed 19 September 2010, available at http://www.wired.com/dangerroom/2009/02/kilcullen-says/; Internet.

Sharkey, N. (2007) Automated Killers and the Computing Profession, *Computer*, 40, pp. 122–124.

Sharkey, N. (2008) *2084: Big Robot is Watching You*, a commissioned report, accessed 19 September 2010, available at http://staffwww.dcs.shef.ac.uk/people/N.Sharkey/Future%20robot%20policing%20report%20Final.doc; Internet.

Singer, P. W. (2009) *Wired for War: The Robotics Revolution and Conflict in the 21st Century* (New York: Penguin Press).

Skinner, B. F. (1953) *Science and Human Behavior* (New York: The Free Press).

Sparrow, R. (2007) Killer Robots, *Journal of Applied Philosophy*, 24(1), pp. 62–77.

Telegraph editors (2010) South Korea Deploys Robot Capable of Killing Intruders Along Border with North, *Telegraph.co.uk*, accessed 19 September, 2010, available at http://www.telegraph.co.uk/news/worldnews/asia/southkorea/7887217/South-Korea-deploys-robot-capable-of-killing-intruders-along-border-with-North.html; Internet.

Turing, A. (1950) Computing Machinery and Intelligence, *Mind*, 59(236), pp. 434–460.

UC Irvine (2009) MURI: Synthetic Telepathy, accessed 19 September 2010, available at http://cnslab.ss.uci.edu/muri/research.html; Internet.

United Nations (1997) Convention on the Prohibition of Use, Stockpiling, Production, and Transfer of Anti-Personnel Mines and on their Destruction, ratified in 1997, accessed 19 September 2010, available at http://www.un.org/Depts/mine/UNDocs/ban_trty.htm; Internet.

United States v. Knotts (1983) 460 US 276, accessed 19 September 2010, available at http://laws.findlaw.com/us/460/276.html; Internet.

US Department of Defense (2009) FY2009–2034 Unmanned Systems Integrated Roadmap (Washington, DC: US Government Printing Office), accessed 19 September 2010, available at http://www.acq.osd.mil/psa/docs/UMSIntegratedRoadmap2009.pdf; Internet.

US Department of Veteran Affairs (2010) VA Stats at a Glance, accessed 19 September 2010, available at http://www1.va.gov/VETDATA/Pocket-Card/4X6_summer10_sharepoint.pdf; Internet.

Walzer, M. (1977) *Just and Unjust Wars: A Moral Argument with Historical Illustrations* (New York: Basic Books).

Wang, V. (2009) Home Security Bots, *Popular Science*, accessed 19 September 2010, available at http://www.popsci.com/gear-amp-gadgets/article/2009-05/home-security-bots; Internet.

Wolfendale, J. (2008) Performance-Enhancing Technologies and Moral Responsibility in the Military, *American Journal of Bioethics*, 8(2), pp. 28–38.

Zamiska, N. (2004) Bionic Knee 'Learns' How to Walk, *Wall Street Journal*, accessed 19 September 2010, available at http://online.wsj.com/ad/article/philips/SB108907039283655627.html; Internet.

Biography

Dr Patrick Lin is the director of the Ethics + Emerging Sciences Group and an assistant philosophy professor at California Polytechnic State University, San Luis Obispo. He is currently an affiliate scholar at Stanford Law School's Centre for Internet and Society and a research fellow at Centre for Applied Philosophy and Public Ethics (CAPPE, Australia). Dr Lin was recently a fellow at the US Naval Academy's VADM Stockdale Center for Ethical Leadership and previously a post-doctoral associate at Dartmouth College, Hanover, New Hampshire.

The Case for Ethical Autonomy in Unmanned Systems

RONALD C. ARKIN[1]
Georgia Institute of Technology, Atlanta, Georgia, USA

ABSTRACT *The underlying thesis of the research in ethical autonomy for lethal autonomous unmanned systems is that they will potentially be capable of performing more ethically on the battlefield than are human soldiers. In this article this hypothesis is supported by ongoing and foreseen technological advances and perhaps equally important by an assessment of the fundamental ability of human warfighters in today's battlespace. If this goal of better-than-human performance is achieved, even if still imperfect, it can result in a reduction in noncombatant casualties and property damage consistent with adherence to the Laws of War as prescribed in international treaties and conventions, and is thus worth pursuing vigorously.*

Introduction

The trend is clear: warfare will continue and autonomous robots will ultimately be deployed in its conduct. Referring to the improving technology of the day and its impact on the inevitability of warfare, Clausewitz stated 'the tendency to destroy the adversary which lies at the bottom of the conception of War is in no way changed or modified through the progress of civilization' (Clausewitz [1832] 1976: 76). More recently, Martin Cook observed 'The fact that constraints of just war are routinely overridden is no more a proof of their falsity and irrelevance than the existence of immoral behavior "refutes" standards of morality: we know the standard, and we also know human beings fall short of that standard with depressing regularity' (Cook 2004). Given this, questions then arise regarding if and how these systems can conform as well or better than our soldiers with respect to adherence to the existing Laws of War. If achieved, this would result in a reduction in collateral damage, namely, noncombatant casualties and destruction of civilian property. The body of research conducted in our laboratory (Arkin 2009;

Arkin and Ulam 2009; Arkin et al. 2009) focuses on this issue directly from a design perspective. As robots are already faster, stronger, and in certain cases (e.g., chess playing) smarter than humans, is it that difficult to believe they will be able to treat us more humanely on the battlefield than we do each other?

This is no simple task, however. In the fog of war it is hard enough for a human to be able to effectively discriminate whether or not a target is legitimate. Fortunately for a variety of reasons, it may be anticipated, despite the current state of the art, that in the future autonomous robots may be able to perform better than humans under these conditions, for the following six reasons:

1. The ability to act conservatively: that is, robots do not need to protect themselves in cases of low certainty of target identification. Autonomous armed robotic vehicles do not need to have self-preservation as a foremost drive, if at all. They can be used in a self-sacrificing manner if needed and appropriate without reservation by a commanding officer. There is no need for a 'shoot first, ask-questions later' approach.
2. The eventual development and use of a broad range of robotic sensors better equipped for battlefield observations than humans currently possess. This includes technological advances in electro-optics, synthetic aperture or wall penetrating radars, acoustics, and seismic sensing, to name but a few.
3. Unmanned robotic systems can be designed without emotions that cloud their judgment or result in anger and frustration with ongoing battlefield events. In addition, 'Fear and hysteria are always latent in combat, often real, and they press us toward fearful measures and criminal behavior' (Walzer 1977). Autonomous agents need not suffer similarly.
4. Avoidance of the human psychological problem of 'scenario fulfillment' is possible, a factor believed partly contributing to the downing of an Iranian airliner by the USS *Vincennes* in 1988 (Sagan 1991). This phenomenon leads to distortion or neglect of contradictory information in stressful situations, where humans use new incoming information in ways that only fit their pre-existing belief patterns, a form of premature cognitive closure. Robots need not be vulnerable to such patterns of behavior.
5. They can integrate more information from more sources far faster before responding with lethal force than a human possibly could in real-time. This data can arise from multiple remote sensors and intelligence (including human) sources, as part of the US Army's network-centric warfare concept (McLoughlin 2006) and the concurrent development of the Global Information Grid (DARPA 2007). 'Military systems (including weapons) now on the horizon will be too fast, too small, too numerous and will create an environment too complex for humans to direct' (Adams 2002).
6. When working in a team of combined human soldiers and autonomous systems as an organic asset, they have the potential capability of

independently and objectively monitoring ethical behavior in the battlefield by all parties and reporting infractions that might be observed. This presence alone might possibly lead to a reduction in human ethical infractions.

Aside from these ethical considerations, autonomous robotic systems offer numerous other potential operational benefits to the military: faster, cheaper, better mission accomplishment; longer range, greater persistence, longer endurance, higher precision; faster target engagement; and immunity to chemical and biological weapons among others (Guetlein 2005). All of these can enhance mission effectiveness and serve as drivers for the ongoing deployment of these systems. But our research (Arkin 2009) focuses on enhancing ethical benefits by using these systems, ideally without eroding mission performance when compared to human warfighters.

Human Failings in the Battlefield

It is not my belief that an autonomous unmanned system will be able to be perfectly ethical in the battlefield, but I am convinced that they can perform more ethically than human soldiers are capable of performing. Unfortunately the trends in human behavior on the battlefield regarding adhering to legal and ethical requirements are questionable at best. 'Armies, armed groups, political and religious movements have been killing civilians since time immemorial' (Slim 2008: 3). The dangers of abuse of unmanned robotic systems, such as the Predator and Reaper, in war are well documented, which occurs even when a human operator is directly in charge (Sullivan 2010; Filkins 2010; Adams 2010). Battlefield atrocities[1] are as old as warfare. 'Atrocity ... is the most repulsive aspect of war, and that which resides within man and permits him to perform these acts is the most repulsive aspect of mankind' (Grossman 1995: 229).

Humanity's propensity to wage war has gone unabated for as long as history has been recorded. One could argue that man's greatest failing is being on the battlefield in the first place. Immanuel Kant asserted 'War requires no motivation, but appears to be ingrained in human nature and is even valued as something noble' (Kant [1795] 1985: 125). Even Albert Einstein, who remained a pacifist well into his 50s, eventually acknowledged 'as long as there will be man, there will be war' (Isaacson 2007: 494). Sigmund Freud was even more to the point: '...there is no likelihood of our being able to suppress humanity's aggressive tendencies' (Isaacson 2007: 382).

In this article, however, we are concerned for the large part with the shortcomings humanity exhibits during the conduct of war (*jus in bello*) as opposed to what brought us there in the first place (*jus ad bellum*). 'The emotional strain of warfare and combat cannot be quantified' (Bourke 1999: 232), but at least there has recently been a serious attempt to gather data on that subject. A recent report from the Surgeon General's Office (Surgeon General 2006) assessing the battlefield ethics and mental health of soldiers and marines deployed in Operation 'Iraqi Freedom' is disturbing. The following 12 findings are taken directly from that report:

1. Approximately 10 percent of Soldiers and Marines report mistreating noncombatants (damaged/destroyed Iraqi property when not necessary or hit/kicked a noncombatant when not necessary). Soldiers that have high levels of anger, experience high levels of combat or those who screened positive for a mental health problem were nearly twice as likely to mistreat noncombatants as those who had low levels of anger or combat or screened negative for a mental health problem.
2. Only 47 percent of Soldiers and 38 percent of Marines agreed that noncombatants should be treated with dignity and respect.
3. Well over a third of Soldiers and Marines reported torture should be allowed, whether to save the life of a fellow Soldier or Marine or to obtain important information about insurgents.
4. Some 17 percent of Soldiers and Marines agreed or strongly agreed that all noncombatants should be treated as insurgents.
5. Just under 10 percent of Soldiers and Marines reported that their unit modifies the ROE (rules of engagement) to accomplish the mission.
6. Some 45 percent of Soldiers and 60 percent of Marines did not agree that they would report a fellow soldier/marine if he had injured or killed an innocent noncombatant.
7. Only 43 percent of Soldiers and 30 percent of Marines agreed they would report a unit member for unnecessarily damaging or destroying private property.
8. Less than half of Soldiers and Marines would report a team member for engaging in unethical behavior.
9. A third of Marines and over a quarter of Soldiers did not agree that their NCOs and Officers made it clear not to mistreat noncombatants.
10. Although they reported receiving ethical training, 28 percent of Soldiers and 31 percent of Marines reported facing ethical situations in which they did not know how to respond.
11. Soldiers and Marines are more likely to report engaging in the mistreatment of Iraqi noncombatants when they are angry, and are twice as likely to engage in unethical behavior in the battlefield than when they have low levels of anger.
12. Combat experience, particularly losing a team member, was related to an increase in ethical violations.

This formal study, although at the very least disconcerting, is by no means the first report of battlefield atrocities. 'Atrocious behavior was a feature of combat in the two world wars, as well as in Vietnam' (Bourke 1999: 163). One sociological study of fighting in Vietnam, pointed out that for all men in heavy combat, 1/3 of men in moderate combat, and 8 percent in light combat had seen atrocities or committed or abetted noncombatant murder (Strayer and Ellenhorn 1975). These numbers are staggering.

Possible explanations for the persistence of war crimes by combat troops are discussed in (Bill 2000; Parks 1976a; Parks 1976b; Danyluk 2000; Slim 2008). These include:

- High friendly losses leading to a tendency to seek revenge.
- High turnover in the chain of command, leading to weakened leadership.
- Dehumanization of the enemy through the use of derogatory names and epithets.
- Poorly trained or inexperienced troops. This lack of training is not just in being a good soldier, but also in understanding the Laws of War.
- No clearly defined enemy.
- The issuance of unclear orders where the intent of the order may be interpreted incorrectly as unlawful.
- Shortage of personnel has been associated in producing stress on combatants that can lead to violations.
- Youth and immaturity of troops.
- An overpowering sense of frustration.
- Pleasure from the power of killing.
- External pressure, for example, for a need to produce a high body count of the enemy.

There is clear room for improvement, and autonomous systems may help. Bourke points out that modern warfare enables violent acts in ways unlike before. Now, 'Combatants were able to maintain an emotional distance from their victims largely through the application of . . . technology' (Bourke 1999: xvii). This portends ill for the reduction of atrocities by soldiers. We now have bombs being dropped in Afghanistan and Iraq by UAV operators from almost halfway around the world in Nevada (Ure 2008). This use of technology enables a form of 'numbed killing'. Bourke further notes that there is now a 'technological imperative' to make full use of the new equipment provided. Although technological warfare has reduced the overall number of soldiers required to wage war, the price is that technology, while increasing the ability to kill, decreases 'the awareness that dead human beings were the end product'. When killing at a maximum range, one can pretend they are not killing human beings, and thus experience no regret (Grossman 1995). This physical distance detaches the warfighters from the consequences of the use of their weaponry.

The psychological consequences on our servicemen and women in Afghanistan and Iraq have reached record levels. In 2007 alone, 115 soldiers committed suicide, up from 102 the previous year; 24 percent of the suicides were by those on their first deployment, and 43 percent were those who had returned from deployment. The suicide rates of active duty soldiers as of August 2008 'were on pace to surpass both last year's numbers and the rate of suicide in the general US population for the first time since the Vietnam war, according to US Army officials' (Mount 2008). This unfortunately was confirmed in July 2010 (Fifield 2010). A statistically significant relationship has been established between the suicide attempts and the number of days spent deployed in Iraq or Afghanistan. To make matters worse, this is coupled with 'a growing number of troops diagnosed with post traumatic stress disorder' (Sevastopulo 2008).

These psychiatric casualties are quite significant and common (Grossman 1995): In World War II alone more than 800,000 Americans were classified unfit due to psychiatric reasons, but an additional 504,000 (approximately 50 divisions) were subsequently rendered unfit as a result of psychiatric collapse after induction; In the 1973 Arab–Israeli War one-third of the Israel casualties were psychiatric in origin, twice the number of dead troops. One World War II study showed that after 60 days of continuous combat, 98 percent of all surviving troops suffer psychiatric trauma of some sort (Swank and Marchand 1946).

These long-term exposures to combat are a recent trend in battle, emerging in the twentieth century. The psychiatric damage can take many forms: battlefield fatigue, conversion hysteria, confusional states, anxiety states, obsession and compulsive states, and character disorders (Grossman 1995). The overall effect on the ability to wage war is obvious, let alone the damage to a nation's surviving citizens.

Creating true warfighters in the first place is a daunting challenge. 'No matter how thorough the training, it still failed to enable most combatants to fight' (Bourke 1999: 61). In World War II most men simply did not kill. In one US Army interview of 400 men, only 15 percent of the men had actually fired at enemy positions (at least once) during an engagement despite the fact that 80 percent had the opportunity to do so (Marshall 1947). There was no observed correlation between the experience, terrain, nature of the enemy, or accuracy of enemy fire on this percentage.

This applied to both land and air forces. One study of the Korean War indicated that 50 percent of F-86 jet fighter pilots never fired their guns and only 10 percent of those had actually hit a target (Sparks & Neiss 1956). During World War II, most fighter pilots never even tried to shoot anyone down, let alone succeeding. Less than 1 percent of the pilots accounted for 30–40 percent of all downed enemy aircraft (Grossman 1995: 31).

One conclusion of this is that human soldiers, although not cowardly, lacked an 'offensive spirit'. One possible reason for this lack of aggressiveness centers on the use of long-distance weapons making battlefields 'lonely' and the feeling the enemy was not real but a phantom. This dehumanization of the enemy also quells guilt in killing (Bourke 1999).

The soldiers in the field are not alone in their complicity. 'Atrocities are the dark secret of military culture' (Danyluk 2000, p.38). 'Servicemen of all ranks were unperturbed by most of these acts of lawless killing' (Bourke 1999: 173). In Vietnam, combat commanders viewed the Laws of War as 'unnecessary' and 'unrealistic' restraining devices which would decrease the opportunity for victory (Parks 1976a: 21). A lawyer, defending one general's decision not to initiate a court martial for suspected war crimes violations, stated 'It's a little like the Ten Commandments – they're there, but no one pays attention to them' (Hersh 1971: 119).

Nonetheless our military aspires to higher ethical performance. General Douglas MacArthur stated:

> The soldier, be he friend or foe, is charged with the protection of the weak and unarmed. It is the very essence and reason for his being. When he violates this sacred trust, he not only profanes the cult, but threatens the very fabric of international society. (Park, 1976a: 18)

In addition, the impact of atrocities on public opinion, as clearly evidenced by the My Lai incident in the Vietnam War, and the consequent effect on troop morale are secondary reasons to ensure that events like these are prevented. Civilians are unfortunately killed during war by other humans for manifold reasons (Slim 2008):

- Genocidal thinking – ethnic or racial cleansing of populations.
- Dualistic thinking – dividing host populations into the 'good guys' and the 'bad guys'.
- Power dominance and subjugation – power lust and to exert force.
- Revenge – emotional striking back for perceived wrongs.
- Punishment and forced compliance – to shape the behavior of civilian populations.
- Utility – it furthers the war strategically.
- Asymmetrical necessity – tactical killing of civilians due to an inferior military position.
- Profit – mercenary and looting activity.
- Eradicating potential – pre-emptive removal of civilians who might otherwise become warfighters in the future
- Recklessness – shooting anything that moves, or other forms of indiscriminate killing.
- Reluctant killing – through human error or accident, collateral damage.
- Collective and sacrificial thinking – killing of groups rather than individuals, they must be sacrificed for a greater good.

These forms of thinking are alien to current artificial intelligence efforts and likely are to remain so. Armed autonomous systems need not nor should be equipped with any of these forms of unacceptable human rationalization or action.

A primary conclusion is that it seems unrealistic to expect normal human beings by their very nature to adhere to the Laws of Warfare when confronted with the horror of the battlefield, even when trained. As a Marine Corps Reserves Captain commented: 'If wars cannot be prevented, steps can be taken to ensure that they are at least fought in as ethical a manner as possible' (Danyluk 2000: 38). One could argue that battlefield atrocities, if left unchecked may become progressively worse, with the progression of stand-off weapons and increasing use of technology. Something must be done to restrain the technology itself, above and beyond the human limits of the warfighters themselves. This is the case for the use of ethical autonomy in unmanned systems.

A Way Forward

Research in our laboratory has provided the motivation, philosophy, formalisms, representational requirements, architectural design criteria, recommendations, and test scenarios to design and construct an autonomous robotic system architecture capable of the ethical use of lethal force (Arkin 2009). These first steps toward that goal, however, are very preliminary and subject to major revision, but at the very least they can be viewed as the beginnings of an ethical robotic warfighter. The primary goal remains to enforce international humanitarian law or the Laws of Armed Conflict (LOAC) on the battlefield in a manner that is believed achievable, by creating a class of robots that not only comply with the restrictions of international law, but in fact outperform human soldiers in their ethical capacity under comparable circumstances. If successful this will result in the saving of noncombatant life and property, ideally without erosion of mission performance. It is too early to tell whether this venture will be successful. There are at least five daunting problems remaining:

- The transformation of International Protocols and battlefield ethics into machine-usable representations and real-time reasoning capabilities for bounded morality using modal logics.
- Mechanisms to ensure that the design of intelligent behaviors only provide responses within rigorously defined ethical boundaries.
- The development of effective perceptual algorithms capable of superior target discrimination capabilities, especially with regard to combatant-noncombatant status.
- The creation of techniques to permit the adaptation of an ethical constraint set and underlying behavioral-control parameters that will ensure moral performance, should those norms be violated in any way, involving reflective and affective processing.
- A means to make responsibility assignment clear and explicit for all concerned parties regarding the deployment of a machine with a lethal potential on its mission.

Hopefully the goals of our limited effort will fuel other scientists' interest to assist in ensuring that the machines that we as roboticists create fit within international and societal expectations and requirements. My personal hope would be that they will never be needed in the present or the future. But mankind's tendency toward war seems overwhelming and inevitable. At the very least, if we can reduce civilian casualties in compliance with applicable protocols of the Geneva Conventions and the ideals enshrined within the Just War tradition, the result will have constituted a significant humanitarian achievement, even while staring directly at the face of war.

Notes

[1] This article is derived principally from a book published in 2009 entitled *Governing Lethal Behaviour in Autonomous Robots* published by chapman & Hall (Taylor & Francis).

[2] Atrocity here is defined as the killing of a noncombatant: either a civilian or a former combatant who has attained *hors de combat* status by virtue of surrender or wound.

References

Adams. J. (2010) US defends unmanned drone attacks after harsh UN Report, *Christian Science Monitor*, 5 June 2010.

Adams, T. (2002) Future Warfare and the Decline of Human Decisionmaking, *Parameters*, 31(4) Winter 2001–2002, pp. 57–71.

Arkin, R. C. (2009) *Governing Lethal Behavior in Autonomous Systems* (Boca Raton, FL: CRC Press).

Arkin, R. C. & Ulam, P. (2009) An Ethical Adaptor: Behavioral Modification Derived from Moral Emotions, *IEEE International Symposium on Computational Intelligence in Robotics and Automation (CIRA-09)*, Dajeon (Korea), December 2009.

Arkin, R. C., Wagner, A. & Duncan, B. (2009) Responsibility and Lethality for Unmanned Systems: Ethical Pre-mission Responsibility Advisement, *Proceedings of 2009 IEEE Workshop on Roboethics*, Kobe (Japan), May 2009.

Bill, B. (Ed) (2000) *Law of War Workshop Deskbook* (Charlottesville, VA: International and Operational Law Department, Judge Advocate General's School, June).

Bourke, J. (1999) *An Intimate History of Killing* (New York: Basic Books).

Clausewitz, C. von ([1832] 1976) *On War*, ed. and trans. M. Howard & P. Paret (Princeton, NJ: Princeton University Press).

Cook, M. (2004) *The Moral Warrior: Ethics and Service in the US Military* (Albany: State University of New York Press).

Danyluk, S. (2000) Preventing Atrocities, *Marine Corps Gazette*, 84(6), pp. 36–38.

DARPA (Defense Advanced Research Projects Agency) (2007) Broad Agency Announcement 07-52, *Scalable Network Monitoring*, Strategic Technology Office.

Fifield, A. (2010) US Army Suicide Rate Exceeds National Average, *Financial Times*, 30 July 2010, available at: http://www.ft.com/cms/s/0/2c662840-9b74-11df-8239-00144feab49a.html; Internet.

Filkins, D. (2010) Operators of Drones are Faulted in Afghan Deaths, *New York Times*, 29 May 2010, available at: http://www.nytimes.com/2010/05/30/world/asia/30drone.html?scp=1&sq=Operators%20of%20Drones%20are%20Faulted%20in%20Afghan%20Deaths&st=cse; Internet.

Grossman, D. (1995) *On Killing: The Psychological Cost of Learning to Kill in War and Society* (Boston: Little, Brown).

Guetlein, M. (2005) Lethal Autonomous Systems – Ethical and Doctrinal Implications, US Naval War College Joint Military Operations Department Paper, February 2005, accessed 22 October 2010, available at: http://oai.dtic.mil/oai/oai?verb=getRecord&metadataPrefix=html&identifier=ADA464896; Internet.

Hersh, S. (1971) A Reporter at Large. The Reprimand, *New Yorker*, 9 October 1971, p. 119.

Isaacson, W. (2007) *Einstein: His Life and Universe* (New York: Simon & Schuster).

Kant, I. ([1795]1985), *Perpetual Peace and Other Essays on Politics, History, and Morals*, trans. T. Humphrey (Indianapolis, IN: Hackett).

Marshall, S. L. A. (1947) *Men Against Fire: The Problem of Battle Command in Future War* (New York: William Morrow).

McLoughlin, R. (2006) Fourth Generation Warfare and Network-Centric Warfare, *Marine Corps Gazette*, 15 September 2006, available at: https://feedback.mcamarines.org/gazette/06mcloughlin.asp; Internet.

Mount, M. (2008) Army Suicide Rate Could Top Nation's This Year, CNN.com, 9 Sept. 2008, available at: http://www.cnn.com/2008/HEALTH/09/09/army.suicides/; Internet.

Parks, W. H. (1976a) Crimes in Hostilities. Part I, *Marine Corps Gazette*, 60(8), pp. 16–22.

Parks, W. H. (1976b) Crimes in Hostilities. Conclusion, *Marine Corps Gazette*, 60(9), pp. 33–39.

Sagan, S. (1991) Rules of Engagement, in: A. L. George (Ed), *Avoiding War: Problems of Crisis Management*, pp. 443–470 (Boulder, CO: Westview Press).

Sevastopulo, D. (2008) US Army Suicide Cases at Record 115, *Financial Times*, 29 May 2008, http://us.ft.com/ftgateway/superpage.ft?news_id=fto052920081802392265; Internet.

Slim, H. (2008) *Killing Civilians: Method, Madness, and Morality in War* (New York: Columbia University Press).

Sparks, B. W. & Neiss, O. (1956) Psychiatric Screening of Combat Pilots, *US Armed Forces Medical Journal*, 7(6), pp. 811–816.
Strayer, R. & Ellenhorn, L. (1975) Vietnam Veterans: A Study Exploring Adjustment Patterns and Attitudes, *Journal of Social Issues*, 31(4), pp. 81–93.
Sullivan, R. (2010) Drone Crew Blamed in Afghan Civilian Deaths, *Associated Press*, 5 May 2010, available at: http://abcnews.go.com/International/wireStory?id=10777234; Internet.
Surgeon General (2006) *IV Operation Iraqi Freedom 05-07, Final Report* (Surgeon General's Office, Mental Health Advisory Team (MHAT), 17 Nov.).
Swank, R. & Marchand, W. (1946) Combat Neuroses: Development of Combat Exhaustion, *Archives of Neurology and Psychology*, 55, pp. 236–47.
Ure, L. (2008) Armchair Pilots Striking Afghanistan by Remote Control, CNN.com, 9 July 2008, available at: http://www.cnn.com/2008/TECH/07/09/remote.fighters/index.html; Internet.
Walzer, M. (1977), *Just and Unjust Wars* (New York: Basic Books).

Biography

Ronald C. Arkin is Regents' Professor of Computer Science and Director of the Mobile Robot Laboratory at Georgia Tech University (Atlanta, GA). He also serves as the Associate Dean for Research and Space Planning in the College of Computing at Georgia Tech since October 2008. During 1997–98, Professor Arkin served as STINT visiting Professor at the Centre for Autonomous Systems at the Royal Institute of Technology (KTH) in Stockholm, Sweden. From June to September 2005, Prof. Arkin held a Sabbatical Chair at the Sony Intelligence Dynamics Laboratory in Tokyo, Japan and then served as a member of the Robotics and Artificial Intelligence Group at LAAS/CNRS in Toulouse, France from October 2005 to August 2006. His research interests include behavior-based reactive control and action-oriented perception for mobile robots and unmanned aerial vehicles, hybrid deliberative/reactive software architectures, robot survivability, multiagent robotic systems, biorobotics, human-robot interaction, robot ethics, and learning in autonomous systems. He has over 170 technical publications in these areas. Prof. Arkin has written a textbook entitled *Behavior Based Robotics*, published by MIT Press in May 1998, co-edited (with G. Bekey) a book entitled *Robot Colonies* published in 1997.

THE CASE AGAINST ROBOTIC WARFARE: A RESPONSE TO ARKIN

Ryan Tonkens

Novel Tech Ethics, Dalhousie University

Semi-autonomous robotic weapons are already carving out a role for themselves in modern warfare. Recently, Ronald Arkin has argued that autonomous lethal robotic systems could be more ethical than humans on the battlefield, and that this marks a significant reason in favour of their development and use. Here I offer a critical response to the position advanced by Arkin. Although I am sympathetic to the spirit of the motivation behind Arkin's project and agree that if *we decide to develop and use these machines they ought to be programmed to behave ethically, there are several major problems with his view as it stands. At present, it is not clear whether such machines would be capable of behaving more ethically than humans. More importantly, to the extent that humans would remain in the context of war, human moral transgressions will continue, especially in the face of complicated ethical challenges accompanying automated warfare. Moreover, even if machines could be more ethical than humans in certain ways and in certain situations, this says nothing about whether warfare that contains these machines would itself be overall more ethical than warfare that does not include them as participants, or whether the inclusion of lethal robots is the best way to guard against human moral transgressions in war.*

Introduction

Semi-autonomous robotic weapons are already carving out a role for themselves in modern warfare. Given the billions of dollars being allocated for artificial intelligence and robotics research by military sources for military applications, if left unchallenged, there is a good chance that this trend will continue in the future, resulting in the increased development and use of more sophisticated military machines.[1] In light of this, it is important that we determine whether the development and use of such autonomous lethal robots is something that we are morally and legally permitted to do, before we (continue to) create and use these machines for this purpose.

One of the leading advocates of the development of such machines is the roboticist Ronald Arkin. In a series of recent works (2009a, 2009b, 2010) Arkin has argued that autonomous lethal robotic systems could be more ethical than humans on the battlefield, and that this marks a significant (indeed, a decisive) reason in favour of their development and use.[2] In this paper I offer a critical response to the position advanced by Arkin. Although I am largely sympathetic to the flavour of the motivation behind Arkin's project (i.e., to make war more ethical and to reduce human casualties), and agree that *if* we decide to develop and use these machines then they ought to be programmed to behave ethically, there are nevertheless several problems with his view.

In what follows I examine the underlying assumptions of Arkin's view, and the case that he advances in favour of the development and use of autonomous lethal robotic systems for military purposes. I argue that, even if we allow Arkin's underlying assumptions (which there is good reason *not* to do), the case that he presents is not as strong as he suggests it to be. At present, it is not clear whether such machines would be capable of behaving more ethically than humans. More importantly, to the extent that humans would remain in the context of automated war, then human moral transgressions will continue to occur, albeit perhaps under different guises, especially in the face of other ethical challenges accompanying automated warfare (e.g., complications surrounding human use of autonomous lethal robotic systems in practice). Moreover, even if machines could be more ethical than humans in certain ways and in certain situations, this says nothing about whether warfare that contains these machines would be overall more ethical than warfare that does not include them as participants, or whether the inclusion of lethal autonomous robots is the best way to guard against human moral transgressions in war, or the inhumaneness of war in general.

By way of giving away the ending: we are not justified in *assuming* that the emergence of automated warfare is inevitable; any case for automated warfare that relies on this assumption is significantly weakened from the start. Moreover, the 'urgency' of lessening human unethical behaviour in war is exaggerated by Arkin, and there are many alternative ways to reduce such behaviour without requiring that we devote scare resources to developing lethal autonomous robotic systems. Even *if* we decide to go down the path of automating warfare, there is good reason to suggest that it will continue to be (just as) unethical, and thus would not alleviate the problem that it was intended to. The important upshot of all of this is that we need a better argument in favour of automating warfare, lest we pursue it in a morally dubious, unjustifiable manner.

Arkin's Case for Automated Warfare

The central thesis of Arkin's case in favour of automated warfare is that, the potential for improved military effectiveness aside, lethal robotic systems *will* be able to perform 'more ethically than human soldiers *are capable of* performing' (emphasis added). This is in part because they would not need to be equipped with the emotional and psychological underpinnings that lead to many of the war crimes and lesser moral violations committed by humans (2009a: 30–1). *If* robots could behave more ethically than human soldiers, argues Arkin, then we should bring such robots onto the battlefield, either *in lieu* of human soldiers (altogether), or as their partners.

At one point, Arkin makes a significantly weaker and more plausible claim: 'There is clear room for improvement, and autonomous systems may help' (2009a: 33). The important difference between this claim and the stronger claim above is the level of conviction – the stronger claim replacing 'may' with 'will'. It seems uncontentious that the current moral calibre of warfare leaves much room for improvement. And, if one goal of developing autonomous lethal robotic systems is to improve the moral calibre of warfare, then, *prima facie* at least, the inclusion of autonomous systems certainly *may* help in this regard. Yet there are many ways in which we could help to improve the current situation, and thus creating autonomous robots is but one of many potential ways to help.

Importantly, even Arkin's weaker, more passive claim depends on the strength of the reasons he advances for accepting his stronger claim (discussed below), including the empirical veracity of the claim that such machines will actually spawn an improvement in this regard. As Arkin repeatedly argues for his stronger claim, and it is the one that is more interesting and philosophically rich, it is that one which I focus on herein. I mention the weaker claim in passing in order to highlight the idea that rejecting *Arkin's* case in favour of automating warfare does not entail a rejection of *all reasons for* automating warfare. Nevertheless, as one of the leading proponents of automated warfare, dismantling his view puts the burden back on proponents to demonstrate anew that the development and use of such machines is a morally acceptable undertaking.

One important point is worth making at the outset. Although it is true that human soldiers are capable of moral transgressions, they are also capable of behaving in line with law and morality, and the great majority of human soldiers that are *capable of* immoral action do not actually behave immorally. Arkin himself (2009a: 35) notes that humans are often reluctant to kill in war, and experience psychological malaise after witnessing others behaving unethically. One implication of this is that increased attention could be placed on strengthening human moral convictions, which is a promising avenue for improving the overall ethical calibre of warfare. Thus, the psychological and dispositional foundations for improving the moral calibre of human warfare are already in place, and may be further strengthened and reinforced through deeper ethical training, for example.[3]

Moreover, humans are also capable of morally praiseworthy and supererogatory behaviour, exemplified by (for example) heroism in battle, something that machines may not be capable of. Something that typically gets overlooked in the debate surrounding automated warfare is the idea that not only will it prove to be quite challenging to actually create robots that are thoroughly sensitive to morality in their programming and behaviour, but that replacing humans with such machines may also eliminate the occurrence of soldiers 'going beyond the call of duty'. Indeed, one may challenge the pursuit of automated warfare on the grounds that it unduly threatens the ability of human soldiers to exhibit morally exceptional behaviour, and undermines important aspects of the military profession. Thus, even if machines could behave more ethically than some humans do, their ceiling for morally good behaviour may not be as high as that of human soldiers; although robots could act more ethically than some human soldiers do, they may also not act as morally excellently as certain human soldiers do. If the inclusion of autonomous robots made it so that humans have less opportunity to demonstrate bravery, or 'selfless sacrifice' in the name of the citizenry that they represent and protect, and if robots were incapable of behaving in such ways themselves, then certain morally praiseworthy aspects of *human* warfare might be lost. Thus making warfare less inhumane by removing humans from the battlefield may also take away certain desirable human-based qualities as well. One upshot of this is that, if there are ways available to improve the moral calibre of warfare that do not mitigate the occurrence of morally praiseworthy and supererogatory action (by humans), then (*ceteris paribus*) this would be good reason to pursue those avenues instead of automating warfare.

Contra Arkin, then, the important issue is not really whether robots could behave better than humans 'are capable of' behaving, but rather whether robots could perform equally as (or more) ethically than humans *actually do* on the battlefield, given that *some* humans commit war crimes and atrocities, under *some* conditions and in *certain* scenarios, *and* given that certain human soldiers perform actions of the highest moral calibre, honour, and bravery. In the face of certain moral transgressions committed by individual human soldiers (and perhaps even robotic soldiers as well), it is prophesized that robots *as a species* will outperform humans as a species, ethically speaking.

According to Arkin, a major aspect of the increased ethicality of robotic systems would stem from their better treatment of non-combatants (e.g., innocent civilians), and their being directly programmed to adhere to the established international laws of war. Although Arkin (2009a: 30–1) admits that autonomous machines may not always behave morally flawlessly, and hence that they will still be imperfect moral entities (i.e., *capable of* immoral actions), rendering the battlefield more moral in this way is thought to be a welcome outcome of developing autonomous military machines; although war may continue to be hellish,[4] making it a more moral context is not a futile enterprise, but rather a welcome one.[5] Arkin takes his project to be a humanitarian one; if the use of lethal robotic systems in war can reduce human casualties, and collateral damage, and reduce the risk of (*certain kinds of*) human moral transgressions, then it would render the context of warfare less inhumane. In essence (and somewhat ironically), the goal is to make war less inhumane by lessening or removing the human element from warfare.[6]

According to Arkin, human dispositions like fear, greed, rage, revenge, stress, and fatigue act as catalysts for misbehaviour in battle. Since we could neglect to program machines with such emotions and psychophysical capacities without thereby sacrificing their military performance, they would not have the same predispositions towards unethical behaviour that human soldiers sometimes fall prey to. In addition to this, as noted above, their ability to behave more ethically would also stem from their programmed (inherent) adherence to the laws of war, which may prove to be more effective than the often limited indoctrination of morality experienced by human soldiers. Even when human combatants are aware of these regulations on their conduct, they have the will to disregard them if they so desire (or in case their emotions get the better of them), something that robotic systems need not be equipped with, despite their otherwise being autonomous.[7] To this extent, the *jus in bello* arm of the international laws of war may be more easily respected by robotic combatants than has traditionally been the case with (some) human combatants.

Arkin (2009a) proposes that autonomous lethal robotic systems could be programmed to possess what he terms an 'ethical governor' and a 'responsibility advisor' that would help ensure to a significant extent that robots behave ethically. Not only would the laws of war be programmed into the very makeup of the robot,[8] but it could constantly survey its own intended behaviour, and advise any human operators or commanders of the ethical consequences of its being used for certain purposes in the given context, prior to its engagement with targets. Although there may still be *some* room for error (due to malfunctioning perhaps, or unforeseen behaviours arising in novel situations), these mechanisms would go towards ensuring that the robotic system performs only those actions that are deemed to be morally permissible.

It is important to note that Arkin is concerned primarily with *autonomous* robots, rather than non-autonomous machines or the sort of semi-autonomous weapons currently in use on the battlefield. Although he recognizes that keeping a human in the operational loop may be beneficial in some cases, he also recognizes several downsides to this more limited autonomy (such as detachment and the dehumanizing effect of long distance and 'remote-controlled' military initiatives [2009a: 35]), and his language suggests that more autonomy in military robots is both desirable (ibid.: 36) and inevitable (ibid.: 29).

With this sketch of Arkin's view in place, I move now to a critical examination of it.

The Inevitability of Automated Warfare

It is important to note that Arkin is not primarily concerned with the larger context of just wars or with human behaviour outside of the context of war. Implicitly at least, he seems to assume an otherwise justifiable context for the use of autonomous lethal robotic systems. Although this starting point is understandable, given that he is not an ethicist, and that his primary concern is with making these military robots safe and moral, the soundness of his case *depends on* the veracity and tenability of these and other implicit assumptions.[9] To use a somewhat inflated example: that concentration camps are used for unjust and inhumane purposes is a crucial fact that needs to be considered prior to construction of such camps. *If* wars that involve autonomous lethal robotic systems are unjust, then ensuring that these robots are safe and effective is morally futile, and unacceptably retrospective.

In this section, however, I would like to highlight and challenge two other assumptions underlying Arkin's view. Arkin's view depends on the veracity and tenability of these assumptions; if there is good reason to reject them, then there is *a fortiori* good reason to reject the underlying reasons for Arkin's project as well.

Despite Arkin's (2009a: 1) explicit recognition that peace is preferable to war, he nevertheless offers the following:

> The trend is clear: Warfare will continue and autonomous robots will ultimately be deployed in its conduct. Given this, questions then arise regarding if and how these systems can conform as well or better than our [human] soldiers with respect to adherence to the existing Laws of War (2009a: 29).

Elsewhere, Arkin concludes:

> My personal hope would be that they will never be needed in the present or the future. But mankind's tendency toward war seems overwhelming and inevitable. At the very least, if we can reduce civilian casualties in compliance with applicable protocols of the Geneva Conventions and the ideals enshrined within the Just War tradition, the result will have constituted a significant humanitarian achievement, even while staring directly at the face of war (2010: 339).

There are two assumptions that Arkin is making here that deserve to be scrutinized:[10] (1) that (automated) warfare will *inevitably* continue in the future, and (2) that the continued development of (more) sophisticated and autonomous lethal robotic systems is itself *inevitable*.[11] For one thing, the language here is much too strong; it is not 'inevitable' that war will continue or that such robots will be developed, especially if this is meant to suggest that it is impossible that things go otherwise. The obvious counter-example here is that we simply do not devote the time and resources to developing such machines (or to waging wars), which are *active* human undertakings. To be sure, pacifists are often dismissed as being naïve and unrealistic, and Arkin's prophecy of the occurrence of (automated) warfare *may* prove to be accurate in the end. However, at this point in time, this outcome is certainly not unavoidable. Moreover, even if our drive towards automated warfare continues, this alone says nothing about whether doing so is morally acceptable. Neither the development nor the use of autonomous lethal robotic systems is inevitable. It may be difficult to curtail the development of these machines, since early versions are already in use and many more are on the drawing board, and it may seem as though these technologies are coming, whether we like it or not. Nevertheless, I am inclined to believe that we still, at present, have a real say in whether this actually occurs. Indeed, even if we do develop autonomous lethal robotic systems, we would still have the unfettered choice *not* to use them in practice. And, we would need a very good reason to both develop and use these robotic systems, beyond the putative inevitability of their development, and the colossal waste of resources if we do not use them once they are available.

Given the rapidly changing nature of modern warfare and the complicated issues that are emerging in the face of increased technological sophistication (Singer 2010), we are presented with an opportunity for critical reflection on whether or not we want to continue to promote the occurrence of (automated) warfare at all. As the nature of warfare continues to morph, there is renewed reason to question our motivations and reasons for allowing war to occur, and to determine whether or not (automated) warfare is something that we want to continue to see happen; as Arkin himself acknowledges (2009a: 5), we have an obligation to determine whether or not we are morally permitted to pursue this project. To this I would add: *prior to actually doing so*. One hopes that if automated warfare is found to be unjust or immoral (as some thinkers have already suggested it to be[12]), outside of whether such robots could behave more ethically than some humans do, then this is something that we would not allow to come to fruition. The only way to ensure this is to ask whether warfare should continue and, if so, whether autonomous lethal robots should be allowed to be deployed in its conduct, rather than to assume that these outcomes are inevitable, no matter what we do. Although this assumption underlying Arkin's view is a common one, it does not speak to any inherent aspect of the world, humanity, or the nature of warfare, and thus needs to be *argued for* insofar as it is to serve as a foundational tenet of our assessment of the ethics of automated warfare. (One way to view my charge here is that, although Arkin recognizes the importance of this issue, he himself does not attend to it, favouring instead issues surrounding the actual development and use of such machines. Yet, by doing this, his view puts the cart before the horse, ethically speaking.) Put differently, it is not enough to recognize, in passing, that we are making the assumptions noted above; the important issue is whether we are *justified* in making these assumptions. Otherwise, we risk begging the question from the very beginning: the development and use of autonomous lethal robotic systems is something that we ought to do because it is something that we assume to be inevitable. If, however, automated warfare turns out to be avoidable, then we need to find other reasons for working towards bringing it about.[13]

I suspect that, to *everyone's detriment*, these assumptions will continue to be made, regardless of whether we are justified in doing so. I discuss them here since their being taken for granted as truthful fuels their putative 'inevitability' – serving as a self-fulfilling prophecy of sorts – and neglects to take seriously the important ethical and legal issues that may otherwise stand in their way. And, to repeat, in order to pursue automated warfare (or not) in a morally justifiable manner, we need good reasons to do so. Removing these assumptions (*pseudo*-reasons) from the foundation of the case in favour of automated warfare significantly weakens its strength and justifiability from the very beginning.

Ethical, Lethal, Autonomous Robots

Arkin (2009a, 2009b, 2010) offers six main reasons why robotic systems will be ethically better than human soldiers are capable of being: (1) with the help of (for example) the

Global Information Grid, robots could be much better than humans at processing relevant information in real time; (2) it is prophesized by some that they will be equipped with superior sensory apparatus for making battlefield observations; (3) robots would not have emotions such as anger, fear, and frustration, which can lead (some) humans to behave unethically; (4) robots need not have faulty psychological dispositions which, again, can lead (some) humans to commit immoral actions; (5) robots will act conservatively and 'self-sacrificially'; and (6) robots could actively monitor the behaviour of (human) soldiers, and report any unethical behaviour they witness. Taken together, these reasons lead Arkin to conclude that robots could behave more ethically than human soldiers are capable of performing in the context of war, at least given certain crucial technological advances in the future (e.g., the establishment of a sophisticated ability for discrimination).

Arkin's language is somewhat misleading here. As it turns out, several of these aspects of heightened robotic ability are not *ethical* in nature; not all of these reasons actually support the idea that robots could be more *ethical* than humans, as Arkin intends them to. For example, claim (2) does not obviously support this thesis, at least not directly; as discussed below, the problems associated with human behaviour cited by Arkin do not stem from a lack of the sensory abilities in human soldiers to make accurate battlefield observations. There is a moral difference between intentional and unintentional wrong-doing, and between culpable and non-culpable ignorance, which Arkin seems to conflate here. Given the conditions of modern warfare, there are cases where human soldiers act responsibly given the information they have, and yet where doing so sometimes leads to *unintentional* wrongdoing (e.g., unforeseen non-combatant collateral damage). It is a stretch to say that human soldiers behave unethically in such situations, even if having more and better information would have lessened the risk of unforeseen, unintentional wrongdoing. To be sure, such collateral damage is undesirable, and if machines have access to more accurate information, then they may be better than humans in many contexts. Yet, it is misleading to suggest that this lack of information automatically renders the human soldier's actions *unethical*. If a lack of sensory abilities results in *unintentional* 'unethical' behaviour by humans, then this is hardly demonstrative of *intentional* unethical activity (however undesirable it may be), and thus is not necessarily something that we *can* guard against (whether in humans or machines). Moreover, it is not clear whether autonomous robots could avoid such problems anytime in the foreseeable future, and thus near-term future robots may be susceptible to the same sort of deficiencies in this regard as human soldiers presumably are. Receiving a sufficient amount of information and processing it properly so as to *eliminate* the prospect of unintentional wrongdoing in its entirety is a tremendous feat indeed.

Perhaps Arkin means to suggest that robots could be better at discriminating than human soldiers are. But, the majority of human moral transgressions are not *unintentional* (e.g., representative of attempted *yet failed* discrimination, or unintentional torture, etc.) and hence are not typically problems of limited sensory abilities leading to poor discrimination. Put differently, misbehaviour stemming from *incorrect* discrimination does not typically represent *immoral* conduct by those human soldiers, although the results are undesirable; the problems that arise here typically stem from a failure or reluctance to discriminate in the first place or, at worst, the misguided belief that targeting, engaging, and killing non-combatants is morally acceptable.[14] Moreover, as Arkin himself repeatedly notes, at present robots are nowhere near as effective at discriminating as humans are. Even if robots could have superior sensory abilities, this does not *necessarily* mean that they would be ethically superior to humans. For one thing, an argument needs to be made to support the claim that this heightened sensory ability, *if* it can be established, would be used ethically; an increased ability to discriminate effectively between combatants and civilians would only be useful to the extent that the robotic (or human) soldier used it to avoid targeting civilians in practice.

Claim (1) also seems misplaced here. It is not clear that human soldiers that violate the laws of war typically do so because of a *lack of* information (about those laws, or the nature of the context, or the status of the target they are engaging, etc.). Perhaps there is a case to be made to suggest that increased information helps soldiers to make better, ethical decisions. And yet, humans (and robots) could use their increased information for ill. Moreover, the information that a soldier needs *in order to be able to behave morally* is quite narrow (e.g., what rules are governing one's behaviour and whether or not the potential target is a legitimate combatant), and thus it seems that even those humans that behave unethically would still have access to this information, even if they choose to ignore it – lacking this information is not the source of their unethical behaviour. (That robots would not ignore this information is certainly an asset, but the point is that increased information does not automatically result in being more ethical, and much of the unethical behaviour of humans does not stem from a culpable lack of information.) Furthermore, any recognized *lack of* relevant information is typically accompanied with a default position of not engaging under such uncertainty, and thus would likely lead to conservatism rather than rash actions, and could be solved by making more information available to human soldiers. At any rate, this potential asset of autonomous robotic systems for enhanced informational capabilities is not obviously *ethically* beneficial, without significant qualification.

Lastly, claim (5) may not support Arkin's thesis either. Arkin (2009a: 31) argues that:

> Autonomous armed robotic vehicles [for example] do not need to have self-preservation as a foremost drive, if at all. They can be used in a self-sacrificing manner if needed and appropriate, without reservation by a [presumably human] commanding officer.

It is not clear how this self-sacrificial disposition would make robots more ethical than humans. It is not typically an ethical breach when a human refuses to sacrifice herself in battle or turns down a suicide mission, for example. Indeed, it is usually morally justifiable for her to do so, although it may often be a sign of heroism or bravery when she accepts such a mission, something lost on the robot that was forced to do so through its very programming.[15]

Perhaps Arkin's point here is that the drive for self-preservation often puts pressure on human soldiers in terms of information processing and decision-making, so that certain sorts of ethical transgressions may emerge when the soldier believes she is acting in ways that are necessary for her self-preservation, ways that are (or turn out to be) unethical, as in cases where she mistakes a civilian for a legitimate threat and erroneously believes her life to be in danger when it is not. The stress, fear, and anxiety that accompany the desire not to be killed in battle (the drive to survive) certainly influence the actions of human combatants, and need not similarly influence robots. Although the sort of scenario described above is unfortunate and undesirable, it is unclear whether the soldier has behaved unethically when she defends herself in that way, and thus whether *this* source of behavioural motivation is ethically problematic, at least in the general case.[16] To be sure, she has made a mistake and harmed someone that she should not have, in the name of her self-preservation. But, at the same time, she was acting out of what she reasonably believed to be self-defence, something that is almost always morally justifiable under (what seems to be) a legitimate threat, and a routine occurrence in the context of warfare. It seems incorrect to suggest that robots could be 'more ethical' than humans in this way – as the human soldier that defends herself has not clearly acted unethically, strictly speaking – but rather that they would not be prone to the same sorts of mistakes, in the rare and unfortunate cases that self-preservation leads to such mistakes (McMahan 1994).[17] Indeed, *unintentional* breaches of the laws of war are not usually taken to be morally problematic and thus she is not found legally culpable (as in cases where a soldier unintentionally and unforeseeably harms a non-combatant), and she all along retains the right to self-defence.[18]

As discussed next, some of Arkin's other points are more plausible, and speak more directly to the ethical aspects of combatant behaviour.

A. *Emotionless drones with minds of steel*: It is true that robots would not need to have the emotions that lead (some) humans to act unethically in battle. At the same time, autonomous robots could be programmed to have certain affective abilities associated with (for example) guilt and remorse, which may make them more effective at learning from their mistakes, and empathizing with illegitimate targets, than they would otherwise be, making them moral from the other direction as well. Worth noting, however, is that this positive affective component only really bears on the morality of such robots in a retroactive manner, after an ethical breach has already occurred; the robot would have nothing to feel guilty about if it had acted morally flawlessly in the first place. Although this affective component may help to prevent further breaches on their behalf in the future, its use is futile unless the robot behaves unethically, *just like some human soldiers do.*

Human users and commanders of autonomous lethal robotic systems would still be susceptible to *their* emotions, which may affect how *they* use robots for their desired ends. Insofar as humans often commit injustices or atrocities even in the face of the threat of being punished (which is a central motivational tenet of Arkin's project as a whole), then the override procedure and ethical advisory role that Arkin suggests should be part of the robot's programming could all along be used for ill, or overruled by its (ethically fallible) human users and commanders. Thus, an important issue that emerges here is *the ethical use of robots by humans*, who would still be susceptible to the emotional, psychological, and physiological catalysts for unethical behaviour. Although the robots themselves may be inclined to behave morally, their human users and commanders would all along be at risk of misusing them; the human element that leads to inhumane behaviour may not have been removed enough. (I return to this line of thought later on.)

To be sure, robots may not commit unethical acts out of emotional shortcomings, since they would not have the emotions necessary for this to be possible. And, as noted above, Arkin is primarily concerned with *autonomous* robots, and hence the extent to which the emotions of human users or commanders could affect the behaviour of these robots may be minimal (although I take this to be an empirical issue). Indeed, he suggests the need for a built-in responsibility advisor that would need to be overruled in order for the human user/commander to succeed in their unsavoury application of the robot's arsenal. Yet, if we assume that humans are as prone to unethical behaviour as Arkin suggests, then this may be little solace and a weak obstacle deterring humans from overruling the ethical prescriptions of the machine. (If humans are not as susceptible to unethical behaviour as Arkin suggests, then *the need for* autonomous lethal robotic systems becomes far less pronounced.)

Other psychological predispositions that Arkin identifies as being correlated with the unethical activity of (some) human soldiers can also be left out of the robot's programming (e.g., limited attention and focus, being prone to mental fatigue and boredom, etc.). Again, Arkin is correct to argue that robots would not need to possess the physical and psychological triggers for wartime atrocities and immoral behaviour. At the same time, we may want to press Arkin a little on this point. For one thing, there are different sources of unethical behaviour other than the flawed emotional or psychological dispositions of human soldiers, including system malfunctioning, vague mission goals or rules of engagement, corrupt command structures, overly strict guiding principles that prevent the soldier (human or robot) from adapting to novel situations in real time, the misuse of robots by humans, etc. Just as human soldiers are prone to certain sorts of imperfections, so too may autonomous robots be similarly flawed (at least in the early stages of their development and use). And, the degree to which human emotion and psychology is the source of the moral transgressions that occur in war is difficult to determine, given that other sources exist. Thus, even if robots could be better than humans in certain ways, the extent to which this would lessen the overall immorality of warfare is one that is open and empirical.

B. *Ethical police*: Arkin is correct to suggest that robots could serve as ethical advisors and 'moral police' on the battlefield. Indeed, one of the most appealing roles for autonomous robots in war would be to evaluate the conduct of human soldiers and the overall justness of wars in general. Interestingly, this role would only be necessary in cases where humans still participated in warfare to a significant extent; in the absence of humans in combat, or in the absence of humans being able to override the recommendations of the robot, then this role for robots may become somewhat superfluous. Thus, the move away from human warfare, as it proceeds closer and closer to human*less* warfare, would eliminate this need and consequently this reason in favour of developing and using these machines in the first place. More importantly, there is no obvious reason why such robotic moral advisors or ethical police would need to have *lethal* capabilities, or actively engage in real military initiatives *as soldiers*. The laudable end of monitoring, improving, and reprimanding unethical activity on the battlefield would not require lethal capabilities in the robot. Thus, an independent reason for making these robotic advisors and moral police capable of autonomous action *and lethal force* is required, something which Arkin does not provide.

Although Arkin provides an important insight here, namely that the inclusion of robots alongside human combatants may change the dynamics of military group behaviour for the better, owing to their role as observers and recorders of what transpires on the battlefield, the role of robotic ethical monitors should not be exaggerated. It seems likely that most instances of human unethical activity would occur without the knowledge of the machine (behind its back or in its absence). In cases where the robot is aware of the immoral activity as it is occurring, the most the robot could accomplish is to arrest the soldier(s) and aid in passing retrospective judgment. But, in such cases, *the moral transgression would already have occurred*. Not even Arkin wants to argue that robots could routinely *prevent* humans from committing war crimes and atrocities, although their presence may act as a deterrent if human soldiers are more aware of the likelihood of their getting caught and being punished thereafter. Thus, whether *human* warfare would necessarily come to be more ethical in practice in the presence of such robots is an important, yet open, empirical issue.

The six reasons that Arkin offers to support the claim that autonomous lethal robotic systems would be more ethical than humans are capable of being are far less definitive than Arkin suggests and, because of this, his case in favour of automated warfare is significantly weakened. Many of them are not ethical in nature (strictly speaking), and thus do not speak directly to improving the moral calibre of human warfare. The remaining reasons are certainly important to consider, and yet overlook the idea that the humans that remained on the battlefield would continue to be able to behave unethically, albeit perhaps in different ways. Although these issues remain inescapably empirical, there is good reason to suggest that Arkin's case is already significantly weaker that it first appeared.

Military Effectiveness and Ethical Superiority

A related part of Arkin's argument in favour of automating warfare is that autonomous robots could be superior to humans in their military effectiveness. As discussed in this section, this relates to the ethical nature of these machines in several ways.

1. *Ethics versus military effectiveness*: Not all weapons or military techniques ought to be used just in case they are more effective than other weapons or techniques. Nuclear bombs and torture are paradigmatic examples here, as their moral repulsiveness (i.e., indiscriminate nature and inhumaneness, respectively) trumps their military effectiveness. Even if we could build autonomous robotic torturers that could render the practice of torture less unethical and inhumane, this is not a good enough reason to build machines for this unethical purpose.[19] Thus, even if it turns out that autonomous robotic systems are militarily effective, i.e., 'faster, cheaper, have better mission success, longer range, greater persistence and endurance, higher precision, faster target engagement, and are immune to chemical and biological weapons' (Arkin 2009a: 30), this does not necessarily entail that they *ought to be* developed or used. To be sure, military effectiveness matters, and being better in this regard *may* bring with it a higher moral acceptability as well (for example, the use of weapons that are more precise may be more ethical than the use of those that are less precise). And yet, if competing moral issues go unattended and unsatisfied, then military effectiveness risks being accomplished at the expense of ethical effectiveness.

2. *Discrimination and proportionality*: Just because robots could, theoretically, behave more ethically than some human soldiers do, does not mean that they will do so in practice; the technological proof will be in the pudding.[20] At the very least, achieving such ethical superiority will be a long process of trial and error in the laboratory and in the field, and will require the realization of presently only prophetical technological advancements, such as the ability for effective real-time discrimination.

Whether or not autonomous military machines could be designed to discriminate effectively and exert proper proportionality in real-world military contexts is a crucial yet open question, although designing a robot with these abilities may not be impossible in principle. Regardless, it seems uncontentious to suggest that the level of autonomy and the ability of machines to act in real-world contexts will increase sooner than our perfecting their ability to exert proper discrimination and proportionality, especially if these latter tasks involve possessing rich mind-reading and folk psychological capabilities (among other things), which are currently only poorly understood.[21]

This is important to recognize since, until autonomous robots can accurately and reliably discriminate between legitimate and illegitimate targets (i.e., to discriminate between combatants and non-combatants, between surrendering combatants and aggressive combatants, between allies and enemies, between weapons and non-weapons, and be able to detect obvious perfidy, and [in the extreme] 'refrain from zapping a little girl who is trying to share her ice cream with the robot' [Sharkey 2008]), then they do not meet a central tenet of the international laws of war. In this case, using such machines in warfare would be legally and morally problematic, and the developers and users of indiscriminate autonomous machines are thereby morally suspect. As mentioned earlier, Arkin is not primarily concerned with this aspect of the ethics of automated warfare. My point here is that he needs to be; even *if* autonomous lethal robotic systems could be more ethical than human soldiers, if their use in warfare is unjust, then improving the moral calibre of war in this way is morally bankrupt.

Once autonomous military robots no longer depend on human operators to discriminate on their behalf and decide for them what constitutes proportionate response, they will need to be effective at doing so for themselves. If they cannot do so, then their participation in warfare is unjust. On the other hand, insofar as humans would continue to be the ones doing the discriminating and assessment of proportional force (on the robot's behalf), then *their* human fallibilities (emotions, psychophysical dispositions, and putative limited sensory abilities) in this regard remain in play, and hence this aspect of 'superior robot ethicality' becomes moot to that extent.

3. *The moral calibre of human soldiers*: As already stressed above, not *all* human soldiers behave unethically, and human soldiers *can* behave more ethically than they currently do on the battlefield. Arkin argues that it is 'unrealistic to expect normal human beings by their very nature to adhere to the Laws of Warfare when confronted with the horror of the battlefield, even when trained' (2010: 338). And yet, (a) we *do* expect this, and justifiably so – why else would we have established the international laws of war and punishments for intentional deviations from them?;[22] (b) there are other ways to work towards changing the conditions of the battlefield so that it could be more conducive to moral behaviour by human soldiers, other than introducing autonomous lethal robotic systems (e.g., lessening fatigue by having shorter tours of duty, improving therapeutic aid available on military bases, tightening up psychophysical screening of soldiers ahead of time, holding more rigorous training on the ethics of warfare and international laws of war, devising clearer rules of engagement, etc.); and (c) despite the existence of certain ethical deviants, *the great majority* of human soldiers do not behave immorally, and hence this problem that Arkin focuses on may be relatively mundane.

Indeed, the statistics that Arkin (2009a) cites about the moral transgressions of humans during warfare are not as overwhelming as he makes them out to be. If ten percent of soldiers report mistreating non-combatants (2009a: 31), then this means that ninety per cent (the vast majority) do not.[23] While it may be argued that ten percent is still a number that we should be uncomfortable with, it is unclear how much the introduction of autonomous lethal robotic systems could reduce this relatively small percentage. The inclusion of drones in certain military contexts has been shown to increase the moral calibre of those contexts (Strawser 2010). And yet, it is way too early to tell if this can be generalized to other contexts, with other (more sophisticated and autonomous) robots, and whether the *overall* immorality of war has decreased (or increased).

Although there is certainly room here for soldiers not to report mistreatment, or to provide faulty reports (in either direction), the above suggests that the great majority of human soldiers, despite their flaws and limitations, do not intentionally mistreat non-combatants. Moreover, although *some* soldiers surveyed believe that torture should be allowed under certain circumstances (2009a: 32), it is not clear that these soldiers actually do participate in torture, even under those circumstances. (Even if some humans do resort to torture in certain contexts, it is unclear how the inclusion of autonomous lethal robotic systems would lessen the occurrence of torture; those humans that desire to torture could continue to do so in the absence of the 'moral machines'.) Thus, although some soldiers may believe that unethical behaviour is sometimes called for, it does not follow that these soldiers have actually dirtied their hands in this way. To be sure, robots would *not* be programmed to be able to torture or rape, and presumably they would not have genocidal tendencies or fascist ideologies. And yet, insofar as humans still take part in warfare, then these moral deviations would continue to be possible *by humans*, despite the presence of ethical robots.

Furthermore, Arkin cites reports that found that 83 percent of soldiers did not agree that civilians should be treated as insurgents (ibid.: 32). Even though this number is promising, it is also unclear what such treatment entailed; acting *as if* an individual is a potential threat until given a reason to believe otherwise does not in itself indicate any *unethical* behaviour; it is the *actual* harm to non-combatants without good reason that would be unethical, not the disposition towards reasonable suspicion. Even though many soldiers reported not being directly informed not to mistreat non-combatants by their superiors (ibid.), this does not mean that those soldiers believed it permissible to, *or actually did*, mistreat non-combatants. *Much more detailed empirical research is required on this important issue* (Sullins 2010). At most, these data may support Arkin's weaker claim noted in Section 1, i.e., that the inclusion of military machines *may* improve things, in certain ways. However, there is not enough here to support Arkin's stronger claim, i.e., that automated warfare would be less immoral than human warfare.

All of this suggests that this putative problem that robots are designed to remedy (i.e., the pervasive unethicality of human soldiers) may not be the calibre of problem that Arkin makes it out to be. The question remains whether or not it is of the level of urgency and significance so as to *require* that we create autonomous lethal robotic systems with the goal of lessening it. The empirical findings that Arkin cites to support his claim of the pervasiveness of the immorality of human soldiers, and to motivate the need for the development and use of autonomous *and* lethal robotic systems, do not seem as worrisome as Arkin suggests them to be. At the very least, the empirical data that Arkin draws upon are open to different and competing interpretations, casting some doubt on their definitiveness. More importantly, the putative abundance of unethical behaviour by human soldiers *alone* does not entail that bringing machines into the battlefield is *the* (or the most appropriate) answer to this problem, even if it is a problem that needs to be attended to, since human unethical behaviour would continue even if robots were present on the battlefield.

Arkin (2009a: 32) also cites findings that suggest that human soldiers sometimes feel unprepared for the situations that they are confronted with in war. Morally salient situations that soldiers are not adequately prepared for is certainly a problem, although it will always be a problem, for both humans and machines, especially as novel and complicated situations arise and as modern warfare changes, and (if) autonomous robots and other futuristic technologies emerge on the battlefield.

Moreover, it is not clear whether machines could perform better in such novel, often morally indeterminate situations. The machine could be programmed to follow the laws of war, at least in principle. However, insofar as novel situations will continue to arise, it is unclear at present whether the robot itself could be adequately prepared (not to suggest that it necessarily would not be); it may be just as (un)prepared as human soldiers report that they are. Presumably, the robot would choose the action that its programming dictates as being the most moral, given the possible alternatives available, in some cases leading to inaction. And yet there would still be room for error here. Moreover, this strategy seems dangerously similar to the fallible 'do the best you can' strategy that human soldiers would predictably adopt in similar cases, something that Arkin wants to guard against. Thus, the introduction of autonomous lethal robotic systems may not remedy this problem that it is intended to remedy.

On the one hand, the unethical activity of human soldiers in the context of war – itself often acknowledged as being an inherently unethical realm of human activity (Walzer 2000, Hawk 2009, McMahan 2009) – seems exaggerated by Arkin. On the other hand, even if the unethical behaviour of some human soldiers is a serious enough problem to deserve sustained and widespread attention, it is not clear that the introduction of autonomous lethal robotic systems would remedy it, or is the most appropriate (ethical, effective, cost efficient, just) way to do so, even if it could. As noted above, alternative strategies are certainly available, including improved moral and legal education of military personnel, the establishment of clearer rules of engagement and international laws of war, more effective deterrents from devious behaviour, increased monitoring of the activities of soldiers (by robots, perhaps), more severe punishments for moral transgressions, and a reduction of the triggers of immoral human behaviour (i.e., stress, fatigue, boredom, confusion, de-sensitization, detachment, anxiety), etc. All of these are reasonable and open options, and may be more cost-efficient than creating and maintaining autonomous lethal robotic systems. Thus, even granting the humanitarian orientation of Arkin's project, he needs to offer independent reasons to justify why he is proposing to develop autonomous lethal robotic systems for the purpose of killing humans, instead of devoting the necessary energy and resources to making human soldiers themselves more ethical. The latter (parsimonious) option would also serve to make warfare less unethical and less inhumane.

A Paradox of Automated Warfare

In light of what has been said above, a paradox of automated warfare begins to emerge. Following Arkin, the goal is to create and use autonomous lethal robotic systems for military purposes. The motivation for this is that robots are expected to be ethically (and militarily) superior to (some) human soldiers, and that automated warfare could presumably lessen the number of human casualties in war.

There are a number of different ways in which automated warfare could manifest itself. Three general manifestations include: (1) *(predominantly) human warfare*, which involves the minimal (or negligible) use of autonomous and semi-autonomous lethal robotic systems by humans in war – the great majority of contemporary military conflicts fall into this category; (2) *human/automated warfare*, which involves a substantial use of autonomous and semi-autonomous lethal robotic systems by humans in war, with a significant number of lethal robotic systems working with or alongside human combatants; and, at the other end of the extreme from human warfare *tout court*, (3) *fully automated (humanless) warfare*, where, outside of the presence of human non-combatants, direct human interaction in combat is negligible, on either or both sides of the conflict.

With respect to (1), it is difficult to see how the inclusion of machines to this minimal extent could have a noticeable and significant impact on improving the moral calibre of warfare overall. In this context, the great majority of combatants are human, and thus come complete with their human fallibilities. Although there has been some research demonstrating that certain kinds of machines already in use have had some impact in this vein, it is unclear at present whether or not such wars are more or less ethical in other ways as well. For Arkin's predictions to hold, and for his project to be worthwhile and well-founded, we would likely need to move to the next stage of automated warfare.

And yet, it is with respect to (2) that *all* of the worries mounted earlier, and other worries surrounding the justness and legality of automated warfare, become most salient. Putting aside the questionable assumptions that I have argued underlie Arkin's rationale, human/automated warfare, of the sort considered throughout this paper, would not clearly be any less unethical (or inhumane) than predominantly human warfare. Even if certain sorts of machines could increase the morality of war in certain ways, other machines may equally detract from this, all in the face of humans continuing to behave immorally on the battlefield. If the worries mounted earlier cannot be overcome, then there is very good reason to resist getting to the stage where this manifestation of automated warfare becomes commonplace. Perhaps, however, removing humans from the context of war *entirely* would be a viable solution.

I see no reason to support this claim, however. Fully automated warfare (i.e., humanless warfare) is undesirable and unproductive, as it would eliminate 'the severity of war', in the sense that human soldiers and the states they represent would no longer have something inherently valuable at stake (i.e., their lives). Indeed, in order for war to make sense, humans will need to remain on the battlefield and be at risk of enemy fire or capture; automated warfare *tout court* is absurd, and hence in order for automated warfare to work, it requires the involvement of humans and human risk in some direct capacity (Krishnan 2009). To be sure, fully automated warfare may be more ethical and humane than the other two manifestations of automated warfare. And yet, machines fighting machines would systematically exclude those things that we are fighting for, and could continue *indefinitely*, given the will and available resources to do so. If such a war came to an end, what would have been settled thereby, or gained by the victor and lost by the defeated? It seems that, *assuming that wars will continue to be waged and fought*, humans will need to continue to be involved in war, whether we like it or not, either alongside robotic systems or as their users and commanders (or both), since wars that do not involve humans and (legitimate) human casualties, and where neither side has anything substantial at stake, cannot be won or lost. In the advent of fully automated warfare, once the machines had defeated the machines of the opposing force, humans would then need to move into the battlefield to settle the dispute themselves.

Yet, insofar as humans are still directly involved in the context of automated warfare (options 1 and 2 above), then these humans would still be at risk of exhibiting the sorts of immoral behaviour identified by Arkin – especially if we devote our attention exclusively to developing autonomous military machines, rather than improving the situation for human soldiers – and hence introducing robots to fight alongside them may not ultimately reduce *human* moral transgressions, or the *overall* immorality of war.[24] Thus, even *if* Arkin is correct to suggest that robots could be more ethical than (some) human soldiers, the presence of autonomous lethal robotic systems in battle may not necessarily reduce the *overall* unethical behaviour performed during war.

It seems that there will need to be a cap on the extent to which autonomous lethal robotic systems should be used, especially if not doing so would mean taking humans out of the context of war altogether. But this means that automated warfare will continue to include human soldiers (if it does continue), complete with all of their emotional, psychophysical, and ethical flaws going more or less uncontested. Doing so entails continuing to sacrifice *many* human lives in battle (both combatants and, collaterally, non-combatants alike), and continuing to endure human moral transgressions, all *in addition to* the ethical challenges that accompany automated warfare (e.g., responsibility for the behaviour of robots, the fairness of remote engagement, etc.). Not only would automated warfare likely continue to witness the sort of moral transgressions already at play in contemporary human warfare, but it also opens up other avenues for unethical behaviour, by both humans and robots, which may actually exacerbate the problem their development and use was intended to remedy.

Conclusion

There may be good reasons to create and use autonomous lethal robotic systems for military purposes. But, the elimination of human unethical behaviour and the overall moral betterment of war are not them, since, as I have argued, introducing these machines into the context of war would in all likelihood not reap these tidings.

In the foregoing I have challenged the assumptions and arguments that underlie Ronald Arkin's view that the development and use of autonomous lethal robotic systems for military purposes is something that we ought to do. Despite certain merits of his view, there are a number of problems with his account that need to be fixed if it is to continue to be accepted. The burden of proof remains on thinkers like Arkin to provide sound argument to show that automated warfare (in any of its three general manifestations) is inevitable. Moreover, even *if* autonomous lethal robotic systems could be more ethical than some human soldiers in certain ways and certain situations, this is insufficient reason for developing and using these machines. And, even if these robots could be more militarily effective, this alone does not entail that they ought to be developed and used for military purposes. Thus, independent motivation is needed, which Arkin does not currently have on offer. In light of this, I take Arkin's case in favour of automating warfare to be a weak one, and, by association, any practice of developing and using autonomous lethal robotic systems that relies on it unconvincing.

NOTES

1. As argued for below, the likelihood of this happening should not be mistaken for the *inevitability* of this occurring.
2. Arkin is not alone in holding this belief. See for example Lin et al. (2008).

3. Something like this is already taking place in Canada. See Canadian Government, Department of National Defence (2010).
4. Walzer (2000).
5. As discussed below, there may be other ways to make war more moral, even aside from the obvious strategy of eliminating war altogether, which would effectively eliminate the potential for *war* crimes in its entirety.
6. *One* way to challenge Arkin's admission here is to note that Arkin is 'criticizing' human soldiers for possessing a similar fallibility that he concedes would be present in machines as well. This raises the question of why we should be including morally flawed (robotic) soldiers as a replacement for other morally flawed (human) soldiers. Indeed, if there are ways to hone in on what makes the *morally solid* human soldiers morally solid, then we may be better positioned to replace morally flawed human soldiers with morally solid *human* soldiers, instead of with morally fallible robotic soldiers.
7. Of course, this will depend on the degree of autonomy that the robot has reached. At a certain level of sophistication, it may turn out that robots would need to be able to disregard orders, in order to meet the requirements for full moral agency. This is a complicated issue, discussion of which is beyond the scope of this paper. See Allen et al. (2000) and Tonkens (2009).
8. Interestingly, because such robots would be programmed to follow the international laws of war, we need to ask – as Sparrow (2007) does – whether their development and use is consistent with those laws in the first place. Indeed, to the best of my knowledge, Sparrow's charges that the use of autonomous 'killer robots' would violate international humanitarian law, and create a morally dubious vacuum of responsibility, remain unanswered by proponents of automated warfare. Without doing so, however, we risk building machines whose very existence goes against the moral and legal codes they are designed to abide by.
9. For a nice discussion on whether automated warfare is consistent with the tenets of received just war theory, see Asaro (2008). See also Sparrow (2007) for a closely related discussion.
10. Arkin is not alone in making these assumptions. See for example Krishnan (2009: 117) and Sullins (2010: 264, 274).
11. Elsewhere, Arkin (2009a: 56) is ambiguous on this point: 'In any case, there is no doubt at the very least that we should *proceed with caution* as these systems *inevitably* move forward toward military use. We must ensure that lethal autonomous systems, *when and if* they are introduced into routine battlefield operations, behave in a manner consistent with international law' (emphasis added). Proceeding with caution suggests that we are (someone is) in control of this enterprise, and thus we can decide the extent to which we pursue this particular path. Yet, if moving in this direction is 'inevitable', then this control may be somewhat illusory. From Arkin's writings as a whole, I believe he is working from the stronger assumptions already noted. If he is *not*, however, then the strength of his underlying motivation comes even further into question; it is only because many people agree that (automated) warfare is inevitable that proponents of automated warfare have been able to get away with *not asking whether we should pursue automated warfare in the first place*.
12. See for example Asaro (2008), Sharkey (2008), and Sparrow (2007).
13. Even if war is avoidable, making it less immoral when it does occur would *prima facie* be a morally desirable task. And yet, without independent reasons for devoting our time

and resources to making *avoidable* wars less immoral (rather than, say, putting those resources into public health and education sectors), we cloak many ethical issues that need to be attended to. See Tonkens (2012).

14. See Slim (2008), especially Chapters 4 and 5.
15. Other issues arise here as well. For example, it may not make much sense to say that a machine could act in a self-sacrificial manner, since it may have no real self to speak of. Moreover, to say that it was acting out of an interest in protecting its fellow (human) soldiers or 'its country' seems to over-extend the meaning of self-sacrifice, as the robot was *merely* following orders. See Vallor (forthcoming) for an interesting discussion in this vein.
16. This example is one of non-culpable ignorance, based on objectively false but subjectively valid belief (McMahan 2009). Similar, less unproblematic cases could also emerge, for example cases of mistaken belief accompanying acts of indiscriminate recklessness. 'Firing at anything with a heartbeat' under some conditions would constitute reckless and negligent behaviour, although in others it may be a reasonable (excusable) response to highly uncertain levels of threat. Thank you to an anonymous reviewer for this point.
17. Elsewhere I have argued that, if machines were to gain a sufficient degree of autonomy (and moral agency in general) then it may be problematic to treat them as if they did not deserve certain rights (e.g., the right to self-preservation). See Tonkens (2011).
18. We can think of cases where, in the heat of battle, human soldiers may be prone to engaging with 'anything that moves'. As robots could be programmed to have a 'higher' standard of assessment and threshold for engagement, in essence championing accuracy and legitimacy over its own personal safety, then such robots would likely not 'fire at anything with a heartbeat'. And yet, it is difficult to imagine a case where a human soldier that opens fire in this way would be behaving *unethically*, assuming that she is being engaged with, and thus has reason to believe that her life is being threatened.
19. For those who disagree that torture is unethical, please see Bufacchi and Arrigo (2006). See also Shunzo Majima's article in this issue of *Journal of Military Ethics*.
20. See for example Sullins (2010: 269), who argues that 'it is unlikely that semiautonomous tele-robotically operated weapons [for example] are enhancing battlefield ethics'. For an opposing view, see Strawser (2010).
21. See Guarini and Bello (2011).
22. One alternative here would be to ground the international laws of war on foundations that humans are capable of adhering to; 'ought implies can', and hence if human soldiers *cannot* abide by the international laws of war (as Arkin suggests), then we may be demanding too much of them, morally speaking. In other words, if acting consistently with the laws of war is practically unrealistic, then (this author is afraid to say) perhaps they ought to be modified accordingly, rather than replacing those who cannot abide by them with autonomous lethal robotic systems.
23. I am somewhat inclined to agree with Arkin that ten percent is a large number when we are talking about the number of human soldiers that behave unethically in the context of war. Two places where we disagree, however, are (1) that this number is high enough to demand serious attention and allocation of an abundance of scarce resources, especially given the fact that most wars are inherently unjust (McMahan 2009), and other sectors of human activity could greatly benefit from such attention and resources (e.g., healthcare, education, welfare); and (2) even *if* this problem deserves our attention and resources,

that creating and using autonomous lethal robotic systems is the *best* way to lower said percentage, given due consideration to *all* the available alternatives.

24. Worth noting is that it may become increasingly unfair to include humans in automated warfare, where their opponents are robots, as it would become exponentially more difficult for them to reign victorious over such opponents, and the robots are not putting similar things on the line (e.g., their lives).

REFERENCES

Allen, Colin, Gary Varner & Jason Zinser (2000) Prolegomena to any Future Artificial Moral Agent, *Journal of Experimental and Theoretical Artificial Intelligence*, 12(3), pp. 251–261.

Arkin, Ronald (2009a) *Governing Lethal Behavior in Autonomous Robots* (Dordrecht: Chapman & Hall).

Arkin, Ronald (2009b) Ethical Robots in Warfare, *IEEE Technology and Society Magazine*, Spring, pp. 30–33.

Arkin, Ronald (2010) The Case for Ethical Autonomy in Unmanned Systems, *Journal of Military Ethics*, 9(4), pp. 332–341.

Asaro, Peter (2008) How Just Could a Robot War Be? in: P. Brey, A. Briggle & K. Waelbers (Eds), *Current Issues in Computing and Philosophy*, pp. 50–64 (Amsterdam: IOS Press).

Bufacchi, Vittorio & Jean Maria Arrigo (2006) Torture, Terrorism, and the State: A Refutation of the Ticking-bomb Argument, *Journal of Applied Philosophy*, 23(3), pp. 355–373.

Hawk, W. J. (2009) Pacifism: Reclaiming the moral presumption, in: H. LaFollette (Ed), *Ethics in practice*, (3rd edition), pp. 735–745 (Oxford: Blackwell).

Government of Canada, Department of National Defence, *Defence Ethics Programme* (2010). http://www.dep.forces.gc.ca/dep-ped/index-eng.aspx. Accessed September 10, 2011.

Guarini, Marcello & Paul Bello (2011) Robotic Warfare: Some Challenges in Moving from Non-civilian to Civilian Theaters, in: Patrick Lin, George Bekey & Keith Abney (Eds), *Robot Ethics: The Ethical and Social Implications of Robotics*, pp. 129–144 (Cambridge: MIT Press).

Krishnan, Armin (2009) *Killer Robots* (Dordrecht: Ashgate).

Lin, Patrick, George Bekey & Keith Abney (2008) Autonomous Military Robotics: Risk, Ethics, and Design. Funded by US Department of Defense/Office of Naval Research; accessed, available at: http://ethics.calpoly.edu/ONR_report.pdf; Internet. Accessed November 11, 2011.

McMahan, Jeff (2009) *Killing in War* (Oxford: Oxford University Press).

Sharkey, Noel (2008) Cassandra or False Prophet of Doom: AI Robots and War, *IEEE Intelligent Systems*, 23(4), pp. 14–17.

Singer, Peter W. (2010) *Wired for War: The Robotics Revolution and Conflict in the 21st Century* (New York: Penguin).

Slim, Hugo (2008) *Killing Civilians: Method, Madness, and Morality in War* (New York: Columbia University Press).

Sparrow, Robert (2007) Killer Robots, *Journal of Applied Philosophy*, 24(1), pp. 62–77.

Strawser, Bradley J. (2010) Moral predators: The Duty to Employ Uninhabited Aerial Vehicles, *Journal of Military Ethics*, 9(4), pp. 342–268.

Sullins, John (2010) RoboWarfare: Can Robots Be More Ethical than Humans on the Battlefield? *Ethics and Information Technology*, 12, pp. 263–275.

Tonkens, Ryan (2009) A Challenge for Machine Ethics, *Minds and Machines*, 19(3), pp. 421–438.

Tonkens, Ryan (2011) Out of Character: On the Creation of Virtuous Machines, *Ethics and Information Technology*, 14(2), pp. 137–149.

Tonkens, Ryan (2012) Should Autonomous Robots be Pacifists? *Ethics and Information Technology*, Online first May 16, 2012. DOI: 10.1007/s10676-012-9292-z

Walzer, Michael (2000) *Just and Unjust War: A Moral Argument with Historical Illustrations*, 3rd edition. (New York: Basic Books).

Ryan Tonkens earned his MA in Philosophy from the University of Windsor in 2006, and a PhD in Philosophy from York University (Toronto) in 2012. His PhD dissertation is an analysis of the ethics of human prenatal genetic alteration, from a character-based perspective. He is currently a postdoctoral research fellow at Novel Tech Ethics, Dalhousie University, where he is conducting research in reproductive ethics and the ethics of parenthood. His primary research interests are in applied ethics, especially ethical issues that accompany advances in biotechnology and artificial intelligence. He has published his work in the *Journal of Medical Ethics*, the *Journal of Medicine and Philosophy*, the *Journal of Applied Philosophy*, *Minds and Machines*, and *Ethics and Information Technology*. *Correspondence address*: Novel Tech Ethics, University of Dalhousie, Halifax, NS B3H 4R2, Canada. E-mail: tonkens@yorku.ca

Moral Predators: The Duty to Employ Uninhabited Aerial Vehicles

BRADLEY JAY STRAWSER
University of Connecticut, Storrs, Connecticut, USA

ABSTRACT *A variety of ethical objections have been raised against the military employment of uninhabited aerial vehicles (UAVs, drones). Some of these objections are technological concerns over UAVs abilities' to function on par with their inhabited counterparts. This paper sets such concerns aside and instead focuses on supposed objections to the use of UAVs in principle. I examine several such objections currently on offer and show them all to be wanting. Indeed, I argue that we have a duty to protect an agent engaged in a justified act from harm to the greatest extent possible, so long as that protection does not interfere with the agent's ability to act justly. UAVs afford precisely such protection. Therefore, we are obligated to employ UAV weapon systems if it can be shown that their use does not significantly reduce a warfighter's operational capability. Of course, if a given military action is unjustified to begin with, then carrying out that act via UAVs is wrong, just as it would be with any weapon. But the point of this paper is to show that there is nothing wrong in principle with using a UAV and that, other things being equal, using such technology is, in fact, obligatory.*

Introduction

Lethal employment of uninhabited aerial vehicles (UAVs) has risen precipitously by a few Western nation-states (most notably the United States) across several theaters of operation (Afghanistan, Pakistan, Yemen, and other locations).[1] The emergence of this technology has sparked widespread debate over the ethical justification of its use. Some claim these drones create a particularly asymmetrical form of warfare that is somehow ignoble or dishonorable. Others contend that UAVs impede certain *jus in bello* principles. Some claim that drones create psychological conflicts for their operators (who are often thousands of miles away) causing unacceptable cognitive dissonance in the mindset of the warrior. Still others raise concerns over drones carrying out targeted killings by non-military government agencies (such as the CIA) and other concerns over their present employment. There is a worry that UAVs could lead to autonomous weapons that make lethal decisions on their own. Finally, some argue that by removing the pilot from the theater of combat a degree of asymmetrical warfare is attained such that the risk threshold for a given state is lowered too far – that it becomes too easy for a state using drones to go to war; thus, their use is ethically pernicious.

In this paper I argue that there is an ethical obligation to use UAVs. Indeed, I hold that, in principle, there is no need for special ethical concern for this weapons system as opposed to any other more standard weapon technology. All of the concerns just listed either miss their mark and do not challenge the ethical obligation to employ UAVs in principle or else do not rise to the level needed to override the principles which form the basis of ethical obligation for UAV employment. I argue that remotely controlled weapons systems are merely an extension of a long historical trajectory of removing a warrior ever farther from his foe for the warrior's better protection. UAVs are only a difference in degree down this path; there is nothing about their remote use that puts them in a different ethical category.

My argument rests on the premise that if an agent is pursuing a morally justified yet inherently risky action, then there is a moral imperative to protect this agent if it possible to do so, unless there exists a countervailing good that outweighs the protection of the agent. Thus, I will contend that, as a technology that better protects (presumably) justified warriors, UAV use is ethically obligatory, not suspicious. After some preliminaries, I will first present the argument for the ethical obligation to use remotely controlled weapons. Then I will walk through the various ethical concerns which are supposed problems for UAV implementation and show how each of these worries is misplaced or fails to adequately counter the ethical obligation for their use.

Remote Weapons as Ethically Obligatory

Media coverage and public debate over the military use of uninhabited remotely controlled weapons is currently *en vogue*.[2] It is surprising then, given such a backdrop, that the case for the ethical obligation to employ UAVs has yet to have been definitively made. That is precisely what I intend to do. First, some distinctions must be made regarding what the target of my claims in this paper will be. Primarily, I am referencing those aircraft presently employed by the United States (and other) militaries commonly known as 'Unmanned Aerial Vehicles' or drones. To avoid unnecessary gender bias I prefer the locution of Uninhabited Aerial Vehicles (UAVs) which I will use throughout.[3] Examples include the General Atomics MQ-1 Predator and the General Atomics MQ-9 Reaper.[4] UAVs have been employed for some time as reconnaissance aircraft, but only fairly recently have such platforms been used for lethal engagement. Critically, when referencing UAVs I only intend those aircraft which are under human control for, at the minimum, any particular lethal action the machine executes. Autonomous weapon systems, which can execute lethal actions apart from a human decision to do so – that can operate 'on their own' – will be addressed below in Objection 1. Finally, my discussion here regarding the ethical obligation to employ UAVs could be applied, with the necessary changes, to any remotely controlled lethal weapon system, including land- or sea-based remotely controlled weapons.[5]

I contend that in certain contexts UAV employment is not only ethically permissible, but is, in fact, ethically obligatory. The basis for this claim rests upon what I call the principle of unnecessary risk (PUR). PUR proceeds as follows: If X gives Y an order to accomplish good goal G, then X has an obligation, other things being equal, to chose a means to accomplish G that does not violate the demands of justice, make the world worse, or expose Y to potentially lethal risk unless incurring such risk aids in the accomplishment of G in some way that cannot be gained via less risky means. That is, it is wrong to command someone to take on *unnecessary* potentially lethal risks in an effort to carry out a just action for some good; any potentially lethal risk incurred must be justified by some strong countervailing reason. In the absence of such a reason, ordering someone to incur potentially lethal risk is morally impermissible. Importantly, PUR is a demand not to order someone to take unnecessary risk *on par* with alternative means to accomplish some goal G. This is what the other things being equal clause is meant to capture. That is, in some cases, the only possible way to accomplish G will be to order Y to undertake a particular means which exposes Y to potentially lethal risk. In such cases, PUR is not directly applicable; whether or not the order is justified must be determined on other grounds. PUR simply demands that no *more* risk than is required for the accomplishment of G (no unnecessary risk) is ordered by X to be incurred by Y.

I take PUR to be uncontroversial. In fact, it is possible that an even stronger form of PUR could be developed that morally bars not only potentially lethal risk, but any risk of bodily harm whatsoever. Further, there may be a reflexive form of PUR available that could entail self-regarding duties not to incur potentially lethal risk unnecessarily. But some may complain that an individual has the moral permission to incur lethal risk in carrying out act X in pursuit of good A even if the risk in no way aids the accomplishment of A (or some other good B) nor is demanded by justice. To avoid such controversy, I employ here the more modest form of PUR as I have developed it. So even if some wish to contend that it is morally permissible for an individual to take unnecessary potentially lethal risks upon his or herself in accomplishing some good, it still seems that PUR holds with no problems, focused as it is upon commanding others to action.[6] That is, if some argue that there are no moral prohibitions against recklessly endangering one's own life for no good reason, certainly morality demands that there is a strong moral prohibition against unnecessarily endangering another's life.[7]

Another important argument can be used for the obligation to employ UAVs over inhabited aerial vehicles. Namely, UAVs are, on par, cheaper to produce and deploy than inhabited planes that accomplish similar missions. Thus, the argument could run, we are obligated to spend as little shared resources as are necessary on any given collective venture (such as a military undertaking), since those resources are scarce and could be used for other worthy goals. A principle of unnecessary waste of scarce resources (PUWSR) could be formulated to capture the normative appeal of such an approach.[8] PUWSR would contend that by not employing UAVs to the greatest extent possible militaries are wasting scarce resources and that UAVs should, therefore, be used in place of inhabited aircraft so as to be better stewards of said shared resources. For, after all, any money not spent on a military venture could be allocated towards other important demands of social justice, such as (say) an egalitarian concern for equal opportunity of welfare.[9] Such a principle, then, could be used to put normative pressure on the financial budgets of Western militaries and demand that efficiency of cost is an important moral issue.

I laud such approaches – and find financial concerns to be particularly relevant in the case of UAV underemployment – but in this paper I set aside such arguments and focus instead on what I see as the stronger normative principle of unnecessarily risking an agent performing a morally justified act. I do this because appeals to principles like PUWSR, while legitimate, are often more easily overridden by other competing normative concerns. That is, even a relatively significant cost difference between two competing methods for carrying out a given act could quickly become moot were there any relevant differences warranting moral concern between the two courses of action. Of course, in this case (UAVs versus inhabited aircraft) whether there are such differences will often be an empirical question. And if, as I assume in this paper, UAVs can carry out similar missions without any significant loss in capability, then concerns over cost would apply just as well. But I still view PUR as a stronger moral claim – one that demands a higher justificatory bar to override – than principles such as the PUWSR; thus, it is upon PUR that I base my central claims in this paper.

Returning then to PUR, an example may help demonstrate its modest moral demands and *prima facie* appeal. Imagine a group of soldiers are fighting in a just war against an unjust enemy. The (unjust) enemy soldiers are, say, invading the just soldiers' country and committing horrific crimes against humanity in the process. In the defensive effort a group of just soldiers, led by Captain Zelda, engage the enemy who are a short 50 yards away. Assume that engaging these enemy soldiers is a just action in pursuit of some good (in this case the good of defending their homes, families, themselves, and other innocents.) Captain Zelda has an idea. She decides to remove her bullet-proof vest, throw down her rifle, and charge the enemies with nothing more than a large rock and chutzpa. She turns to the troops under her command and orders them to do likewise. Set aside whether or not such an action is morally permissible for Captain Zelda to pursue individually. Also assume that charging the enemy in this fashion would in no way aid in accomplishing the good of successfully attacking the enemies yet would dramatically increase the lethal risk her troops incur. PUR says that it is morally impermissible for her to order her fellow troops in her squad to take off their bullet-proof vests, throw down their rifles, and charge the enemy with only a rock since there is no good reason to do so. PUR holds that it is morally impermissible for Captain Zelda to endanger the lives of her troops any more than is necessary for the accomplishment of good A. My argument below for the moral obligation to employ UAVs rests on PUR as a sound moral principle.

Note that such an action as Captain Zelda's planned foolhardy charge *may* contribute to some other thing, Q, which she takes as a good, such as an adrenaline rush or perceived valor gained by taking such inordinate risks. In such a case, one could try to argue that the act passes PUR since it aims at some other (purported) good. But PUR is not in the business of determining whether or not certain goals are goods worthy of being sought. It is a structural principle that functions on permissible commands to others only after it has been determined what the morally proper good to pursue should be. So, granting a proper good, PUR demands that one commands others to incur lethal risk (or increased lethal risk) only in pursuit of that good (or some equal or greater good) if it is necessary in the way defined. That is, the risk one orders another to incur must track exactly with the necessity of that risk in relation to the accomplishment of the purported good. In this case, we are agreeing that the good sought (or the good that *should* be sought) is the successful attacking of the enemy; hence, Captain Zelda's actions are impermissible by way of PUR for they do not aid in the accomplishment of the proper good nor are they demanded by justice or some other good. If Captain Zelda engages in reckless warfare and orders others to do likewise not because of necessity but because of some personal pleasure gained by the excitement of risk-taking and combat, then we would conclude her actions to be morally impermissible for other reasons outside PUR. That is, it may very well be that Captain Zelda orders her troops to make such a brash charge in the pursuit of something she takes as a good. In that case, the reason her action is wrong is not due to PUR but because she is mistaken that this is a good worthy of being sought and ordering others to seek (the adrenaline rush of risky combat, say). But, if we agree that the good that *should* be sought is attacking the enemy, her orders are impermissible via PUR because her commands in no way help in this aim even though they cause her troops to incur (greater) lethal risk.

Granting PUR then, consider the following claim, OP:

> (OP) For any just action taken by a given military, if it is possible for the military to use UAV platforms in place of inhabited aerial vehicles without a significant loss of capability, then that military has an ethical obligation to do so.

I argue that OP is true. It could, of course, very well turn out that OP is only vacuously true because the antecedent is false. This paper will not primarily be arguing for or against the truth of OP's antecedent, but instead assume it is true and argue that the normative consequent follows.[10] The antecedent of OP could be false for any number of reasons. First, it could turn out to be technologically infeasible to transition some military aircraft into remotely piloted vehicles without a significant loss of capability, for various reasons.[11] Or it could be near impossible to do so due to budgetary constraints.[12] Further, it could be that the antecedent of OP is false because remotely controlled weapon systems cannot practice target discrimination as effectively as inhabited vehicles can; and this would constitute a significant loss of capability. Or it could turn out that for some as of yet unforeseen reason remotely piloted weapon systems are not as capable in some other manner as inhabited vehicles. In any such case, the antecedent is false and OP is vacuously true.

There are very good reasons to believe, however, that the antecedent of OP could be true and even likely, as will be mentioned at points below. The central aim of this paper, however, is to establish that the normative consequent follows if the antecedent is true. Further, the antecedent of these claims is an empirical question – one that can be tested for its veracity. What I am investigating is whether there is any principled reason for *not* employing UAVs. I contend that there is not, and further (based on PUR) that there is a strong moral obligation to use them in place of inhabited aircraft. If there is such an obligation, then OP follows.[13]

Note that the 'in place of' criterion of OP is crucial for its derivation from PUR. A given commander in a combat context is obligated by PUR to order her troops to use weapon Z in place of W if and only if Z reduces the risk placed on that soldier in comparison with and as an alternative to W. It is the risk differential between options Z and W that is the source of the obligation. To put it another way, because Z exists and is presently available for the commander to order her troops to use *in place of* W, the commander is obligated *not* to order the use of W so long as Z is a viable alternative that meets the other criteria (such as not violating the demands of justice). That is to say, the ordering to use Z is (presumably) permissible in a just warfighting context; it becomes obligatory only as an *alternative* to W. But, if only W exists (or is the only option for other reasons, such as the demands of justice), then it could very well be permissible to order the use of W. Both W and the less-risky Z must be viable options for the obligation to use Z to instantiate via PUR.

To build the case for OP's consequent, consider the following scenario. Two small towns, Prudentville and Recklessville, each have a local police force that includes a highly trained 'bomb squad.' Each bomb squad has been very successful in disarming and disposing of malicious explosive ordnance throughout the years with few (but some) casualties. Recently, both towns acquired remotely controlled robots that can be used to disarm explosives while being operated from afar. Under the control of a trained explosive ordnance disposal (EOD) technician, these robots are just as capable at disarming bombs as the EOD technicians are themselves working 'hands on.' And with the robots, of course, the EOD technicians are not at any risk of injury or death. After some initial experimentation to ensure use of the robots did not cost them any bomb-disarming capability, Prudentville decides to have their bomb squad use the robots in any situation where it was possible to do so. They viewed the decision as a 'no-brainer': saving the life of even one bomb-technician would be well worth the cost of using the robot. Recklessville decides not to have their EOD technicians use the robots, even though they have them available and are capable of doing so. Thus, they put their bomb technucians at risk for no reason (or no *good* reason, at any rate) and violate PUR.

Take the above story as a guiding normative analogy for claim OP.[14] If it is possible for the bomb squad to use a robot to defuse the bomb remotely, with no significant or relevant loss of capability, then via PUR the bomb squad has a clear ethical obligation to use the robot in place of a human handling the bomb directly. The situation is relevantly analogous with the current and future use of remotely controlled military aircraft. That is, if it is possible for a state to have its military use remotely controlled weapon systems to carry out combat missions instead of inhabited weapon systems, with no significant or relevant loss of capability, then via PUR (assuming military missions carry potentially lethal risks) the state has a clear ethical obligation to do so. This is simply because by operating at a much greater distance from combat, the operator of the weapon system is exposed to significantly less risk. And if there is no compelling reason to expose a soldier to risk, then it is wrong to do so. Hence, OP.

One important caveat: The justification of remotely controlled weapons in war here assumes that their employment is done as part of a fully justified war effort meeting both *jus ad bellum* and *jus in bello* criteria. Thus, if the military in question is justified in a particular military strike in the first place, they should protect the just warrior carrying out the action to the greatest extent as is possible – up until protecting the warrior impedes his/her ability to behave justly in combat, as will be argued below. Granted, if a given military action is unjustified, then it is unjustified whether it is done by a pilot flying an aircraft remotely or otherwise. That is, my argument that the employment of UAVs is ethically obligatory follows out of PUR in that a given military action in question must be a proper good in the first place. If the act is morally unjustified to begin with, then it is morally impermissible for other reasons outside of the scope of PUR. Notice, for example, that this leaves open the possibility that universal pacifism may be the correct moral outlook towards warfare and yet OP still holds (although vacuously, because a pacifist will hold that there simply are no justified military actions).

A related point is that some may here object that my analogy between a bomb squad and a military force fails for the bomb squad is trying to disarm a bomb, and thereby *prevent* the loss of life, whereas a military strike is attempting to *take* life. Yet the point of connection for the analogy is not what, specifically, the given action is attempting to carry out (be it disarming a bomb or delivering a bomb), but simply that a particular action is justified and aiming towards *some* worthy good combined with being inherently risky to the agent engaging in the action. Again, the analogy to UAV use rests on a presumption that a given military strike employing a UAV is justified to start with – if it is not, then the UAV strike is morally impermissible, of course.[15] So the case with the bomb squad is intended to focus on the moral principle of unnecessary risk in the execution of *some* good. The bomb squad, commanded by their town, undertakes morally justified but risky action F aiming to accomplish good goal G. If G can be accomplished just as effectively but with less risk to the bomb squad by a means other than F (such as by using a robot), then there is a moral obligation to use the robot.

The same reasoning applies, with the necessary changes, for a given military force. A military, commanded by their state, undertakes morally justified but risky action F aiming to accomplish good goal G. If G can be accomplished equally as effectively but with less risk to the military members (such as by using an uninhabited drone), then there is a moral obligation to use the drone. That the good G for the bomb squad case is the protection of life while in the UAV case G is the taking of life is not a relevant difference for the analogy. What matters is that G is a good worthy of pursuit.

To put the position another way still: ordering a warfighter to take on risk in any activity must be justified. If a given action can be equally well accomplished via two different methods, one of which incurs less risk for the warfighter's personal safety than the other, then a justification must be given for why this safer method is not used. If there is no good reason not to use it, then we are obligated to employ the safer method. For all cases of ordering a warfighter to undertake any given risky action, there should be a reason that demonstrates why the risk is necessary for the accomplishment of the given objective. If one grants that removing a pilot from the theater of combat by using a UAV instead of an inhabited weapon platform greatly reduces the risk to that pilot, then there should be a presumption for using a UAV (or any remote weapon) whenever it is possible to do so in a way that does not compromise the capability of a given warrior to behave justly. The burden of proof, then, is on those who argue that we should not employ UAVs or similar remote technology. Such a position needs to justify why we should have pilots take on such risk. As mentioned above, there are a variety of objections usually offered as to why UAV employment is ethically suspicious. I shall now review each of these in turn and show why they fail to overcome the claim that UAVs are, in principle, ethically obligatory.[16]

Objection 1: The Move to Independent Autonomous Weapons Systems

Some worry that UAVs lead us down a road toward independent autonomous weapons (IAWs); robots that make lethal decisions on their own.[17] Where to draw the line when a weapon system is 'autonomous' is notoriously difficult.[18] For simplicity's sake here, I refer to any weapon that makes a decision on its own accord to launch a particular lethal attack as 'independently autonomous' (or 'fully' autonomous as is sometimes used). Thus, a cruise missile that guides itself to a target would not be an IAW because a human agent made the decision to launch it and for it to go attack the given target, but a Predator drone programmed so as to make the particular decision to fire on a specific target of its own accord would become an IAW. So long as there is a 'human in the loop' (to use the common military parlance) for each particular lethal decision, I consider it non-autonomous for the purposes of this paper. That is, so long as a human agent makes the decision whether or not to employ lethal force, the weapon is not an IAW as I use the term. The argument against the employment of UAVs runs like this: IAWs are morally impermissible. UAV development will lead to IAWs. Therefore, UAV development is impermissible.

Response

As an objection against UAV usage goes, this fails to counter the moral obligation for their employment. In fact, we can grant the first premise (that

'IAWs are morally impermissible') but dispatch the objection by arguing that its second premise is presently unsubstantiated (that 'UAV development will lead to IAWs'). One could agree with the objection that we should not develop IAWs and that we should not allow development of UAVs to lead us down the road towards IAWs. Indeed, it is plausible that it could be difficult to stop such a progression, but it is not true that the development of UAVs will *necessarily* lead to the development of IAWs. Thus, we need empirical evidence to show that this is the case. The objection is a kind of slippery slope objection because it assumes that the development and employment of UAVs must lead to the development and deployment of IAWs. Slippery slope objections are problematic because they fail to acknowledge a plausible middle ground stopping point. Namely, this objection misses the possibility of maintaining the employment of UAVs while at the same time working for the banning of IAWs (something I recommend Western nation-states do). Thus, at present, this objection fails as an argument against the ethical obligation to employ (and develop) UAV technology.

I raise this objection first so as to make an important distinction for the scope of this paper. In this paper I am only arguing for the moral obligation to use remote weapons that are explicitly non-autonomous, at least regarding any lethal decisions. On my view, the distinction between IAWs and non-autonomous remote weapons is of paramount importance in this debate and is often neglected. One reason it is so important is that if this distinction is neglected and, even more importantly, if this distinction is not enforced and efforts to develop IAWs are not stopped, then objection 1 stands (assuming that one grants its first premise). That is, to be clear, it is entirely possible that the use of UAVs will in fact lead to the use of IAWs. If this can be shown to be the case and if it cannot be stopped, then, since I do grant the first premise, I see it as a legitimate objection against the employment of UAVs. But my hope is that the development of IAWs can be stopped even while UAVs are employed and developed. I do not here have space to argue against the moral permissibility of IAWs – that has been done effectively elsewhere (see Sparrow 2007; Asaro 2006, 2007; Himma 2007).[19]

Some may object that my acceptance of the premise that 'IAWs are morally impermissible' is inconsistent with my use of PUR to ground the moral obligation to use UAVs. The objection would contend that many weapon systems which could (arguably) be considered IAWs offer far better protection of a just warfighter and are thereby obligatory via PUR. Examples could be weapons systems such as the Phalanx Close In Weapon System or the SeaRAM employed by the US Navy when they are used in fully autonomous mode. Without such weapon systems many sailors would potentially be at unnecessary risk, or so this objection claims.[20] But this objection fails to appreciate that PUR, although a strong at first view moral principle, can be overridden by a strong enough countervailing normative reason. In this case, although I do not argue for it here, I find the principled objections to IAWs to be sufficiently strong such that they override the moral demands of PUR. That is to say, it is perfectly compatible and in no way logically inconsistent to hold (as I do) that some non-autonomous weapon systems (such as UAVs) are obligatory via PUR and at the same time hold that IAWs are impermissible on grounds specific to their autonomous nature which overrides PUR. In any case, regardless of whether or not one accepts the first premise of objection 1, the objection on the whole fails because it is a slippery slope argument that is inadequately substantiated.

Objection 2: UAV Limitations Lead to *jus in bello* Violations

Some grant that remotely controlled weapons better protect the just warfighter but argue that they do so at the cost of a decreased ability to discriminate combatants from noncombatants and other *jus in bello* compromises.

Response

Certainly, if an UAV operator engaging the battlefield from thousands of miles away through a video feed is unable to properly adhere to the *jus in bello* principles of discrimination and proportionality, then such drones should not be used. Indeed, if using a UAV in place of an inhabited weapon platform in anyway whatsoever decreases the ability to adhere to *jus in bello* principles, then a UAV should not be used. This is consistent with OP since adhering to principles of discrimination and proportionality are key aspects of a weapon system's capability. And the just warrior's increased protection (which a UAV provides) should not be bought at an increased risk to noncombatants. Martin Cook (2004) makes this point effectively when he discusses the 1999 NATO air campaign waged in Kosovo. It seemed to some that by conducting missions at a minimum of 15,000 feet, NATO was more concerned with force-protection than noncombatant discrimination (see Cook 2004: 126–127). Had the combat missions been flown at a lower altitude they would have put the pilots at more risk but would have been significantly better at discriminating between, say, an ambulance and a military transport. It is the duty of the just warfighter, I contend, to take additional risk upon him/herself if such risk is required in order to better shield innocents from harm.[21] Thus, in arguing for OP, part of the assumption of the antecedent is that the use of UAVs does not hamper the warfighter's (technical) ability to discriminate between combatants and noncombatants nor make judicious decisions of proportionality. Such a technical weakness would constitute a 'significant loss of capability.'

However, there is good reason to think just the opposite is true: that UAV technology actually *increases* a pilot's ability to discriminate. For example, the Israeli government-owned Rafael Armament Development Authority claims that with the new Spike Extended Range precision missile, which is designed to be used by UAVs, they have achieved 'urban warfare precision' (Rafael Advanced Defense Systems 2010). The missile can be launched in a fire, observe, and update mode (as opposed to a 'fire and forget' mode) that 'allows the UAV operator to update the missile, aim, point, or steer the missile off course if the intended target turns out to be a civilian' (Rafael Advanced

Defense Systems 2010). The report goes on to quote an Israeli pilot who has used the weapon system: 'The beauty of this seeker is that as the missile gets closer to the target, the picture gets clearer...The video image sent from the seeker via the fiber-optic link appears larger in our gunner's display. And that makes it much easier to distinguish legitimate from non-legitimate targets' (Rafael Advanced Defense Systems 2010).[22]

And recent studies bear out that UAVs appear to have, in fact, greater technical capabilities at making determinations of combatant status. Avery Plaw (2010) has recently compiled a database combining reports from a variety of sources on the results of United States UAV attacks carried out in Pakistan from 2004 to 2007. This data shows that UAV strikes were far better at noncombatant discrimination than all other methods used for engaging Taliban fighters in the region. For example, the UAV strikes resulted in a ratio of over 17 to 1 of intended militant targets to civilian deaths compared with a 4 to 1 ratio for Pakistan Special Weapons and Tactics Teams team offensives or a nearly 3 to 1 for Pakistan Army operations in the same region during the same time period. Or, compare the 17 to 1 ratio for the UAV employment to the shocking 0.125 to 1 militant to civilian casualty ratio estimate for all armed conflict worldwide for the year 2000 (Plaw 2010).[23] If these numbers are even close to accurate, it seems that there is strong evidence which directly contradicts the central premise of objection 1. That is, UAVs are better, not worse, at noncombatant discrimination.

Regardless, however, whether or not UAVs are as technically capable of making determinations of proper target discrimination is an empirical question. If it turns out that UAVs are not as capable, then OP's antecedent is false and the claim is vacuously true. At present, however, all available evidence points strongly towards there being no reduction in the technical ability of UAV pilots to discriminate as opposed to inhabited aircraft pilots' ability. But, this being an empirical matter, there is no in-principle objection here to UAVs being ethically obligatory for military use.

Objection 3: Cognitive Dissonance for UAV Operators

This objection worries that the use of drones leads to psychological conflicts for their operators causing cognitive dissonance in the mindset of the warrior. The worry can manifest two separate ethical concerns, first that it is wrong to do this to UAV operators – for them to kill the enemy from their 'desk' at work and then go home to dinner and their child's soccer match – that this places an unjust psychological burden on them. The second and greater concern is that this cognitive dissonance will weaken the operator's will to fight justly in several ways (e.g. the operators not taking the warfare as 'real' or serious enough but instead viewing it as a video game; the operators suffering mental problems and post traumatic stress disorder which, because of their distance from the battlefield, could go untreated and unrecognized, causing further problems and leading to inappropriate decisions; and so forth.)[24]

Response

The argument that the ethical justification for UAVs is threatened if UAV operators are more likely to behave unjustly in their combat actions due to this cognitive dissonance is unsound. First, it can be argued that the temptation for the warfighter to commit *jus in bello* violations would actually lessen, perhaps significantly so, once the warfighter is not at risk. The remote pilot can take more time in evaluating a target before firing – to ensure that target is an enemy combatant – than they would be able to otherwise; for in the worst case scenario a machine is lost, not a human pilot. Returning to the bomb squad analogy, in using a robot the EOD technicians do not experience the same level of stress because there is no danger to themselves; thus, they are not as nervous and, presumably, more successful. The same could hold true for UAV pilots making judicious decisions in combat. Once fear for their own safety is not a pressing concern, one would assume the operator would be more capable, not less, of behaving justly.

But perhaps this is not the case. Maybe the distance and disjunct of this level of remote weaponry does create a significant and genuinely new kind of stress on warfighters that might compromise their abilities to behave justly. There is significant empirical work here yet to be done. But even if we grant that displaced combat harms UAV pilots' abilities, first note that there are means of overcoming this problem and, second, that this issue is not a knock against the ethical justification of UAVs themselves. If necessary we could, for example, move all UAV operators much closer to the theater of combat; forcing them to live in a deployed environment, along the same time-zone as the combat, and under more standard battlefield conditions and stresses.[25]

Further, note that all UAV action has the ability to be recorded and monitored. By default since it is remotely controlled, whatever data feed a UAV pilot received can easily be overseen by many others simultaneously and later for review and critique. This added level of accountability could be used to get, if necessary, further added layers of scrutiny over lethal decisionmaking – even demanding more than one officer agree to a kill, for example. Indeed, an entire team of officers and human rights lawyers could oversee every single lethal decision made by a UAV, if desired or deemed necessary. The point is that there are a variety of ways to overcome any concerns that the pilots of UAVs would be somehow less judicious on average than inhabited weapon systems would be. All of this argues against this cognitive dissonance problem as being somehow insurmountable, much less negating the ethical obligation for UAV use in principle. Moreover, even if there is some psychological harm done to UAV pilots that we cannot overcome, it certainly seems that such harm would be less damaging than the expected harm that could come about via inhabited flights.

Objection 4: Targeted Killing by UAVs

Recent media coverage has raised concerns over the use of UAVs for targeted killings, particularly as is currently being done by the Central Intelligence

Agency (CIA) in Pakistan, Yemen, and other theaters of operation.[26] The specific objection is that assassinations fall outside the bounds of acceptable just-war theory/practice and that UAVs somehow make this practice too easy or contribute to it in some unacceptable manner.

Response

Although I will not argue for the position here, I wholeheartedly share the ethical concerns over assassinations.[27] I further share the underlying concerns regarding a non-military government agency carrying out independent lethal operations in a foreign theater. But none of these concerns are restricted in any significant way to remotely controlled weapon systems. The CIA could be carrying out these same missions with a trained sniper or an inhabited aircraft. It is this particular *policy* that is of proper ethical concern here, not UAV technology or use in general.

Some might argue, however, that the UAV makes targeted killing of this sort particularly pernicious because, first, an aerial vehicle flying over airspace is in some principled way different than sending in a ground special forces unit. Second, the objection claims that the battle for the 'hearts and minds' of local nationals in a given theater is significantly worsened by what they view as ignoble warfare; UAVs are thought to be 'cowardly.' And, third, the objection continues, there are some ways in which UAV technology makes such policies easier to execute because of the abilities unique to current UAV platforms.

As to the first concern, this is admittedly an interesting case that could appear to be peculiar to UAVs. Namely, if a nation-state sends a UAV over another sovereign nation-state's airspace they have not sent an actual person or agent over the airspace. This could perhaps leave room for a contrived argument that because no actual person crossed a border no infringement of national sovereignty occurred. Although intrigued by this distinction for UAV weaponry, I do not find it persuasive. For a UAV strike in terms of sovereignty issues is analogous to a long-distance artillery shell fired across the border or other forms of attack that do not involve an agent crossing an actually geographic border such as cyber-warfare.[28] In such cases, yes, no actual person violated the territorial integrity of the sovereign state in question, but, of course, all nations would still (rightly) view such acts as a direct violation of their sovereignty. So, contra the worry, UAVs do not create a special class of weapons that can escape traditional just-war theory scrutiny or respect for territorial integrity and national sovereignty through an odd loophole.

As for the second concern, two points are in order. First, I would argue that it is at least possible that if UAVs are used in line with the rules of warfare, and civilian causalities are not increased (and perhaps even lessened) due to their usage, then there might be no *greater* resistance from a local populace than would be encountered for more conventional weapons. There is some empirical evidence (albeit limited) to back up this possibility (Plaw 2010). Further, the possibility has some intuitive plausibility when we note that the majority of hostile responses to UAVs by local populaces have come, as usual, when they have inadvertently hit civilian targets – but we have seen this same response in other conflicts when similar strikes were delivered from (say) a B-52 bomber flying at altitude dropping munitions. Again, this seems to point to the possibility that the particular platform dropping the bomb (inhabited or uninhabited) is not what generates a hostile response from the people below, but whether the attack was justified and hits legitimate targets.

But perhaps this response fails. There is, admittedly, some strong empirical evidence suggesting just the opposite: that local populaces' particular resistance to UAVs is precisely due to the fact that they are uninhabited. But so be it. For even if the first response fails, recall that my argument for the ethical justification of UAVs requires that there be no reduction in just warfighting capability. So even if it *does* turn out that in a given theater of operation UAVs do, in fact, cause significantly greater resistance from the local populace as compared to the resistance that similar inhabited vehicles would generate (perhaps because the population thinks they are cowardly or some similar response), then they should not be used on OP grounds. Such a limitation would clearly fall under the 'significant loss of capability' clause of OP. And, of course, this is an empirical question, not an in-principle objection to UAVs.[29]

The third concern – that UAV technology makes such actions easier to carry out – similarly does not offer a principled objection to the moral obligation to use UAVs.[30] It is true that the extended ability of platforms such as the Predator to 'hover' and stay in a localized area for long hours, even days, and observe targets, is a clear combat advantage. Many inhabited aircraft do not have such capacities. And, further, some of the remote areas where such strikes are carried out by UAVs are such that they would be inaccessible to similar inhabited weapon platforms. But these facts about the superior capabilities of UAVs do not count against OP. Just as the advent of airpower brought with it many new and often superior ways warfighters could engage in combat (both justified and not), such advantages do not imply anything inherently wrong with airpower as airpower. Further, the mere existence of such advantages does not force policymakers to misuse these capabilities. Certainly, it would be impossible to drop bombs on innocent civilians if planes did not exist. But that some drop bombs on innocent civilians does not make airplanes morally suspicious, but rather those who so use them to drop bombs. The same holds true for the new capabilities brought about by UAVs.

Thus, there is nothing peculiar to UAVs in regards to the ethical concerns over their present use in targeted killings around the globe. It is the morality of the United States' recent policy of targeted killings we must debate here; not the ethical justification of UAVs.[31]

Objection 5: UAVs Create Unjust Asymmetry in Combat

This objection normally runs as follows: The use of remotely controlled weapons by one force against another force that does not have similar technology crosses an asymmetry threshold that makes the combat inherently ignoble. That is, the extent of asymmetry in the combat abilities between two opposing sides becomes so great when one side employs remote weapons that the fight is intrinsically unfair and that, in turn, makes the use of said remote weapons morally impermissible. This position is usually held because in such circumstances one side literally does not take *any* life-or-death risks whatsoever (or nearly so, since its warfighters are not even *present* in the primary theater of combat) whereas the opposing side carries all the risk of combat.

Response

As an objection against the ethical justification for remotely controlled weapons broadly, and UAVs in particular, this commonly heard argument fails. First, if someone holds that justified combat should be a 'fair fight' between sides, at least to some degree, then I wouldd argue that contemporary military engagements crossed that threshold long ago. How fair is the present fight between an F-22 pilot flying at altitude delivering a precision missile and a tribal warrior wielding a rocket-propelled grenade? If there is a moral problem here due to asymmetry, it seems to have occurred long before UAV implementation and is not endemic to them. But, second, even *if* the actual removal of the warrior from the theater of combat represents a truly new level of asymmetry in combat (and perhaps it does), this alone is still no argument against doing it.[32] This is because if one combatant is ethically justified in their effort, and the other is not, then it is *good* that the just warrior has the advantage and is better protected.[33]

Here I am following Jeff McMahan's recent work rejecting the moral equality of combatants (see McMahan 2009). That is, the warrior fighting for a just cause is morally justified to take the life of the enemy combatant, whereas the unjust fighter is not justified, even if they follow the traditional principles of *jus in bello* such as only targeting combatants and the like, to kill the justified fighter. Thus, there is no chivalrous reason for a just combatant to 'equal the playing field' or 'fight fair.' If combatant A fights under a just cause, while combatant B fights for an unjust cause, combatant A owes nothing to combatant B by way of exposing his/herself to some minimal threshold of risk. Thus, it is *right* for combatant A to reduce the risk in an engagement with the unjust enemy.

But even if one disagrees with McMahan's position and the rejection of the MEC, there are still no grounds to object to the protecting of a soldier under the 'fair fight' objection. A MEC advocate would still presumably agree that armed forces pursuing a justified action as part of a just war is justified to do all they can to protect their soldier so long as that protection does not hinder the soldier's ability to follow *jus in bello* principles. The only difference is that a MEC advocate will think the unjust aggressor state enjoys the same allowance to protect their warfighter similarly.[34] That is, even if one thinks that soldiers enjoy a symmetrical position of the right to individual defensive measures in a given conflict, this in no way prevents either side from maximizing their personal defense so long as it is not at the cost of *jus in bello* precepts; indeed, such precepts (under MEC) would explicitly allow it.[35]

Thus, again, the argument for a 'fair fight' fails on two counts. First, it is already overcome by earlier technological advancements because present military operations are already far from fair even without the asymmetry of UAV weapon systems and thus the issue here is not with UAVs properly speaking. And, second, the desire for a 'fair fight' is simply a weak claim in the first place; something akin to an archaic demand of military commanders in eighteenth century warfare to line up their troops across from one another for a 'dignified battle.' There is simply no normatively compelling reason to think a justified military force need have a fair fight anymore than we would think a police force ought not use bullet-proof vests to offer dangerous criminals a fair fight.[36]

But perhaps this still does not give the objection its due. Paul Bloomfield once remarked that simply the idea of 'being killed by remote control' is powerful and disturbing.[37] The intuition seems to be that killing someone in such a manner is profoundly disrespectful; that a human being deserves to be able to at least point at his or her killers (and condemn them, if they are unjust) even if his or her killers are cruising 20,000 feet above in a plane. The thought is that at least a human being in a plane high above is less of a 'faceless' death wrought upon someone than a robot being operated remotely would be. Or consider the sentiment Uwe Steinhoff raises in discussing remote weaponry generally and how the odd risk asymmetry it creates (making the targets of attack 'defenseless') does not feel like 'honorable' warfare:

> To be sure, I do not deny that there is something fishy about attacking the defenseless. What is fishy about it might be captured very well in this passage: 'The pilot of a fighter-bomber or the crew of a man-of-war from which the Tomahawk rockets are launched are beyond the reach of the enemy's weapons. War has lost all features of the classical duel situation here and has approached, to put it cynically, certain forms of pest control' (Steinhoff 2006: 7).[38]

It must be admitted that there does appear something ignoble or dishonorable in such a vision of warfare as 'pest control' that Münkler's quote describes. Perhaps it is that such distance makes warfare seem too clinical or cold-hearted.[39] Many will have sympathy with such a sentiment when envisioning UAV warfare – myself included. But whatever this sentiment is, it does not amount to a normative argument; such a 'feeling' does not constitute a moral reason for rejecting UAV use. Something being disturbing does not by itself make it wrong. This sense of the ignobility must be elucidated into a coherent and compelling ethical argument against using UAVs; mere displeasure at imagining their employment does not help us. As Steinhoff writes,

> Judged from a traditional warrior's code of honor, a code that emphasizes, among other things, courage, there is nothing honorable in killing off defenseless enemies (whether it is therefore already *dis*honorable is yet another question). But honor and morality are not the same, and honor and the laws of war are not either. In short, the prohibition of

> assaults upon the defenseless is neither an explicit nor an implicit principle of the laws of war or of just war theory (Steinhoff 2006: 8).

Steinhoff is certainly right in this. I would add that a crucial element in how one 'feels' about imagining such warfare depends on whether or not the precision missile strike in the picture envisioned is justified or not. Is it a military strike as part of a fully justified defense against an aggressing, unjustified, destructive enemy force? Is the strike hitting a legitimate and morally culpable target? If it is, such factors temper our view of the strike considerably and move us away from the 'pest control' picture. In such a case, we should desire that the just warrior be well protected from any possible threat that this enemy might proffer – protection that the UAV affords.

Objection 6: Reduction of the *jus ad bellum* Threshold

The worry here is that the asymmetry in combat abilities created by the advanced technology of UAVs, and in particular by the massive reduction of risk to the UAV pilot, makes it too easy for the nation employing UAVs to go to war.[40] That is, the asymmetry created by UAVs lowers the *jus ad bellum* threshold such that more unjust wars might be conducted because the risks of war to a nation-state could become so minimal.[41]

Response

This objection, on first glance, may appear to be the strongest objection to the implementation of UAVs. The worry that it will be easier to go to war if we have technology X, and thus more tempting to enter into unjust wars (making more unjust wars more likely), is intuitively plausible. But this kind of argument ultimately fails for the objection does not succeed in negating the present moral imperative to use UAVs as derived from PUR. To see why this is, consider two possible worlds, Alpha and Beta. In Alpha, nation-state Zandar has developed the technology to make bullet-proof vests for its military members to wear in combat which significantly decreases the risks they incur in battle. Zandar, in accordance with PUR, produces these bullet-proof vests and has its military members wear them. In world Beta, nation-state Zandar has developed the same technology and has the bullet-proof vests available. However, it reasons that if it uses these bullet-proof vests, war would 'cost' it less in terms of risks to its own troops and, thus, be easier (and thus more tempting) to wage. In such circumstances, Beta-world-Zandar worries, more unjust wars are more likely. So it decides not to use bullet-proof vests in order to make war more costly to wage (by intentionally increasing the risk to its soldiers) in the hopes that this will lessen the likelihood that Zandar will engage in an unjust war in the future. Aside from this one decision, the worlds Alpha and Beta are identical.

Let us assume that it turns out Beta-world-Zandar's reasoning was correct. That is, going forward from this juncture there does, indeed, end up being some greater number of unjust wars waged in world Alpha than in world

Beta. The use of the bullet-proof vests in some way lowered the threshold for going to war for Alpha-world-Zandar enough that it made a positive difference on the total number of wars fought – which included some unjust wars. I still contend that Beta-world-Zandar's decision was morally impermissible. This is because the normative force of PUR upon present actions is too strong to overcome such weak predictive calculations of future possibilities.[42]

I will show why this is shortly, but first note that the scope of this issue far exceeds UAVs and bullet-proof vests, of course, but strikes at *any* asymmetry in military technological development whatsoever. Any improvement to a given military's capabilities that gives it an advantage over its potential enemies will face the same objection offered here against UAVs. But that would mean that this objection could be used to block the development and implementation of *any* military technology that creates any asymmetry. Further, the objection could actually be employed to work backwards: that current militaries should intentionally reduce military capabilities in order to make war more costly to them since doing so would place their soldiers at greater risk. Following this logic could even lead to the conclusion that a state should have their militaries throw away their weaponry and all defensive technology, for certainly a neutered military would be less likely to engage in unjust wars in the future.

I grant that this worry about asymmetry created by improvements in military technology making it easier to go to war may well be a legitimate concern. But it is a logic that quickly runs to demanding no military technology whatsoever in the hopes of avoiding future unjust wars. Perhaps this is correct. Perhaps there should be no militaries. But notice that we are now a far cry from arguing over UAV technology. We are arguing over the existence of any military weaponry or advancement whatsoever. If so, then this is not actually an objection *specific* to UAVs in principle. Moreover, if objection 6 is correct in this way, then OP still stands – it is just a vacuous claim: as would be *any* claim about the possibility of justified use of military weaponry of any kind.

But the problems with the objection run even deeper. As I alluded to above, the reasoning by Beta-world-Zandar not to use the vests, notice, rests on epistemically dubious calculations that are predictive about *themselves* doing something wrong in the future ('we might be more likely to do wrong action X down the road') over epistemically solid calculations to protect their own just warfighters presently ('our soldiers will be safer *today* if they wear the vests.') Notice what odd moral reasoning would be occurring were objection 6 to work: *because* we will most likely behave unjustly in the future, we should behave unjustly in the present (by violating PUR in choosing not to protect our warriors as best we can) in order to try to prevent ourselves from acting unjustly in the future. If that holds, we have a strange account of moral epistemology at work, to say the least. We should forego taking presently morally correct action A in order to help restrain our future selves from the likelihood of committing morally wrong action B. In other words, we should do something wrong now in order to (hopefully) better stop ourselves from doing something wrong in the future.

This seems odd, although there could perhaps be cases where such decisions are the right actions – the lesser of two evils, perhaps. But notice that the Beta-world-Zandar decision is not a straightforward case of present self-restraint to limit future potential wrongdoing for, presumably, *usual* cases of present self-restraint are not acts that are themselves impermissible. For example, imagine a man, Tom, who knows he tends to get very angry and do intemperate things. Tom decides he should lock up his gun in a safe and give the key to it to a trusted friend. Tom does this present act to restrain his future self from doing something wrong. But Tom's locking up his gun is not an impermissible act viewed on its own. Violating PUR by not protecting just warfighters is a presently impermissible act viewed in isolation. Thus what makes the reasoning of Beta-world-Zandar's decision so strange: they are intentionally putting their soldiers at greater risk now (which would be considered impermissible in isolation) in order to restrain themselves from doing something impermissible in the future. The comparison back to Tom would be if Tom decides to punch his friend Sam now (which is impermissible) because it will help him not do something worse in the future (such as kill Bob). If that is actually the case, then this *could* be the right thing to do. But notice that we would require a rather high level of epistemic certitude for Tom's knowledge of this scenario – and that there is no other means to avoid killing Bob – in order to deem his act justifiable. That is, if Tom has near *certainty* that the only way to prevent himself from killing Bob in the future is by punching Sam now, then perhaps it is a justified act. But one wonders how Tom could *ever* have such epistemic certitude predicting future acts. The same is true for Beta-world-Zandar.

But perhaps it is still possible that such a decision is justified. This is because, one could argue, I am here equivocating on the moral weight of the present wrong of failing to protect just warfighters and the future potential wrong of more unjust wars. If they are of vastly different moral significance and consequence, then perhaps it is justifiable to do a lesser wrong now in order to increase even the slightest chance of avoiding a much greater wrong in the future. I grant this possibility. Indeed, one could argue that such a decision is directly in accord with PUR since the good of avoiding future wars is a greater good than the present protection of just soldiers (that is, some would argue that such a calculus is demanded by justice).

The trouble with applying this to our present case is the high degree of epistemic uncertainty we have in predicting future states of affairs, particularly future decisions to go to war. That is, even if the wrong of sacrificing the protection afforded to just soldiers is a lesser evil than the possibility of future unjust wars, we have complete confidence in the *present* wrong occurring but we would be far less than certain of the future wrong occurring. In other words, the *odds* of that future wrong occurring will matter and it is unclear how we could reliably predict such odds. The odds need not be equal between the wrongs, of course. If the greater evil was great enough, we could perhaps need only a relatively small chance of its occurrence to outweigh even the certainty of a lesser evil. But, again, it is unclear if we can even have *that* small level of epistemic confidence that a given weapon technology (be it bullet-proof vests, M-16 rifles, or UAVs) would lead to greater instances of unjust wars; or at least not the level of confidence we would need to trump the demand not to commit the present wrong.

So even if it turns out that future worlds would be ones with less war were we to intentionally limit our own military technological development, we cannot have enough epistemic certainty in knowing this presently to overcome the demands to protect the just warfighter. Short of a crystal ball, I cannot imagine how we could ever have the level of certainty for predictive knowledge claims of future group behavior that would be necessary to claim that the future *possible* good should outweigh our present moral duty not to unnecessarily risk others. Perhaps it can be done, but this must be demonstrated before we intentionally bring unnecessary risk upon others. That is, one would have to show how we can have such epistemic confidence that we are not violating PUR (via the demands of justice override) in not presently protecting just warfighters. The burden of proof will be on those claiming that we must presently undertake an act that we would usually consider impermissible in isolation in order to avoid a future evil that we do not have complete confidence in. Hence, although it is certainly possible that use of UAVs could lower the costs of going to war for a given state and, thereby, lower the threshold for going to war such that a state might have an increased likelihood of engaging in a war that is unjust, such predictions cannot be the basis for demanding an intentional violation of PUR given our present epistemic limitations.

This is an unhappy conclusion. While I have great sympathy for the worry, it seems PUR is too strong to overcome with such shaky future predictions as to the unethical decisions a future state would make. And, again, if we allow this block against UAVs it would set in motion a moral principle that would not stop at UAVs but encompass all military technology – not just its future development but retroactively demand that present military technology creating force asymmetry be intentionally reduced.[43] If this is sound we could eventually be back at demanding that Captain Zelda *should* be required to fight with no bullet-proof vest, no rifle, and with only a rock in order to make war 'cost us more' so that we would be less likely to engage in an unjust one. But this is absurd. If a war is just, we are obligated to protect the just warfighters engaging in it. UAVs do precisely that.

Conclusion

UAVs will have an increasingly large presence in military operations on all levels; this much appears increasingly inevitable. Here I have made the case that any new technology that better protects the just warfighter is at least a *prima facie* ethical improvement and is morally required to be used unless there are strong countervailing reasons to give up that protection. I have argued that if using UAVs (as a particular kind of remote weapon) does not incur a significant loss of capability – particularly the operators' ability to engage in warfare in accordance with *jus in bello* principles – then there is an ethical obligation to use them and, indeed, transition entire military inventories to UAVs anywhere it is possible to do so. In fact, I endorse the stronger claim that such a proposed transition would not only be feasible without a significant loss of capability but would actually *increase* weapons systems capability *and* the ability to fight justly.[44] All of the concerns regarding UAVs presently on offer do not negate this ethical obligation to use uninhabited weapon systems and should be properly viewed instead as indictments against mistaken policy decisions and specific instances of force application – not as principled objections against UAVs themselves for none of the concerns are endemic to UAVs in any significant way.

Finally, I note that this paper is in the odd position of arguing for the ethical obligation to use UAVs for a putatively just military action in the current context wherein much, if not all, *actual* UAV employment is part of military actions that are morally questionable or outright impermissible. The particular contemporary circumstances and misuses of UAVs, however, do not trump the moral principles underlying the ethical obligation to employ UAVs for *just* actions. Indeed, this highlights the central point well: the first question for the morally permissible use of *any* weapon technology is, of course, whether the military action itself is morally justified. If it is not a justified undertaking in the first place, then it does not matter if it is carried out via a crossbow, a sniper rifle, or a UAV; it is morally impermissible regardless. If the act is morally justified, however, we are obliged via the demands of PUR to protect the agent ordered to carry out that action as best we can; be it a police officer apprehending a dangerous criminal, an EOD technician disarming a bomb, or a just warrior fighting an unjust enemy. Hence, the ethical obligation to employ UAVs.

Acknowledgements

The author would like to thank Jeff McMahan, Martin Cook, Uwe Steinhoff, Avery Plaw, Paul Bloomfield, Shawn Kaplan, Andrew Ely, Alexis Elder, Donald Joy, Matt Hallgarth, John Sherman, Casey Johnson, Abbilynn Strawser, and Bill Rhodes for helpful discussions, comments, and editorial advice on this paper throughout its various stages. Special thanks are owed to Stephen Kershnar, in particular, for his thorough criticism and help. Thanks as well to an anonymous reviewer for the *Journal of Military Ethics* for his or her very helpful comments. Thanks are also due to the participants at the 2010 International Society of Military Ethics Conference and the participants of the 7th Global Conference on War and Peace where some early sections of this paper were presented. Finally, the author is indebted to the members of the UConn Ethics Summer Reading Group for a thorough discussion of the paper.

Notes

[1] And there are other locations where it is presumed (although not verified) that lethal UAV employment has taken place, such as in Gaza by the Israel Defense Forces. In this paper I will refer to these remotely controlled weapon systems primarily as UAVs (Uninhabited Aerial Vehicles) and occasionally

as drones. UAVs that are used for lethal combat purposes are sometimes referred UCAVs (Uninhabited Combat Aerial Vehicles), but I will not use that locution here. See below for further clarification of these terms and some important distinctions. For a far more in-depth classification of various types and kinds of remote weapon systems see Sparrow (2009).

[2] Singer (2009) gained wide press and much of his work discusses these ethical concerns over drones. Recent events such as the potential *jus in bello* violations wrought by Predator drones in Afghanistan – have received international media attention. Various human rights watchdog groups raised alarm over recent Israeli strikes in Gaza using the Predator platform supposedly against noncombatants. In the past year alone, publications such as *The New Yorker*, *The Atlantic*, the *Washington Post*, *Scientific American*, the *New York Times*, and media outlets such as National Public Radio and Public Broadcasting Systems, have all had substantial reports and several highly critical opinion pieces on the use of UAVs. Additionally, there are many in the United States military community itself who do not question the efficacy of UAV usage but rather have principled worries concerning their use such as those mentioned above. It is currently a 'hot topic' at professional military ethics and development conferences.

[3] Occasionally, such aircraft are instead referred to as Remotely Piloted Vehicles (RPVs). This is especially the case in present US Air Force usage which could be due to public relations concerns over worries regarding autonomous weapon systems. That is, 'RPV' emphasizes that these vehicles are still controlled by human pilots. That we see a move away from the UAV moniker back to the RPV idiom in common discourse is telling of the felt need by some in the military community to emphasize that these aircraft still require human pilots (particularly the pilot community in the US Air Force). For more on this phenomenon see Fitzsimonds and Mahnken (2007). See Sparrow (2009) for more fine-grained distinctions on the kinds of uninhabited weapon systems and their classifications. Also see Sparrow (2007).

[4] See all of the following for good expositions of the historical trail leading to present day lethal UAVs: Singer (2009), Card (2007), Mustin (2002). For a good overview of the planned future of UAVs, see Office of the Under Secretary of Defense (2006). Note that technically speaking, UAVs are not individual aircraft but weapon systems involving several aircraft and ground control stations.

[5] Such as the land-based Foster-Miller's Special Weapons Observation Remote Direct-Action System (SWORDS) or Qiniteq's Modular Advanced Armed Robotic System (MAARS) weapons. See Arkin (2009) for a good overview of such weapons.

[6] Some argue that PUR as presented here is false due to the possibility of someone entering into a foolish agreement with another. Imagine if X signs a contract with Y to follow his commands no matter how stupid or irrational they may be. In such a case, if Y orders X to incur potentially lethal risk for no good reason, then X cannot claim that his right is infringed. Assuming X entered into the agreement with informed consent and was not under coercion or exploited, one could argue that Y's order is not morally impermissible. (Thanks to Stephen Kershnar for this objection.) If one wish to grant this is possibility, then PUR can be amended to reflect those relationships where X enters under the authority of Y on the assumption that Y will not order him to take on risk for no good reason; that is, X assumes Y will follow PUR. At any rate, I think most military members in Western militaries implicitly expect their commanders not to risk their lives unnecessarily.

[7] That is, in my view there may very well be a self-regarding duty to oneself that is entailed by PUR, but I set aside the possibility here to avoid libertarian objections and paternalism concerns. I am strongly inclined to think, however, that there is a self-regarding form of PUR that could hold up against many libertarian objections so several initially apparent counter-examples to PUR would not actually hold. For example, some may offer the activity of skydiving as a morally permissible act to undertake even though it involves incurring potentially lethal risk. But PUR would allow for the moral permissibility of this action for, presumably, the lethal risk involved in skydiving is actually a necessary part of the good sought by the action. In this case, the good is the thrill and excitement of the act of jumping out of an airplane. Thus, the 'rush' sought after by such skydivers (among other possible goods they seek when undertaking the activity) requires taking on the risk. So, according to a self-regarding PUR, it would be morally permissible for an agent to take on the risk of skydiving because that risk directly contributes to the good sought. Whether or not seeking out death-defying activities for the sake of an adrenaline rush is a good that should be sought (or a good at all) is another matter. The PUR does not resolve disputes over what is and is not a good to seek, but rather only the moral demand not to unnecessarily incur lethal risk in seeking an established *putative* good.

[8] Thanks to an anonymous reviewer for suggesting such a principle.

[9] Indeed, I think there are strong arguments that can be made against resource expenditures for military ventures in general when it is shown the plethora of other good ways such resources could be alternatively spent. It is an empirical question, certainly, but it is not implausible in the least to imagine that, for example, the roughly $750 billion dollars the United States has spent thus far on the war in Iraq could not have been spent in other ways that would have done far more good in the world. But that is a debate for another paper. For an interesting presentation of the various trade-offs military expenditures impose on a populace, see the National Priorities Project (2010) for the literal financial cost of war.

[10] Notice that, if true, then OP carries with it a corollary to pursue the development of and transition to an all-UAV military force: (OPT) For any given state, if it is possible to transition its entire military inventory of inhabited aerial vehicles to UAVs without a significant loss of capability, then that state has an ethical obligation to do so.

[11] Although it certainly appears to be technologically possible since there are already UAVs in operation. See below for some discussion on this.

[12] Although, granting the ethical obligation to protect the just warfighter I lay out below, this would have to be a truly astronomical cost – particularly in relative comparison to the amount presently spent on defense budgets. Of course, it's entirely possible that morality demands resources be spent on other things entirely outside of defense costs, such as education, development, and the like. But that is another matter.

[13] Notice that although this paper is focused on lethal UAVs, the corollary claim of OPT (see note 10) would hold for all aircraft. That is, even cargo planes and the like (even those used to transport soldiers), should be transitioned to UAVs. The idea is simple: risking one less person on the flight (the pilot or pilots) is better than risking them if not necessary. However, of course, it is quite possible that troops would refuse to fly on a plane without a present pilot. If that is the case, then that would be a 'significant loss of capability' and so, perhaps, UAVs are not equally capable as inhabited aircraft in the case of troop transporting cargo planes. An example of such platforms for small cargo loads is the Mist Mobility Integrated Systems Technology (MMIST) CQ-10A SnowGoose which is already operationally capable.

[14] Singer (2009) also discusses the use of bomb-disarming robots and the connection between their use and the use of remote robots and weaponry more broadly by military forces.

[15] Thus, while it is true that UAV usage is ethically impermissible in such instances, so too would *any* kind of strike via *any* kind of weapon system (inhabited or uninhabited) be impermissible; the impermissibility does not derive from UAV-specific employment. That is, it is not the UAV *qua* UAV that makes such a strike impermissible.

[16] I will note briefly that I am not even airing objections that claim UAV employment is 'weak' or somehow not 'tough enough' or 'cowardly.' These responses against UAVs derived from some type of wrong-headed machismo are certainly common but, one hopes, are not taken seriously by any military policy decisionmakers. See below for discussion on how the perception of the UAV by enemy forces, however, could have an impact on determining its capability.

[17] Some worry over this development and others laud it. Arkin (2009), for example, sees the development of UAVs as in-line with the development of IAWs and focuses his work on how to develop such autonomous systems to follow the Laws of Armed Conflict and various Rules of Engagement (how to give the machines an 'ethics upgrade'). Arkin contends that such developments are moral improvements and should be pursued with vigor.

[18] For some helpful efforts to this end, see Sparrow (2007).

[19] Indeed, precisely because I do agree with the first premise of objection 1, I argue that now is the time to institute policies that would block IAW development even while we develop UAV and other remote weapon systems that are human controlled.

[20] Many thanks to an anonymous reviewer for this objection.

[21] This position is controversial and widely discussed. Or, better, the entire notion of who should bear risks in any conflict is greatly contested, but most of the debate hinges on questions of liability, debates over the doctrine of the moral equality of combatants (MEC), and distinction issues. If a just warrior was fighting in a truly just war and the innocents in question were truly innocent and in no way liable, some will argue their moral status (the just combatant and the innocent) is equal. And, thus, while just warriors should do all they can to shield the innocent from harm, they should not treat the

noncombatants' worth as above their own safety. And once the doctrine of the moral equality of combatants is disposed of (see below on McMahan), it can become difficult to sustain my view (that the just warrior ought to bear the burden for shielding innocents from harm) for all cases, particularly once strict boundaries between combatants and noncombatants are questioned. Seth Lazar has argued that McMahan's position becomes untenable in precisely this way because liability will extend to far too many noncombatants in ways that should make them justifiable targets in McMahan's rubric. See Lazar (2010). Steinhoff (2008) has challenged McMahan on his rejection of MEC by using claims from within McMahan's own theory. See McMahan (2008) for a thorough response.

[22] *Nota bene*: In this objection I am focusing solely on the *technical* ability of UAVs to discriminate properly; I will consider the impact on the psychology of the warfighter (and subsequent *in bello* worries therein) below in objection 2.

[23] And see Plaw (2010) cited above for all the various references used in creating the database. Regarding the estimate for the global causality ratio, see Osiel (2009: 143) and Kaldor (1999: 8).

[24] Many thanks to several UAV pilots for firsthand accounts and discussion of these phenomena (their identification is withheld by request). See Wallach and Allen (2009) and Singer (2009) where the cavalier attitude of treating UAV operations like a video game is discussed.

[25] Militaries could even go so far as to force UAV operators to live in bunkers and set off fake mortar rounds and so forth around the compound in order to make it feel more 'real' if such effects were shown to help overcome this supposed problem of cognitive dissonance caused by being too far from the battlefield.

[26] See, for example, Mayer (2009). Notice that nowhere in this long article does the author ever discuss the ethical justification of UAVs themselves. As nearly all recent discussions of UAVs in the public square do, it goes into great detail regarding the ethical concerns raised by targeted killings, sovereignty issues regarding operations in Pakistan, worries over military functions being carried out by a non-military department of the government (the CIA), and so forth, without ever discussing the ethical justification of UAVs in principle.

[27] For an interesting discussion of the justification of assassination see Kaufman (2007). See also Kershnar (2004). For an argument that many leaders should not be treated with non-combatant immunity but are legitimate targets, see Kershnar (2005). Presumably, the UAV-targeted leaders of Al-Qaeda and the Taliban would fit as such targets under Kershnar's argument. Gross (2006) gives the argument that targeted killings cannot fit into a proper moral category. If they are an extension of law enforcement, they fail due process, and assassination as self-defense seems implausible, or so Gross argues.

[28] For a helpful account of the future of this new form of warfare, see Dipert (2010).

[29] I will note, however, that such an empirical question would be very difficult to determine in many contexts. That is, the question is not if a given local populace would display resistance to munitions being dropped by UAVs (that is likely). Rather the empirical question that would have to be determined is if that resistance to UAVs is significantly greater than what resistance *would be* encountered via an inhabited aircraft. And that would be hard to determine, to say the least.

[30] This kind of objection is raised by Phythian (2010).

[31] For a helpful discussion of this point see Kolff (2003).

[32] A further reason to think that the removal of the warrior from the theater of combat in itself is nothing particularly new is the existence of Intercontinental Ballistic Missile operators and the like who are certainly removed from the theater of combat where their munitions would be delivered. (Thanks to an anonymous reviewer for this point.)

[33] An important point to note here is that the operators of UAVs would be considered combatants under the traditional just-war theory rubric. (Thanks to Uwe Stienhoff for raising this point.) I'll remain neutral on this point (if for no other reason than that I reject the moral equality of combatants thesis), but note that whatever one's division of combatants, the UAV pilots would certainly still qualify. This has the (perhaps odd) result of meaning the UAV operators would be legitimate targets under most just-war accounts, even though they would be conducting operations from their office thousands of miles away (at places like Nellis Air Force Base, Nevada.) So be it.

[34] Interestingly, McMahan holds that those who fight without a just cause cannot, in principle, ever satisfy the *jus in bello* principle of proportionality. See McMahan (2004). But a traditional advocate of MEC would disagree, of course. For an attempt to defend some of the traditional elements entailed by MEC, see Benbaji (2008).

[35] For a good discussion of some of the complexities and difficulties of symmetrical rights to personal defensive measures enjoyed by soldiers on the traditional just-war theory model (contra the asymmetry of a right to personal defensive measures in individual self-defensive cases where culpability is included in such determinations) see Emerton and Handfield (2009). For a good note on this issue of liability and defense see McMahan (2005: 10) and McMahan (2009).

[36] Additionally, there is a different concern with the asymmetry created by advanced technology which worries that the asymmetry makes it too easy to go to war and thus lowers a nation's *jus ad bellum* threshold too far. I will address this separate concern below.

[37] Paul Bloomfield, personal correspondence, 12 July 2010 and discussions held at the University of Connecticut ethics reading group, summer 2010.

[38] Steinhoff is quoting from Münkler (2003: 234), Steinhoff's translation. In the original paper Steinhoff is discussing the general lack of bravery involved in attacking the defenseless as part of the debate over torture. The quote is referencing an attitude against any remote weaponry where the warrior has effectively removed his/herself from risk, and, as such, can easily be applied to UAV usage. Note that it further affirms the point above that if there *is* some kind of dishonorable fight for UAVs due to the asymmetry of the combat, the threshold was crossed long before UAVs (as in Münkler's referencing Tomahawk missiles) and is therefore not endemic to them, in particular, but to modern warfare across the board. The use of the phrase 'pest control' to describe what is seen as this particularly non-courageous form of warfare was discussed in my panel at the 7th Global Conference on War and Peace, 2010, Prague, Czech Republic. Many thanks to Uwe Steinhoff for permission to use the remark and directing me toward Münkler's work.

[39] Bloomfield argues that the root of the 'pest control' worry is our aversion to being treated as pests ourselves were we to be attacked via remote control. If UAVs were used by our enemies against us, we would think they are wronging us in some way to kill us in this manner (wronging us over-and-above the killing itself, that is.) Thus, the thought runs, we should extend this respect for all people into all contexts, even against the unjust enemy: that is, all humans deserve the respect of not being killed via such remotely controlled 'pest control' measures.

[40] Notice I say the 'massive reduction in risk' not total reduction for, presumably, on most just-war theory accounts, the UAV operators would still be considered liable targets for attack since they would most certainly be combatants. For more on UAV operators being legitimate targets, see Singer (2009: 386).

[41] Thanks to the audience at the 2010 International Society of Military Ethics Annual Conference in San Diego for the thorough discussion of this objection.

[42] It could be objected here that this analogy does not hold because of different obligations that arise from 'purely' defensive military technologies (such as bullet-proof vests) as opposed to those offensive weapons that serve to increase defensive capabilities (such as UAVs). The distinction between offensive versus defensive military capabilities is contentious for many reasons. For one, anything (such as a vest) that increases a soldier's defensive abilities will thereby increase that person's va;ue as an offensive force. But, I will not argue this point here. If one is convinced that my story regarding Zandar does not apply to UAVs due to the offensive/defensive distinction, then the entire thought experiment could be re-cast with the use of eighteenth century muskets versus the use of contemporary M-16 rifles as the competing choices. The muskets (clearly offensive weapons) would reduce the troops' defensive capabilities because they take longer to load, are less accurate, etc. After replacing vests with M-16s, the results of this thought experiment, *mutatis mutandis*, would be the same. Thanks to Donald Joy for helping to develop this point.

[43] I should note, of course, that there is a long history of arguments for disarmament that proceed precisely along these grounds; particularly for nuclear weapons. Notice, however, that the strongest arguments of this type are advanced against particular technologies that are viewed as ethically problematic in principle in isolation from other evils. That is, there is something wrong with (say) nuclear weapons or landmines in principle (the inability to discriminate, etc.) that provides an impetus for banning them in the first place, wholly apart from what future harms they could make more likely. *Additionally*, by reducing them we reduce the future chance of their unjust use. But this is precisely because any future use of them would be unjust so we can have a *certainty* that if they were ever used in the future, such use would be unjust. This is entirely different for UAVs, which *can* be used justly in some circumstances.

[44] Although I have not fully argued for this stronger claim here. Again, see Plaw (2010) cited above. Of course, if this stronger claim is true, it would press an even greater ethical obligation to employ UAVs and transition military inventories to all-UAV forces.

References

Arkin, R. (2009) *Governing Lethal Behavior in Autonomous Robots* (New York: Chapman &Hall).

Asaro, P. (2006) What Should We Want From a Robot Ethic? *International Review of Information Ethics*, 6, pp. 9–16.

Asaro, P. (2007) How Just Could a Robot War Be? Paper presented at 5th European Computing and Philosophy Conference, Twenthe, NL, June.

Asaro, P. (2007) Robots and Responsibility from a Legal Perspective. Paper presented at IEEE International Conference on Robotics and Automation, Rome.

Benbaji, Y. (2008) A Defense of the Traditional War Convention, *Ethics*, 118, pp. 464–495.

Card, J. (2007) Killer Machines, *Foreign Policy*, May, p. 92.

Cook, M. (2004) *The Moral Warrior* (Albany, NY: SUNY Press).

Dipert, R. (2010) The Ethics of Cyberwarfare. Paper presented at International Society of Military Ethics Annual Conference, San Diego, CA, January.

Emerton, P. & Handfield, T. (2009) Order and Affray: Defensive Privileges in Warfare, *Philosophy and Public Affairs*, 37, pp. 382–414.

Fitzsimonds, J. R. & Mahnken, T. G. (2007) Military Officer Attitudes Toward UAV Adoption: Exploring Institutional Impediments to Innovation, *Joint Forces Quarterly*, 46, pp. 96–103.

Gross, M. (2006) Assassination and Targeted Killing: Law Enforcement, Execution, or Self-Defence? *Journal of Applied Philosophy*, 23(3), pp. 323–335.

Himma, K. (2007) Artificial Agency, Consciousness, and the Criteria for Moral Agency: What Properties Must an Artificial Agent Have to be a Moral Agent? Paper presented at 7th International Computer Ethics Conference, San Diego, CA, July.

Kaldor, M. (1999) *New and Old Wars* (Palo Alto, CA: Stanford University Press).

Kaufman, W. R. P. (2007) Rethinking the Ban on Assassination, in: M. W. Brough, J. W. Lango & H. van der Linden (Eds), *Rethinking the Just War Tradition* (Albany, NY: SUNY Press).

Kershnar, S. (2004) The Moral Argument for a Policy of Assassination, *Reason Papers*, 27, pp. 45–67.

Kershnar, S. (2005) Assassination and the Immunity Theory, *Philosophia*, 33(4), pp. 129–147.

Kolff, D. W. (2003) Missile Strike Carried Out with Yemeni Cooperation' – Using UCAVs to Kill Alleged Terrorists: A Professional Approach to the Normative Basis of Military Ethics, *Journal of Military Ethics*, 2(3), pp. 240–244.

Lazar, S. (2009) Responsibility, Risk, and Killing in Self-Defense, *Ethics*, 199, pp. 699–728.

Lazar, S. (2010) The Responsibility Dilemma for Killing in War: A Review Essay, *Philosophy and Public Affairs*, 38(2), pp. 180–213.

Mayer, J. (2009) The Predator War: What are the risks of the CIA's covert drone program? *The New Yorker*, October, accessed 26 October 2010, available at: http://www.newyorker.com/reporting/2009/10/26/091026fa_fact_mayer; Internet.

McMahan, J. (2004) The Ethics of Killing in War, *Ethics*, 114(4), pp. 708–718.

McMahan, J. (2005) Just Cause for War, *Ethics and International Affairs*, 19(3), pp. 1–21.

McMahan, J. (2008) Debate: Justification and Liability in War, *Journal of Political Philosophy*, 16(2), pp. 227–244.

McMahan, J. (2009) *Killing in War* (New York: Oxford University Press).

Münkler, H. (2003) *Die neuen Kriege* (Reinbek at Hamburg: Rowohlt).

Mustin, J. (2002) Future Employment of Unmanned Aerial Vehicles, *Air and Space Powe Journal*, 16(2), pp. 86–97.

National Priorities Project (2010) *Cost of War*, accessed 1 September 2010, available at: http://www.nationalpriorities.org/costofwar_home; Internet.

Office of the Under Secretary of Defense (2006) *Unmanned Aircraft Systems Roadmap: 2005–2030* (Washington, DC: Office of the Under Secretary of Defense).

Osiel, M. (2009) *The End of Reciprocity* (Cambridge: Cambridge University Press).

Phythian, M. (2010) Ethics and Intelligence: The Implications of the Rise of the Armed Drone. Paper presented at 7th Global Conference on War and Peace, Prague, 30April.

Plaw, A. (2010) Sudden Justice. Paper presented at 7th Annual Global Conference on War and Peace, Prague, 1 May.
Rafael Advanced Defense Systems (2010) *Spike ER Precision Missile*, accessed 1 June 2010, available at: http://www.rafael.co.il/marketing/sip storage/files/0/600.pdf; Internet.
Singer, P. W. (2009) *Wired for War: The Robotics Revolution and Conflict in the 21st Century* (New York: Penguin Press).
Sparrow, R. (2007) Killer Robots, *Journal of Applied Philosophy*, 24(1), pp. 63–77.
Sparrow, R. (2009) Building a Better Warbot: Ethical Issues in the Design of Unmanned Systems for Military Applications, *Science and Engineering Ethics*, 15, pp. 169–187.
Steinhoff, U. (2006) Torture: The Case for Dirty Harry and against Alan Dershowitz, *Journal of Applied Philosophy*, 23(3), pp. 337–353.
Steinhoff, U. (2008) Jeff McMahan on the Moral Inequality of Combatants, *Journal of Political Philosophy*, 16(2), pp. 220–226.
Wallach, W. & Allen, C. (2009) *Moral Machines: Teaching Robots Right from Wrong* (New York: Oxford University Press).

Biography

Bradley Jay Strawser is currently pursuing his Ph.D. in Philosophy at the University of Connecticut. He specializes in ethics and political philosophy, particularly just-war theory, military ethics, various applied ethics issues, and issues pertaining to autonomy and liberalism. He also works in metaphysics and other areas. His paper, 'Rea's Revenge and the Persistent Problem of Persistence for Realism,' was recently accepted for publication in *Philosophia*. Formerly, Strawser served as a Captain in the United States Air Force. His service therein included work in Anti-terrorism, Force Protection, and Manpower, among other roles. He also was an Instructor of Philosophy at the US Air Force Academy where he taught ethics, logic, and other courses. He earned his MA in Philosophy from the University of Connecticut in 2005 under the Air Force Institute of Technology civilian institution fellowship program. He separated from military service in early 2008.

Saying 'No!' to Lethal Autonomous Targeting

NOEL SHARKEY
Department of Computer Science, University of Sheffield, UK

ABSTRACT *Plans to automate killing by using robots armed with lethal weapons have been a prominent feature of most US military forces' roadmaps since 2004. The idea is to have a staged move from 'man-in-the-loop' to 'man-on-the-loop' to full autonomy. While this may result in considerable military advantages, the policy raises ethical concerns with regard to potential breaches of International Humanitarian Law, including the Principle of Distinction and the Principle of Proportionality. Current applications of remote piloted robot planes or drones offer lessons about how automated weapons platforms could be misused by extending the range of legally questionable, targeted killings by security and intelligence forces. Moreover, the alleged moral disengagement by remote pilots will only be exacerbated by the use of autonomous robots. Leaders in the international community need to address the difficult legal and moral issues now, before the current mass proliferation of development reaches fruition.*

Introduction

Robots in warfare are an inevitable stage in the evolution of weapons development designed to separate fighters from their foes. Weapons technology has evolved to enable killing from ever increasing distances throughout the history of warfare. From stones to pole weapons to bows and arrows, from cannon to aerial bombing to cruise missiles, killing has become ever easier. Battlefield robots may not change the character of war but they will change the way that wars are fought.

In 2010, the ultimate distance weapons in the sights of military organisations throughout the world are armed robots. To be clear, the robots discussed here are mobile platforms that can carry multiple weapons systems, bombs, cameras, sensors or other payloads. They can be recalled, refuelled and re-weaponised. Operationally, they can be sent into the theatre on a fishing expedition without a preset goal. It is this type of flexibility that separates them from other robots such as cruise missiles and torpedoes.

Of course there are some exceptional systems such as loitering munitions that fall between the cracks. For example, the Israeli Harpy and Harop made by Israel Aircraft Industries are a cross between a missile and an Unmanned Aerial Vehicle (UAV). They are launched like a missile but hover over an area, select their own targets (hostile radar signatures) as they arise and then impact them like a missile. Thus they have some mission flexibility but not reusability.

Many thousands of robots have been used during both the Iraq and Afghanistan conflict (up from 150 in 2004) for tasks such as bomb disposal and cave clearance on the ground and for surveillance from the air. The undisputed success of UAVs (now more commonly known as Remote Piloted Aircraft, RPA, or Drones) for gathering intelligence in the Iraq/Afghanistan conflicts has made them a showcase for military powers everywhere. There is an almost insatiable military demand for UAVs. Troops do not like to move without them. This same demand is reflected in more than 40 other countries that are either developing or purchasing military robots. These include Russia, China, Israel, Pakistan, Korea, India and Iran.

The focus of this paper is on ethical concerns about the application of armed robots in areas with mixed combatant and civilian populations. It is unclear whether International Humanitarian Law (the Laws of War) will eventually need to be amended to accommodate emerging technologies. But for now we need to scrutinise the mapping between the applications of the new technologies and the current laws to ensure that they are preserved and followed.

The main ethical concerns expressed here are about the planned use of the armed autonomous 'decision making' robots that have appeared in all of the US military's roadmaps since 2004. However, before examining the issues surrounding autonomous lethal force in detail it will be useful to look at some of the ethical problems that have arisen for man-in-the-loop or Remote Piloted Drones. There are lessons to be learned that will highlight some of the dangers of fully autonomous armed robots.

Armed 'Man-in-the-Loop' Systems

The United States leads the field with its armed drones. The Predator MQ-1 was first armed by the Central Intelligence Agency (CIA) in 2001 with its now standard two Hellfire missiles. Nowadays 'pilots' of the 432nd Air Expeditionary Wing fly official military missions over Iraq and Afghanistan from Creech Air Force Base in the Nevada desert, thousands of miles away from the action. The Royal Air Force 39th Squadron also participates at Creech. Each plane has a two-person crew: a remote pilot and a sensor operator who sit at what looks like games consoles making decisions about the application of lethal force. The planes are flown around the clock in shifts and it is easy for pilots to take a break from 'battle' at any time or even go home for dinner with their families.

In October 2007, the Predator fleet was joined by the larger and more powerful MQ-9 Reaper with its larger payload capacity of 1,700kg – up to

14 Hellfire missiles or a mixture of missiles and bombs. The number of Reapers flying in operations doubled after their first year and General Atomics have reported difficulties in keeping up with the demand. By March 2009, the number of armed flying drones in the field was reported to be 195 Predators and 28 Reapers. The number has grown dramatically since.

Anticipated increases can be seen in the training figures for Predator crews. In 2005, 32-person crews consisting of a pilot and a sensor operator were being trained per year, whereas in 2008, 160 crews were trained and in late 2008, an additional $412m was added to the budget for training more non-pilot operators. In 2009, the number of remote pilot operators trained outnumbered the number of conventional pilots. Figures for the size of the current fleet of Predators and Reapers were not publicly available at the time of writing but in May 2009 the Obama administration budgeted $5.5 billion to add to the fleet in 2010. The military demand for the armed UAVs is highlighted by the more than a million mission hours flown by the MQ-1 by 2010 (up from 250,000 in June 2007).

Moral Disengagement

Flying a UAV from thousands of miles away from the battlefield clearly alleviates one of the two fundamental obstacles that war-fighters must face – fear of being killed. As Daddis (2004) pointed out, fear is one of the greatest obstacles to a soldier's effectiveness in battle. It is obvious that the greater the distance from the enemy, the less fear will play in the action. Many battles throughout history have been lost by men bolting in panic as fear swept through the ranks (Holmes 2003).

Certainly, the remote pilots of attack planes like the Reaper and the Predator have no need to worry about their personal safety. Sitting in cubicles thousands of miles away from the action they can give a mission their full attention without worrying about being shot at. They are in highly secure home ground where no pilot has ever been safer.

The second fundamental obstacle is *resistance to killing*. Grossman (1995) argues that 'not firing' is not cowardice but really a compulsion of ordinary men not to kill. In his book *Acts of War*, Holmes (2003) argues that the *hit rates* in a number of historical battles show that many soldiers were not prepared to fire directly at the enemy. This is also borne out by the analyses and findings of Marshall ([1947]2000]) from his interviews of soldiers after World War II.

Similarly, Royakkers and van Est (2010) discuss the emotional and moral disengagement of UAV pilots, or 'cubicle warriors' as they calls them, from the consequences of their actions in the field. The crux is that the type of control system used encourages a 'Playstation' mentality that makes an operator more careless about decisions to kill. Royakkers and van Est suggest that new recruits may have been playing videogames for many years and may not see a huge contrast with that of being a cubicle warrior. They provide examples from Peter Singer's book *Wired for War* in which a young pilot is reported to have said, 'It's like a video game. It can get a little bloodthirsty. But it's fucking cool' (Singer 2009b: 308–309). And in a further example, a young pilot says, 'The truth is, it wasn't all I thought it was cracked up to be. I mean, I thought killing somebody would be this life-changing experience. And then I did it, and I was like "All right, whatever."... Killing people is like squashing an ant. I mean, you kill somebody and it's like "All right, let's go get some pizza."' (ibid.: 391–392).

In contrast to this view, there are reports of remote pilots suffering from different types of stress than onboard pilots. After an attack the operator gets to see the aftermath, using high-resolution cameras. This is more than a conventional fighter pilot ever sees and, at least on the surface, it would seem that remote operators would be less morally disengaged than the crew of a high altitude bomber.

However, for an article appearing in the *Air Force Times* (Lindlaw 2008), five remote pilots interviewed said that they had had not been particularly troubled by their missions although they can sometimes make for a strange existence. Colonel Chris Chambliss, a commander of the 432nd Wing at Creech, said that on only four or five occasions had sensor operators gone to see the Chaplain or their supervisors and that this was a very small proportion of the total number of remote operators.

All of this is anecdotal evidence and proper scientific research needs to be conducted to find out what kind of stresses, if any, the operators are facing. It is well known that making good decisions is more difficult under stressful conditions. Longitudinal studies would be required as the operators may start out being stressed by graphic scenes but then become inured with sufficient exposure. This is a bit like the process trainee surgeons and nurses go through. They are known often to faint when they start out but quickly get habituated to horrific medical events.

Killing distance does not have to be physical distance. It can be psychological as well. Grossman (1985) cites Clausewitz and the French Colonel Ardant du Picq (1821–70) for expounding at length that the vast majority of deaths in battle occurred when the victors chased the losing side in retreat. Du Picq suggests that Alexander the Great lost fewer than 700 men over all his battles because his army was never pursued by a victorious enemy. Although UAV operators have high-resolution images on large screens, the cameras are looking down from above and it can be very hard to make out the faces of their victims. Is this so different from attacking retreating troops?

One type of stress that keeps getting mentioned in the media as if to counter the game mentality notion is that of remote pilots not getting a chance to wind down together and bond in their mutual operations. They return home to their families after their shifts have finished and even go to parents evening at school. The problem is that they have no debriefing about their day on the battlefield and cannot talk to their families about it. This is a very different issue from whether or not they are morally engaged or disengaged from the killer strikes.

Targeted Killings

Up until now the discussion has centered on conventional forces using armed drones to support troops on the ground in the war zones of Afghanistan and Iraq. But there is a second armed UAV 'Air Force' controlled by the CIA out of Langley in Virginia. The first ever US kill with a drone was in Yemen on 3 November 2002 when a CIA-operated MQ-1 killed a known Al-Qaeda leader and five other men travelling in the same vehicle.[1] This was controversial at the time but was considered, by Department of Defense lawyers, to be a legitimate defensive pre-emptive strike against Al-Qaeda. Since then, the use of drones for targeted killings or 'decapitation strikes' in states that are not at war with the US has been rising at an accelerating pace. It is what the *Asia Times* has called 'the most public "secret" war of modern times' (Turse 2010).

According to estimates published on the New America Foundation website,[2] the US drone attacks in Pakistan have risen from nine between 2004 and 2007 up to 85 by mid-October in 2010. The estimates are shown in Table 1. The number of civilian deaths has been very difficult to estimate and have ranged from as few as 20 to more than a thousand. There are many different counts and no objective way to decide between them. The New America Foundation gives a figure of 1 in 3 civilian to insurgent ratio while the Brookings Institute study in 2009 indicated that only 1 in 10 of those killed was an insurgent.

The legality of targeted killing is at best questionable. The word decapitation is often used to mean cutting off the leaders of an organisation or nation fighting a war from the body of their war-fighters. The aim of the current spate of aerial decapitation strikes was initially to target Al-Qaeda and Taliban leaders without risk to US military personnel. The idea is to eventually leave only replacements from the shallowest end of the talent pool that will be ineffective and easy to defeat. With less risk, the targeted leaders do not have to be at the highest level. This explains the increasing number of death-delivering attacks in Pakistan using Predators and Reapers.

These individually targeted killings come despite the banning in the US of all politically motivated killing of individuals since the famous Church commission report on the CIA political assassinations in 1975. In 1976, President Gerald Ford issued a presidential executive order that 'no person employed by or acting on behalf of the United States Government shall

Table 1. US drone attacks in Pakistan 2004–2010

		Numbers killed		
Year	estimated strikes	high estimate	low estimate	leaders killed
2004–2007	9	109	86	3
2008	34	296	263	11
2009	53	709	413	10
2010*	86	719	426	10

*Until 10 October 2010
Source: See endnote 2.

engage in, or conspire to engage in, assassination'. This is not codified in the US law but is an executive order that the president can change at will and without public notice. President Ronald Reagan issued Executive Order (EO) 12333, which expanded the ban to include persons acting on *behalf* of the US government and all subsequent presidents have supported it. However, EO 12333 does not limit lawful self-defence options against legitimate threats to the national security of US citizens. (Parks 1989)

During wartime, a combatant is considered to be a legitimate target at all times. If a selected individual is sought out and killed it is not termed an assassination. According to a Memorandum on EO 12333, which is said to be consistent with United Nations (UN) Charter 51, 'a decision by the President to employ clandestine, low-visibility, or overt military force would not constitute assassination if US military forces were employed against the combatant forces of another nation, a guerrilla force, or a terrorist or other organization whose actions pose a threat to the security of the United States'.

But the real legal question is, does the CIA have a right to assassinate *alleged* insurgent combatants without due process. Seymour Hersh (2002), whose writings were one of the main motivations for the Church commission, complained that, 'the targeting and killing of individual al-Qaeda members without juridical process has come to be seen within the Bush Administration as justifiable military action in a new kind of war, involving international terrorist organizations and unstable states'. The insurgents have been redefined as combatants, but without receiving the rights of prisoners of war and without being given the chance to surrender or to face trial. This move, in combination with Article 51, provides legal cover for the right to assassinate insurgent combatants.

The legality of the targeted killings has also been challenged at the UN General Assembly meeting in October 2009 by Philip Alston, UN special rapporteur on Extrajudicial Killings. He made a request for US legal justification about how the CIA is accountable for the targets that they are killing. The US turned down the request by refusing to comment on what it said were covert operations and a matter of national security.

A rebuttal by Harold Koh (2010), Legal Adviser, US Department of State, stated, 'US targeting practices including lethal operations conducted by UAVs comply with all applicable law including the laws of war.' However, there are no independent means of determining how the targeting decisions are being made. It remains unclear as to what type and level of evidence is being used to sentence non-state actors to death by Hellfire missile without right to appeal or right to surrender. It is also unclear what other methods were exhausted or attempted to bring the targets to justice. The whole process sits behind an unjustified cloak of national secrecy.

The CIA use of drones was also questioned at a Senate Hearing in a prepared statement by US law Professor Kenneth Anderson (2010). Anderson said that Koh 'nowhere mentions the CIA by name in his defense of drone operations. It is, of course, what is plainly intended when speaking of self-defense separate from armed conflict. One understands the hesitation of senior lawyers to name the CIA's use of drones as lawful when the official position of the US government, despite everything, is still not to confirm or deny the CIA's operations.'

However, the Director of the CIA, Leon Panetta (2009), has been very open about the CIA use of drones. He told the Pacific Council on International Policy in 2009 that, 'it's the only game in town in terms of confronting and trying to disrupt the al-Qaeda leadership.' Revealing the CIA's intentions on the expansion of targeted drone kills, Panetta went on to say of Al-Qaeda that, 'If they're going to go to Somalia, if they're going to go to Yemen, if they're going to go to other countries in the Middle East, we've got to be there and be ready to confront them there as well. We can't let them escape. We can't let them find hiding places.'

This expansion of targeted killing is just what was of concern to the UN special rapporteur on Extrajudicial Killings. A subsequent report by Alston (2010) to the UN General Assembly discusses drone strikes as violating international and human rights law because both require transparency about the procedures and safeguards in place to ensure that killings are lawful and justified: 'a lack of disclosure gives States a virtual and impermissible license to kill'. Some of Alston's arguments revolve around the notion of 'the right to self-defence' and whether the drone strikes are legal under Article 51. He also examines some of the complex legal issues surrounding proportionality the different conditions that pertain when the military action is called 'self-defence'.

Panetta and many others have made the case that armed UAVs are more accurate and will kill less civilians than a B-52 bomber in attacking the tribal regions in Pakistan. But as a former CIA operative told me, there is no way that Pakistan or other state actors not at war with the US could 'turn a blind eye' to the bomber strikes as they do now for drones. It can be argued that it is their perceived precision and accuracy that allows them to penetrate areas and kill people in ways that would not previously have been available without major political and legal obstacles.

Attacking with a remote piloted vehicle is no different under the Laws of War than attacking with a manned helicopter gunship or even with a rifle. The worry is that the nature of an unmanned vehicle with no risk to military personnel, an ability to hover over an area for very many hours, and its perceived accuracy may lead to considerable expansion of potential targets.

I qualified 'accuracy' with the word 'perceived' here both because the attacks are often on buildings where the targets cannot be verified and, if the Brookings or New America Foundation figures discussed above are anything to go by, the accuracy is not as high as some might imagine. One of the oft-cited targeting methods of the CIA is to determine the locations of people through their cell phones; switch on your phone and you are dead. But a recent copyright lawsuit between two companies sheds doubt on the accuracy of this targeting method (Stein 2010). A small company called Intelligent Integration Systems alleges that one of its client companies, Netezza, reverse-engineered their software, Geospatial, on a tight deadline for the CIA. The court heard that the illegal version of the software could produce locations that were out by as much as 40 feet and that the CIA had knowingly accepted the software.

But even if targeting was 100 per cent accurate, how can we be sure that alleged insurgents are 'guilty as charged'. Information about target identity and their role and position is heavily dependent on the reliability of the intelligence on which it is based. There are lessons that should have been learned from the Vietnam War investigations of Operation 'Phoenix' in which thousands were assassinated. It turned out that many of those on the assassination list had been put there by South Vietnamese officials for personal reasons such as erasing gambling debts or resolving family quarrels

Philip Alston (2010) reports that during a mission to Afghanistan he found out how hard it was for forces on the ground to obtain accurate information. 'Testimony from witnesses and victims' family members showed that international forces were often too uninformed of local practices, or too credulous in interpreting information, to be able to arrive at a reliable understanding of a situation.' He suggests that, 'States must, therefore, ensure that they have in place the procedural safeguards necessary to ensure that intelligence on which targeting decisions are made is accurate and verifiable.'

From *in*-the-Loop to *on*-the-Loop to *out-of*-the-Loop

Since 2004 all of the roadmaps and plans of the US forces has made clear the desire and intention to develop and use autonomous battlefield robots. Fulfilment of these plans to take the human out of the loop is well underway. The end goal is that robots will operate autonomously to locate their own targets and destroy them without human intervention (Sharkey 2008a). The Committee on Autonomous Vehicles in Support of Naval Operations (2005) wrote: 'The Navy and Marine Corps should aggressively exploit the considerable war-fighting benefits offered by autonomous vehicles (AVs) by acquiring operational experience with current systems and using lessons learned from that experience to develop future AV technologies, operational requirements, and systems concepts.' We can only hope that the 'lessons learned' will include ethical and moral lessons. The ethical problems will only be amplified and added to by the use of autonomous robots.

Before delving into these problems it is important to clarify what is meant by 'robot autonomy' here. This is often confused with science fiction notions of robots with minds of their own with the potential to turn on humanity. The reality is very different. The autonomous robots being discussed for military applications are closer in operation to your washing machine than to a science fiction *Terminator*. The way the term 'autonomy' is used in robotics should not be confused with how the term is used in philosophy, politics, individual freedom or in common parlance. It is more related to the term automatic. An automatic robot carries out a pre-programmed sequence of operations or moves in a structured environment. A good example is a robot arm painting a car.

An autonomous robot is similar except that it operates in open or unstructured environments. To do this, the robot is controlled by a programme takes inputs from its sensors and adjusts the speed and direction of its motors and actuators as specified. If the goal of a robot is to avoid objects and the sensors detect an object, the programme would adjust the motors so that the robot moves to avoid it. For example if the left-hand sensors detect the object, the robot would move right, and if the right-hand sensors detect the object, the robot would move left.

This leads us to another often-misunderstood topic, the robot decision process. This should not be confused with human decision making except by analogy. A computer decision process can be as simple as, IF object on left, THEN turn right OR IF object on right, THEN turn left, ELSE continue. Alternatively, the activity on a sensor may activate a different sub-programme to help with the decision. For example, to get smoother passage through a field laden with objects, a sub-programme could be called in to calculate if a turn to the left would result in having to negotiate more obstacles that a turn to the right.

Programmes can become very complex through the management of several sub-programmes and decisions about which sub-programme should be initiated in particular circumstances. But the bottom line for decision making by a machine, whether is it using mathematical decision spaces or artificial intelligence reasoning programmes, is the humble IF/THEN statement.

The other important thing to say about autonomy is that a system does not have to be exclusively autonomous or exclusively remote operated. The US Army, Navy and Air Force have all at some point discussed the classification of robot control on a continuum from totally human operated to fully autonomous. Each has separate development programmers and each has its own operational definitions of the different levels of robot autonomy. The Army has ten levels while the Air Force has four. The Navy goes for a characterisation in terms of mission complexity but points to three different classes of autonomous robot vehicle: (i) *scripted* in which the robot carries out a pre-planned script of the 'point, fire and forget' variety; (ii) *supervised* in which some or all of the functions of planning, sensing, monitoring and networking are carried out with the help of human operators to make decisions; (iii) *intelligent* in which attributes of human intelligence are used in software to make decisions, perceive and interpret the meaning of sensed information, diagnose problems, and collaborate with other systems.

The first of the Navy classifications, *scripted,* is the closest to 'automatic' and includes cruise missiles, torpedoes and automated weapons such as the Phalanx Close In Weapon System. The *supervised* category refers to the both the current 'man-in-the-loop' systems such as the Predator and Reaper control and the newly proposed 'man-on-the-loop' systems. And the *intelligent* category refers to fully autonomous operation although frankly I am not sure exactly what they mean by 'attributes of human intelligence in software'. Perhaps there is a little slippage into wishful thinking here.

There are at least four reasons that make an autonomous robot militarily desirable including: (i) remote operated systems are more expensive to manufacture and require many support personnel to run them; (ii) it is possible to jam either the satellite or radio link or take control of the system.

(iii) one of the military goals is to use robots as force multipliers so that one human can be a nexus for initiating a large-scale robot attack from the ground and the air; (iv) the delay time in remote piloting a craft via satellite (approximately 1.5 seconds) means that it could not be used for interactive combat with another aircraft. At a press briefing in December 2007, Dyke Weatherington, deputy director of the US Department of Defense's Unmanned Aerial Systems Task Force, said, 'Certainly the roadmap projects an increasing level of autonomy . . . to fulfill many of the most stressing requirements. Let me just pick one for example. Air-to-air combat – there's really no way that a system that's remotely controlled can effectively operate in an offensive or defensive air combat environment. That has to be-the requirement of that is a fully autonomous system.'[3]

The most recent United States Air Force Unmanned Aircraft Systems Flight Plan 2009–2047[4] opens the strategy for a staged move from current remote piloted systems to fully autonomous systems. It represents a push to shrink the role of the man in the loop. To begin with, autonomous operation will be mainly for tasks such as take-off, landing and re-fuelling. As unmanned drones react in micro- or nanoseconds the 'humans will no longer be "in the loop" but rather "on the loop" – monitoring the execution of certain decisions. Simultaneously, advances in AI [artificial intelligence] will enable systems to make combat decisions and act within legal and policy constraints without necessarily requiring human input' (ibid.: 41).

The idea of a human on-the-loop means that the human will be in executive control overall to call in or call off the robots rather than being in control of each individually. In other words, the robots will be essentially autonomous: 'SWARM technology will allow multiple MQ-Mb aircraft to cooperatively operate in a variety of lethal and non-lethal missions at the command of a single pilot' (ibid.: 39). Such a move will require decisions being made by the swarm, human decision making will be too slow and not able to react to the control of several aircraft at once. With the increasing pace of the action and with the potential of several aircraft to choose targets at the same time, it will not be possible to have the human make all of the decisions to kill.

The main ethical concern is that allowing robots to make decisions about the use of lethal force could breach both the Principle of Distinction and the Principle of Proportionality as specified by International Humanitarian Law (Sharkey 2008b). Currently and for the foreseeable future no autonomous robots or artificial intelligence systems have the necessary properties to enable discrimination between combatants and civilians or to make proportionality decisions.

Under the Principle of Distinction only combatants/warriors are legitimate targets of attack. All others, including children, civilians, service workers and retirees, should be immune from attack. The same immunity covers combatants that are wounded, have surrendered or are mentally ill.[5] The Principle of Proportionality applies in circumstances where it is not fully possible to protect non-combatants in an action. It requires that the loss of life and damage to property incidental to attacks must not be excessive in relation to the concrete and direct military advantage.

The discrimination between civilians and combatants is problematic for any robot or computer system. First, there is the problem is the specification of 'civilianness.' A computer can compute any given procedure that can be written as a programme. We could, for example, give the computer on a robot an instruction such as, 'if civilian, do not shoot.' This would be fine if and only if there was some way to give the computer a precise specification of what a civilian is. The Laws of War do not help. The 1949 Geneva Convention requires the use of common sense to determine the difference between a civilian and combatant while the 1977 Protocol 1 essentially defines a civilian in the negative sense as someone who is not a combatant:

1. A civilian is any person who does not belong to one of the categories of persons referred to in Article 4 A (1), (2), (3) and (6) of the Third Convention and in Article 43 of this Protocol. In case of doubt whether a person is a civilian, that person shall be considered to be a civilian.
2. The civilian population comprises all persons who are civilians.
3. The presence within the civilian population of individuals who do not come within the definition of civilians does not deprive the population of its civilian character.

Protocol 1 Additional to the Geneva Conventions, 1977 (Article 50)

Even if a clear definition of civilian did exist, it would have to be couched in a form that enabled the relevant information to be extracted from the sensing apparatus. All that is available to robots are sensors such as cameras, infrareds, sonar, lasers, temperature sensors and ladars etc. While these may be able to tell us that something is a human or at least animal, they could not tell us much about combat status. There are systems that can identify a face or a facial expression but they do not work well on real time moving people. And even if they did, how useful could they be in the fog of war or from the air? British teenagers beat surveillance cameras just by wearing hooded jackets.

In a war with non-uniformed combatants, knowing who to kill would have to be based on situational awareness and on having human understanding of other people's intentions and their likely behaviour. In other words, human inference is required. Humans understand one another in a way that machines cannot. Cues can be very subtle and there are an infinite number of circumstances where lethal force is inappropriate. Just think of children being forced to carry empty rifles or of insurgents burying their dead.

The problem has been put clearly by Major Daniel Davis (2008), a combat veteran of Iraq 1991 and Afghanistan 2005, who writes, 'Suggesting that within the next 12-plus years technology could exist that would permit life-and-death decisions to be made by algorithms is delusional. A machine cannot sense something is wrong and take action when no orders have been given. It doesn't have intuition. It cannot operate within the commander's intent and use initiative outside its programing. It doesn't have compassion and cannot extend mercy.' Davis also quotes Colonel Lee Fetterman who has a high regard for remote controlled robots:

> The function that robots cannot perform for us — that is, the function we should not allow them to perform for us – is the decide function. Men should decide to kill other men, not machines. This is a moral imperative that we ignore at great peril to our humanity. We would be morally bereft if we abrogate our responsibility to make the life-and-death decisions required on a battlefield as leaders and soldiers with human compassion and understanding. This is not something we would do. It is not in concert with the American spirit. (ibid.)

Turning to the Principle of Proportionality, there is no way for a robot to perform the human subjective balancing act required to make proportionality decisions. No clear objective methods are provided for calculating what is proportionate in the Laws of War (Sharkey 2009). The phrase 'excessive in relation to the concrete and direct military advantage expected to be gained' is not a specification. It is also practically impossible to calculate a value for 'the actual military advantage'. What could the metric be for assigning value to killing an insurgent relative to the value of non-combatants, particularly children who could not be accused of willingly contributing to insurgency activity? Proportionality calculations should be based on the likely differences in military outcome if the military action killing innocents had not been taken (Chakwin et al. 2002).

Commanders have to weigh the circumstances before making a decision but ultimately this is a subjective matter. Clearly the extremes of wiping out a whole city to eliminate even the highest value target, say Osama bin Laden, is out of the question. There has to be circumstantial and subjective estimates about just how many innocent people are equal to the military value of the successful completion of a given mission. A computer system could be used to assist commanders to determine which weapons would cause the minimum of collateral damage. However, after the calculation is completed, a decision must still be made about whether the minimal number of civilian casualties and damage to civilian property is proportional to the military advantage gained.

In the heat of battle, both the principles of discrimination and proportionality can be problematic. Humans do make errors and can behave unethically but they can be held accountable. Armed autonomous robots could violate both of these principles but could not be held accountable (Sharkey 2008c). There is no way to punish a robot. We could threaten to switch it off but that would be like telling your washing machine that if it does not remove stains properly you will break its door off. This leaves the question about who is responsible along the long causal chain that includes: the manufacturer for the description they gave, the programmer, the designer, the Department of Defense, the generals or admirals in charge of the operation and the operator. Can we hold someone accountable? If not, then, legally speaking, these robots cannot be used (Sparrow 2007).

Some Lessons to Learn

It is quite likely that autonomous robots will come into operation in a piecemeal fashion. Research and development is well underway and the fielding of autonomous robot systems may not be far off. However, to begin with they are likely to have assistive autonomy on board such as flying or driving a robot to a target destination and perhaps even selecting targets and notifying a human (the Israeli autonomous ground robot, Guardium, already has this functionality). This will breed public trust and confidence in the technology – an essential requirement for progression to autonomy (Dahm 2010). The big worry is that allowing such autonomy will be a further slide down a slippery slope to give machines the power to make decisions about whom to kill.

Apart from the legal and ethical problems of discrimination and proportionality for fully autonomous armed robots, there are lessons to be learned from the current application of remote piloted vehicles. We discussed some of the controversy about the issue of moral disengagement of the operators and the Playstation mentality (Royakkers & van Est 2010). While this is an interesting and plausible narrative, it is not possible to be conclusive because the evidence, outside of the anecdotal, is just not available. It must be noted, nonetheless, that the power of graphic detail to desensitise should not be underestimated. The narrative about moral disengagement is even more compelling when it comes to autonomous robot warriors. As Cumming (2006) puts it, 'developing technologies of this sort also have the potential to provide for the creation of moral buffers that allow humans to act without adequately considering the consequences'.

Another lesson can be learned from the use of the remote piloted armed aircraft by the CIA for targeted killings. At best, such targeted killings are legally questionable in terms of the Laws of War as discussed above. Apart from problems with verifying intelligence about targeting and the lack of a transparent judicial process, there is concern that the use of drones has greatly expanded the range of targets that can be picked off by the CIA. Autonomous aircraft that can hover on their own for extended periods and select their own targets could greatly exacerbate the problems. The Defense Advanced Research Projects Agency Vulture programme has just awarded a contract to Boeing (teamed with QinetiQ) to develop a heavier-than-air platform capable of keeping 1,000 lb of payload with 5 kilowatts of power aloft for five years. These developments need to be watched very carefully.

It is difficult to design and develop these new technologies but they are not so difficult to copy. We have already seen the rapid proliferation of drones in response to their perceived military success in the Iraq/Afghanistan conflicts. If armed autonomous robots are fielded in the same way with or without a human in the weapons loop, it will not be long before there is similar proliferation. If it turns out to be militarily advantageous to keep a human out of the loop – for example in circumstances when it is not possible to maintain a radio or satellite link – then they will be used by all.

Notes

[1] Israel may have been using armed drones for longer but denied it for several years despite eyewitness testimonies. It cannot be verified here.

[2] The Year of the Drones, New America Foundation website: http://counterterrorism.newamerica.net/drones, accessed 15 October 2010; Internet.

[3] DoD Press Briefing with Mr Weatherington from the Pentagon Briefing Room, US Dept. of Defense, 18 December 2007, www.defenselink.mil/transcriptstranscript.aspx?transcriptid=4108; Internet.

[4] United States Air Force Unmanned Aircraft Systems Flight Plan 2009–2047, Headquarters of the United States Air Force, Washington, DC, 18 May 2009.

[5] But see also Ford 1944.

References

Alston, P. (2010) Report of the Special Rapporteur on Extrajudicial, Summary or Arbitrary Executions, Philip Alston to the UN Human Rights Council, 14th session, A/HRC/14/24/Add.6, 28 May 2010.

Anderson, K. (2010) Written Testimony to the US House of Representatives Committee on Oversight and Government Reform, Subcommittee on National Security and Foreign Affairs, Subcommittee Hearing: 'Drones II', 28 April 2010, accessed 3 November 2010, available at http://papers.ssrn.com/sol3/papers.cfm?abstract_id=1619819; Internet.

Chakwin, B., Voelkel, D. & Enright S. (2002) Leaders as Targets, Joint Forces Staff College, Seminar # 08, accessed 3 November 2010, available at http://www.stormingmedia.us/13/1361/A136124.html; Internet.

Committee on Autonomous Vehicles in Support of Naval Operations National Research Council (2005) *Autonomous Vehicles in Support of Naval Operations* (Washington DC: The National Academies Press).

Cumming, M. L. (2006) Automation and Accountability in Decision Support System Interface Design, *Journal of Technology Studies*, 32(1), pp. 23–31.

Daddis, A. (2004) Understanding Fear's Effect on Unit Effectiveness, *Military Review*, July–August, pp. 22–27.

Dahm, W. J. A. (2010) 'Technology Horizons': A Vision for Air Force Science & Technology, Volume 1, AF/ST-TR-10-01, accessed 14 October 2010, available at http://www.airforce-magazine.com/SiteCollectionDocuments/TheDocumentFile/Strategy and Concepts/TechnologyHorizonsVol1_2010.pdf; Internet.

Davis, D. (2007) Who Decides: Man or Machine? Armed Forces Journal, 11, accessed 3 November 2010, available at http://www.armedforcesjournal.com/2007/11/3036753; Internet.

Ford, J. S. (1944) The Morality of Obliteration Bombing, *Theological Studies*, 5, pp. 261–309.

Grossman, D. (1995) *On Killing: The Psychological Cost of Learning to Kill in War and Society* (Boston: Little, Brown).

Hersh, S. M., (2002) Manhunt: The Bush Administration's New Strategy in the War against Terrorism, *New Yorker*, 66 (December), accessed 3 November 2010, available at http://www.newyorker.com/archive/2002/12/23/021223fa_fact; Internet.

Holmes, R. (2003) *Acts of War: The Behaviour of Men in Battle* (London: Cassell).

Koh, H. (2010) Speech to The American Society of International Law, Washington, DC. March 25, 2010, accessed 3 November 2010, available at http://www.state.gov/s/l/releases/remarks/139119.htm; Internet.

Lindlaw, S. (2008) UAV operators suffer war stress, *Air Force Times*, 8 August 2008, accessed 10 October 2010, available at http://www.airforcetimes.com/news/2008/08/ap_remote_stress_080708/; Internet.

Marshall, S.L. A. ([1947] 2000) *Men against Fire: the Problem of Battle Command* (Oklahoma City: Oklahoma University Press).

Panetta, L. E. (2009) Director's Remarks at the Pacific Council on International Policy, 18 May 2009, accessed 15 October 2010, available at https://www.cia.gov/news-information/speeches-testimony/directors-remarks-at-pacific-council.html; Internet.
Parks, H. W. (1989) Memorandum on Executive order 12333, Reproduction Department of the Army, office of the Judge Advocate General of the Army, Washington DC 20310–220.
Royakkers, L. & van Est, R. (2010) The Cubicle Warrior: the Marionette of Digitalized Warfare, *Ethics and Information Technology*, 12, pp. 289–296.
Sharkey, N. (2008a) Cassandra or the False Prophet of Doom: AI Robots and War, *IEEE Intelligent Systems*, 23(4), pp. 14–17.
Sharkey, N. (2008b) Grounds for Discrimination: Autonomous Robot Weapons, *RUSI Defence Systems*, 11(2), October, pp. 86–89.
Sharkey, N. (2008c) The Ethical Frontiers of Robotics, *Science*, 322, pp. 1800–1801.
Sharkey, N. (2009) Death Strikes from the Sky: The Calculus of Proportionality, *IEEE Science and Society*, Spring issue, pp. 16–19.
Singer, P. W. (2009) *Wired for War: The Robotics Revolution and Conflict in the 21st Century* (New York: The Penguin Press).
Sparrow, R. (2007) Killer Robots, *Journal of Applied Philosophy*, 24, pp. 62–77.
Stein, J. (2010) CIA Drones Could be Grounded by Software Suit, *Washington Post*, SpyTalk, 11 October 11, 2010, accessed 14 October 2010, available at http://blog.washingtonpost.com/spy-talk/2010/10/cia_drones_could_be_grounded_b.html; Internet.
Turse, N. (2010) Drone Surge: Today, Tomorrow and 2047, Asia Times, 26 January 2010, accessed 15 October 2010, available at http://www.atimes.com/atimes/South_Asia/LA26Df01.html; Internet.

Biography

Noel Sharkey is a Professor of Artificial Intelligence (AI) and Robotics and a Professor of Public Engagement at the University of Sheffield (Department of Computer Science), UK. His research and teaching span academic disciplines, including engineering, philosophy, psychology, cognitive science, linguistics, artificial intelligence, computer science and robotics. Previously he worked at Yale in AI, Stanford in Psychology, Essex in Linguistics and Cognitive Science and Exeter in Computer Science. He is on the editorial board of several journals and is Editor-in-Chief of the *Journal of Connection Science.* His work has ranged from computational modelling of psychological processes, to neural computing and machine learning, to monitoring safety critical systems, to robotics. He appears regularly on television, radio and in the press.

As well as more than 150 academic articles and books, Professor Sharkey writes for national newspapers and magazines and has created thrilling robotics museum exhibitions and mechanical art installations. His core research is now on ethical applications of robotics in domains such as the military, child care, elder care, policing, medicine and crime. He is cofounder of the International Committee for Robot Arms Control and has been awarded a two-year Leverhulme Research Fellowship (September 2010) to conduct an ethical and technical appraisal of Robots on the Battlefield.

ADDITIVE MANUFACTURING AND ITS IMPLICATIONS FOR MILITARY ETHICS

John Mark Mattox

National Defense University Center for the Study of Weapons of Mass Destruction, Washington, DC, USA

Recent advances in so-called 'additive manufacturing' pose significant, new challenges, in scope if not in kind, for military ethicists. While the problem of dual-use technologies (i.e. technologies that can be used for both good and malevolent purposes) is not new, the possibility of the rapid, uncontrolled replication of highly sophisticated tools of violent action – tools that heretofore have been largely inaccessible to laymen – could vastly expand the number of persons able to commit violent acts, or even wage war, far beyond traditional boundaries. This article explains the nature of additive manufacturing and identifies the challenges it poses for militaries and governments with the de facto *responsibility to keep war-making tools out of the wrong hands. In light of the industrial revolution occasioned by the advent of additive manufacturing and the revolution in military affairs that it portends, it proposes a research agenda for military ethicists. In particular, it argues that military ethicists must now expand their scope of inquiry in a way that accords due prominence to the nexus between these technologies and* jus ante bellum *issues of conflict avoidance.*

Introduction

Whenever a copy is made of anything, there arises a set of easily predictable normative questions, to include:

- What should one be allowed or forbidden to copy?
- Who should or should not be allowed to make, see, or otherwise use the copy?
- For what purposes should or should not the copy be allowed to be used?

Indeed, there exist entire bodies of national and international law, in the form of copyright and patent laws and treaties, that deal with these and similar issues. Hence, the ethical issues involved may seem to be essentially settled matters and, to that extent, uninteresting.

However, with the advent of the most recent copying technologies, ethicists of all kinds, to include military ethicists, must now seriously consider the ramifications of what it means to copy things that, in the wrong hands, could prove monumentally disastrous. Such is the ethical challenge associated with the emerging technology known as 'additive manufacturing'.

A New Industrial Revolution

In order to appreciate the magnitude of the ethical issues potentially associated with additive manufacturing, it is necessary first to appreciate the magnitude of the difference between additive manufacturing and traditional manufacturing techniques. For several millennia, and in increasingly refined form, traditional manufacturing has relied principally on two major kinds of processes: 'subtractive manufacturing' and 'forming' (Williams & Campbell 2011).

- Subtractive manufacturing involves starting with a large mass of some medium – say, a block of steel or aluminum – and removing portions of that mass until the desired shape is all that remains – not unlike what happens when a sculptor chisels away at a rough-hewn block of stone until all that remains is the smooth image of a human figure.
- 'Forming', the other traditional approach to manufacturing, is accomplished through processes such as 'casting', 'stamping' or 'injection molding'.
    - 'Casting' involves pouring a molten liquid into a mold that contains a hollow cavity of the desired shape, and then allowing the liquid to solidify. This technique is used frequently to make complex shapes – for example, an automobile engine block – that would be difficult or uneconomical to make by other methods (see DeGarmo et al. 2003: 277).
    - 'Stamping' includes a range of manufacturing processes in which a medium is manipulated to produce a new shape or design. For example, the metal kitchen sink found in many western homes was once a flat piece of metal, which was bent into the shape of a sink and into which holes were then punched to accommodate the drain and faucets. Coins are another example of something produced by stamping.
    - 'Injection molding' involves the creation of a mold that can be separated into two parts, into which a liquid medium is injected and from which, after the liquid solidifies, the mold can be removed. For example, many plastic objects – like resealable food containers or cell phone cases – are formed in this way.

In contrast to these traditional manufacturing methods, additive manufacturing actually 'adds' materials together so as to form a larger object. This technique starts with a three-dimensional (3D) computer-aided design model and uses a computer algorithm to 'slice' the 3D model into very thin cross-sections. The computer images are then passed into a 3D printer, which builds the object one minute layer at a time.

The implications associated with the ability to construct things in this manner are nothing less than staggering. Additive manufacturing has already been incorporated, in one form or another, into large segments of the manufacturing economy, to include consumer products, motor vehicles, medicine and dentistry, industrial and business machines, and aerospace. It has also found its way into architecture, academe, government and military.[1] As one observer has noted, the first industrial revolution harnessed steam power and mechanization and defined the nineteenth century; the second industrial revolution combined standardization and mass production to define the twentieth century; additive manufacturing marks the beginning of the third industrial revolution and is already defining the twenty-first century (Pierpoint 2012). Indeed, there appears to be no theoretical limit to the array of objects that can be produced using additive manufacturing – even objects that there is no physical way to create by any other means. This includes, incredibly, the possibility of fabricating replacement human body organs or parts on demand, using the genetic specifications of the intended recipient so that the recipient's body will not reject it.[2] The only limit to the size of the object to be printed in three dimensions is the size of the 3D printer. To date, there exist 3D printers large enough to 'print' whole car bodies or large building construction components.[3] Conversely, micro-scale objects have also been printed in 3D. For example, additive manufacturing technology has been used by a Japanese university to produce the world's smallest sculpture – 10-μm long and 7-μm high – the size of a red blood cell (Iga & Kokubun 2006: 175).

Likewise, there seems to be no theoretical limit to the media that can be used in additive manufacturing processes. Although, 'there are only a few "multi-material" printers currently available', still a much broader array of manufacturing possibilities clearly are feasible,[4] and there is no reason to believe that the coming decades will not include breakthroughs in additive manufacturing that would astonish laypersons living today. Moreover, because additive manufacturing represents not merely an advance but indeed a quantum leap in the development of industrial processes, the industrial revolution of which it is the foundation brings with it the potential for a revolution in military affairs in which laymen acquire the ability to produce highly sophisticated items for violent or even war-making purposes as readily as they could produce an utterly benign item for routine household use.

Implications for Military Ethics

However, even if one grants that additive manufacturing likely represents the advent of a new industrial revolution, one may be tempted to conclude that, ethically speaking, there really is nothing new here. After all, traditional technologies, to include both subtractive manufacturing and forming, have been used both for peaceful and violent, good and evil purposes for millennia, and so the fact that one and the same technology can be put to contrary purposes hardly seems, *prima facie*, to generate a new or even broadened set of moral-philosophical concerns. However, precisely because additive manufacturing introduces the prospect of small groups or even individuals operating on small budgets being enabled to produce war-making materiel that, up to the present, only governments or large commercial firms possessed the wherewithal to produce, the vast scope of the regulatory problem presented by additive manufacturing is unprecedented in the history of technology. Indeed, one is reminded of words frequently attributed to Joseph Stalin: 'Quantity has a quality all its own.' The idea of ubiquitously available weaponry previously available only to nation-states has profound implications on a host of accounts, including military ethics. Consider the vast extent of the problem set:

Small Arms

At first blush, small arms might seem to be unlikely candidates for additive manufacturing, since militaries typically purchase items like these in large quantities, and at least at present, additive manufacturing technologies are not optimized for mass production. That intuition, however, may be irrelevant in the case of extra-national paramilitaries and terrorist organizations. What a perfect way to replicate weaponry that cannot be traced by serial number or by any other traditional accounting means! As the sophistication of the technology, and with it the ability to produce small arms in large quantities matures, paramilitary or terrorist organizations could look far more like regular armies than they now do because of their ability to equip themselves. Whether these regular-army-like organizations would feel themselves bound by long-standing normative undertakings relative to the conduct of war is questionable indeed. However, the problem of regulating small arms is not restricted to organizations. Already, individuals are able to produce restricted automatic weapon components using additive manufacturing – an issue that has not gone unnoticed in the current gun control debate in the USA (Farivar 2013). That same manufacturing capability could be employed by individual uniformed military personnel, particularly those in specialized units, to equip themselves at will with personal kit that could transgress a legal or moral line.[5] Couple these concerns with what appears to be an insatiable global small arms market and a commercial business ethic that is all too often trumped altogether by the profit motive, and what results is a milieu in which the problem of illicitly copied small arms may become as ubiquitous as the problem of illicitly copied music for iPods. While the proliferation of pirated music may not be something that the military ethicist spends a lot of time thinking about, the completely uncontrolled and unlimited ability to produce small arms on demand is something the military ethicist ignores at the risk of significant consequences.

Major Military Hardware

Additive manufacturing is ideal for the reproduction of complex and even custom geometries and low-volume-production replacement parts, and it can produce these shapes in a way that greatly reduces assembly requirements. Take, for example, a duct for an F-18 fighter jet. The conventional duct fabricated using traditional processes consists of 16 parts, including nuts and bolts, to form an assembly that still requires being glued together. Using additive manufacturing technologies, the same duct can be made as one complete piece – no assembly required. Soon, major components of military hardware also will be produced in this way. In 2012, it was reported that 'Airbus is planning to print out a wing for an airliner in the near future. They are not planning to print out a plan for the wing or a picture of the wing. They are going to print the actual wing itself – a wing that will someday be part of an aircraft flying at 500 miles per hour at 40,000 feet altitude' (Pierpoint 2012).

The ability to manufacture repair parts – including major assemblies – in this way means that a whole host of issues relative to foreign military sales and export regimens for replacement parts for so-called 'major-end items' of military equipment (such as fighter jets and tanks) must be carefully rethought. Why? Because, hitherto, a certain modicum of control could be exerted upon purchasers of major-end items by regulating their access to the specialized replacement parts required by these systems. This limitation has long served both the political and moral purpose of regulating purchasing nations' bellicose conduct and of keeping otherwise unobtainable military hardware out of malevolent hands. However, as the ability of purchasers to create their own replacement parts dramatically increases, the ability to regulate violent action by regulating the supply of replacement parts commensurately decreases. This creates a greatly expanded array not only of political problems but also of moral ones, because it requires that nations make new normative decisions about, and weigh more carefully than ever before, the question of to whom and under what circumstances it is morally permissible to sell military equipment abroad and when and to whom it is not. Failure to do so might result in enabling non-state actors to wage war on a much larger scale than they otherwise would have been able to do, because of their enhanced ability to supply their own replacement parts.

New Concepts in War-Making Equipment

An even bigger problem than the prospect of additive manufacturing technologies making replacement parts readily available, or even that those technologies could be used to replicate the major-end items themselves, is that additive manufacturing could render altogether obsolete the big, bulky machines of the kind that have typified military equipment from the Renaissance to the end of the twentieth century. For example, additive manufacturing is already well on its way to producing, *in toto*, unmanned aerial vehicles of the kind that have proliferated in the first decade of the twenty-first century and which have, in their own right, become objects of *jus in bello* inquiry. In 2011, a university in the UK designed an entire unmanned aerial vehicle, printed it using additive manufacturing technology, and flew it – all in one week (Tate 2011).

Catastrophic Weapons

The problem likely to be of greatest concern, however, is that which ongoing advances in additive manufacturing could pose relative to the production of weapons of mass destruction. Up until now, chemical and biological weapon production often has required expansive industrial complexes. However, emerging advances in nanotechnology coupled with emerging additive manufacturing technologies may make possible the production of chemical or biological warfare agents using micro-reactors. It could be, therefore, that additive manufacturing technologies may be just the solution to the production of highly toxic substances, very small quantities of which could produce large numbers of human fatalities. Nuclear weapon production also historically has required vast industrial complexes. (Indeed, one of the secret cities associated with the Manhattan Project, which produced the world's first nuclear weapon, consumed 10 per cent of all electricity produced in the USA at the time.[6]) Although up until now, additive manufacturing processes have not involved heavy metals like uranium or plutonium, if (or when) the technological barriers to their use in additive manufacturing are breached, one cannot discount the prospect of nuclear weapons components being manufactured by something that looks far more like a 'Mom and Pop' operation with a fancy copying machine rather than by something that looks anything remotely like the enormous industrial centers associated with nuclear weapons production by the superpowers during the Cold War. Indeed, one can barely imagine more catastrophic consequences than those that attend the prospect of additive manufacturing technologies being applied to produce weapons of mass destruction. They do not require production in quantity; indeed, the ability to produce them at all is sufficient to shift balances of military and political power globally. Imagine additive manufacturing technology applied to weapons of mass destruction in the hands of persons like the 9/11 terrorists. This could mean that highly sophisticated technological means to perpetrate not merely monstrous criminal acts, but indeed acts of *war* could find their way into the hands of terrorist cells bent on wholesale destruction. One might be tempted to dismiss the problem by observing that would-be mass destroyers would have to obtain special nuclear materials – weapon-grade uranium or plutonium – before they could manufacture a device capable of producing a nuclear yield. While this is true, it is also the case that these materials exist in abundance in the aftermath of the Cold War and that some nation-states continue to manufacture them. Moreover, some possessors of these materials have either questionable ties to terrorists or are suspect in their ability to maintain strict accountability of, or control over, their weapons-grade nuclear materials. Similarly worrisome is the specter of disposable bioreactors, kits for conducting gene-splicing experiments, or chemical micro-reactors – all of which are part and parcel of the unfolding drama of additive manufacturing, and which, in the wrong hands, could allow private citizens to manufacture inherently dangerous materials in the privacy of their own homes.[7]

New Imperatives for Military Ethics

The challenges posed by additive manufacturing have implications for all aspects of military ethical inquiry. Let us begin, for example, with *jus in bello* – the most thoroughly developed domain of the military ethicist. While the world has had – particularly in the last decade – extensive experience in dealing with non-state actors whose organizations have looked and acted very much like armies, these 'armies' have had to rely on sympathetic entities to supply them with manufactured military equipment. What if they were able to make their own equipment – not crude devices like roadside bombs, but highly sophisticated devices of the kind traditionally available only to nation-states? Such a capability would further blur the already blurred distinction between combatant and common criminal, and along with it, questions surrounding which rules of engagement, which standards of evidence governing targeting criteria, and which safeguards on civil rights should be applied to whom. For example, is an additive manufacturing capability in the possession of entities sympathetic to perpetrators of violent action a proper military target? Should persons who use the equipment, even if not for military purposes, be regarded as combatants?

With respect to *jus ad bellum*, if advances in additive manufacturing enable the waging of *de facto* war (as opposed to, say, merely terrorist action) by non-state entities, the whole question of what constitutes a war and the moral boundaries for responding to it with military as opposed to law enforcement means may have to be reconceived. In a world full of non-state entities able to self-equip for war, what new kinds of aggrievements, if any, might constitute 'just cause' to go to war against them? What might count as a 'last resort'? In what way and to what extent should nation-states be held accountable for the malevolent acts of *de facto* armies operating within their boundaries but independent of state support? Of course, questions like these are not, *prima facie*, novel. However, they heretofore have occurred in a context in which the role of non-state actors was considerably more peripheral than it may become with the aid of the emerging technologies described herein.

Given the proposition that, while it is good to mitigate the evil effects of war, it is far better to avoid them altogether, the military ethicist of the twenty-first century might, in fact, find the intellectual tools of the profession best applied to *jus ante bellum*. Until recent history, this domain could be left to diplomats, lawyers and policy makers of different sorts, and, of course, these will continue to play a vital role. However, in a world where wars with non-state entities frequently occur but are rarely declared, and in which these entities can equip themselves with high-tech military hardware, moral strictures more thoroughly focused on war prevention rather than war regulation may become essential. That is not by any means to suggest that *jus ad bellum* and *jus in bello* considerations would become obsolete, but merely to acknowledge that, perhaps far more than any time in the past, *jus ante bellum* considerations calculated to prevent violent conflict could assume an unprecedented level of prominence. Indeed, it may be argued that the challenge presented by additive manufacturing provides precisely the impetus needed to extend *jus ante bellum* discussions into a broader framework that includes not only anticipations of war but also war and violence prevention. This is so because, clearly, the issues involved in the military application of additive manufacturing technologies – especially as they could one day apply to weapons of mass destruction – invite an ethical discussion on how to restrict easy access to these tools (and thereby restrict easy access to the incentives) to engage in violent action in the first instance, whether in the form of traditional warfare or terrorist action. *Jus ad bellum* discourse has always observed a strong presumption against war – that wars, if they are to be fought at all, should be fought only as a last resort. The logical extension of that premise is that, if the set of circumstances that constitute justified resort to war *can be* constrained, then morality requires that they *should be* constrained. In that respect, the advances now evident in additive manufacturing technologies provide ample justification for an expanded view of *jus ante bellum* that includes questions on how to constrain access to emerging technologies applied to violent purposes. In short, while, until the present day, a sharp, perhaps even exclusive, focus on *jus in bello* as traditionally conceived may have been appropriate for military ethicists, problems of the kind now posed by additive manufacturing may make it so that it is no longer sufficient for military ethicists to focus exclusively on dissecting questions about who can or cannot kill whom when. On the other hand, if the military ethics community is willing to expand the scope of its discourse beyond its present circumscription, it may find itself uniquely positioned to provide an essential intellectual service to the emerging world of twenty-first-century conflict and violence.

In particular, military ethicists might profitably seek a voice in the formulation of new global norms. Since it may be both physically impossible and, indeed, undesirable, to prevent altogether the proliferation of these emerging technologies – especially since, when used properly, they offer tremendous benefits to humankind –it may be that the best safeguard against their malevolent use is the establishment of new international norms.[8] A host of international undertakings (e.g. the Nuclear Nonproliferation Treaty, the Chemical Weapons Treaty, the Biological and Toxins Weapon Convention, and the Conventional Forces in Europe Treaty) represent well-established attempts to regulate access to weaponry. However, such regulatory efforts historically have focused upon nation-states. Since additive manufacturing technology could open the door to the possibility of small groups or individuals availing themselves of weapons that states are not allowed to possess (or if allowed to possess, only within strict parameters), the result could require a radical re-conceptualization of what 'warfare' entails. For example, if an act of war traditionally associated with nation-states is committed instead, and in similar magnitude, by a self-arming non-state entity, should it be treated as an act of war – and if so, to what end? (Note that a low-tech and small-scale version of what the future could bring is already evident in the challenges associated with fighting a *war* with the Taliban in Afghanistan.) Regardless of the form that new international understandings to govern the coming revolutions in manufacturing and military affairs might take, the formulation of appropriate norms is a matter in which military ethicists should seek to have a voice. In a related vein, it may be necessary for military ethics to re-tool its thinking with respect to the problem of terrorism and of what constitutes a terrorist. Not only could additive manufacturing make guns, bullets, bombs and so on cheaper and more accessible, it also could facilitate their disguise, such that improvised explosive devices (IEDs) could be made to look identical to non-weapons (Williams & Campbell 2011); terrorists could altogether lose their dependency upon sympathetic nation-states for their high-tech manufactured war materiel; and the entire complexion of organized warfare could become radically different.

A Proposed Research Agenda for Military Ethics

Much of the history of ethics is a history of reacting to extant dilemmas. While this is understandable, the perils of the twenty-first century are such that it may no longer be sufficient for ethicists to content themselves with what might be termed 'reactive reflection'. In light of the possibilities presented by additive manufacturing, it now becomes necessary for ethicists to become *proactively* reflective and to consider dilemmas to which technology will surely give birth with ever-diminishing lead time. The slate of military ethical problems associated with this rapidly emerging and truly transformative technology suggests a research agenda that includes a host of specific questions, of which the following are representative:

- What morally justifiable approaches can be taken perhaps along the lines of the proposed Fissile Material Cutoff Treaty, to restrict access either to stockpiled special nuclear materials or the production of these materials, or to precursors to chemical and biological weapons? In light of the emerging threat, it may be that states unable or unwilling to prevent access to these materials by small groups or individuals should be held accountable – if necessary, with the appropriate application of force. If so, an expanded understanding of acceptable *jus ad bellum* parameters may be required.
- In a related vein, what justifications can be given for violent interventions beyond the boundaries of one's own nation-state's borders, to include pre-emptive action, to prevent the use of emerging technologies for malevolent purposes – particularly as they might be brought to bear with respect to weapons of mass destruction? Implicit in this question as well is a requirement to explore expanded understandings of *jus ad bellum*.
- What moral justification can be given to insist that some entities are entitled access to special nuclear materials and that others are not? While this has long been an issue, emerging additive manufacturing technologies could add new impetus to it. Already, states such as Iran and North Korea appeal to moral arguments to justify possession of special nuclear materials outside the established international legal frameworks. The ability of small groups or individuals to possess these materials would fundamentally transform the nature and scope of the regulatory mechanisms required.
- To what extent can persons be morally required to surrender claims to rights of privacy in order to ensure that their use of additive manufacturing technologies is not malevolent? The enormous complications associated with this question are reflected in the perennial gun control debate in the USA. One can easily imagine the following variation on an already famous argument: 'Additive manufacturing – even if used by private persons to produce deadly weapons – doesn't kill people; people kill people.' Instead of the current preoccupation with the question 'What should private individuals be allowed to buy?', the more pressing normative question might become 'What should private individuals be allowed to *make*?'
- Indeed, is there any such thing as a right to privacy when it comes to the production of weapons of mass destruction? (An interesting question – especially in light of the fact that both nation-states and non-state actors have always shrouded their production of such weapons in utmost secrecy).
- How should additive manufacturing as it relates to the manufacture of large and small conventional war materiel be regulated?

Conclusion

These are not easy questions. Moreover, the ethical problems that they present – not to mention those persons whose actions give rise to these problems – will not wait for ethicists to decide when it is professionally interesting or intellectually stimulating to address them. These problems are quickly heading our way. Indeed, additive manufacturing technologies are already being used to produce out of plastic fully functioning weapons and weapon components that are presently made out of metal. Just over the temporal horizon lies the prospect of manufacturing small arms and many other weapon-related items out of multiple media simultaneously. Only a bit farther over the horizon lies the production of not only repair parts, but also whole major-end items themselves.

Past discussions over the control of military wherewithal have not necessarily been informed by voices from the military ethics community. However, the community can no longer responsibly avoid participation in that discussion – especially in light of the prediction that, in a future day, individuals will have 3D printers in their homes and, instead of having manufactured items shipped to them, they simply will download a design file and manufacture the item themselves with a 3D printer on their kitchen counter (Pierpoint 2012). If questions related to the possibility of laypersons being empowered to create weapons and wage unrestricted war are not proper questions for military ethicists, then there *are no* proper questions for military ethicists.

The sweeping technological revolution on our immediate horizon could change war fundamentally, and the revolution is occurring so rapidly that, unless we are vigilant, we could find ourselves without the ethical tools necessary to deal with the change. The impending changes will profoundly affect both soldiers on the battlefield and private citizens attacked as noncombatants. Indeed, unless these ethical problems are dealt with, every additive manufacturing problem may become a military ethics problem.

ACKNOWLEDGEMENTS

The author wishes to acknowledge his debt of gratitude to his colleagues, Dr Christopher Williams and Dr Thomas Campbell of the Virginia Polytechnic Institute and State University, for introducing him to the world of emerging manufacturing technologies, as well as for providing insights without which this paper would not have been possible. Thanks are also due to the referees for this journal. All views expressed herein are the responsibility of the author and do not necessarily reflect the position of any US government entity.

NOTES

1. See, for example, http://www.wohlersassociates.com/index.html, accessed 23 September 2013.
2. Making a Bit of Me: A Machine that Prints Organs is Coming to Market, *The Economist*, 18 February 2010, accessed 23 September 2013, available at: http://www.economist.com/node/15543683; Internet.
3. See, for example, 3D printing a supercar, the Citroën GT, accessed 23 September 2013, available at: http://i.materialise.com/blog/entry/3d-printing-a-supercar; Internet.
4. Christopher Williams, personal correspondence to the author, 29 May 2012.

5. The author is indebted to Mr Forrest Waller, Senior Research Fellow at the National Defense University Center for the Study of Weapons of Mass Destruction, for this insight.
6. See The Manhattan Project, accessed 23 September 2013, available at: http://www.paducahvision.com/manhattan-project, p. 12; Internet.
7. The author is indebted to Mr Forrest Waller, Senior Research Fellow at the National Defense University Center for the Study of Weapons of Mass Destruction, for this insight.
8. The author is indebted to Dr Richard J. Danzig, Chairman of the Board, Center for a New American Security, Washington, DC, for this insight.

REFERENCES

DeGarmo, E. Paul, Black, J. T. & Kohser, Ronald A. (2003) *Materials and Processes in Manufacturing*, 9th edn. (Hoboken: Wiley, Inc.).

Farivar, Cyrus (2013) 'Download this Gun': 3D-printed Semi-Automatic Fires over 600 Rounds, accessed 23 September 2013, available at: http://arstechnica.com/tech-policy/2013/03/download-this-gun-3d-printed-semi-automatic-fires-over-600-rounds; Internet.

Iga, Kenichi & Kokubun, Yasuo (Eds) (2006) *Encyclopedic Handbook of Integrated Optics*, s.v. 'Nanophotonics' by Satoshi Kawata, Yasushi Inouye and Hong-Bo Sun (Boca Raton: CRC Press).

Pierpoint, Paul (2012) Why Industry Needs to Embrace Additive Manufacturing and the New Industrial Revolution, accessed 23 September 2013, available at: http://www.lvb.com/article/20120517/LVB01/120519887/0/FRONTPAGE/Why-industry-needs-to-embrace-additive-manufacturing-and-the-new-Industrial-Revolution; Internet.

Tate, Paul (2011) Additive Manufacturing: Can It Bring Manufacturing Home?, accessed 31 May 2012, available at: http://www.manufacturing-executive.com/community/leadership_dialogues/game-changing_technologies/blog/2011/12/02/additive-manufacturing-can-it-bring-manufacturing-home; Internet.

Williams, Christopher & Campbell, Thomas (2011) Additive Manufacturing (3D Printing): State of the Art, Potential and Implications, lecture given at National Defense University, Washington, DC, 19 August 2011.

John Mark Mattox is a Senior Research Fellow at the National Defense University Center for the Study of Weapons of Mass Destruction and Director of the Countering Weapons of Mass Destruction Graduate Fellowship Program. He served on three continents as a field artillery officer and is a veteran of the First Gulf War. He holds a BA from Brigham Young University, Master of Military Arts and Sciences from The United States Army War College, and MA and Ph.D. degrees from Indiana University. He has served on the faculty of The United States Military Academy at West Point, The University of Maryland, Missouri State University, and The U.S. Army Command and General Staff College, where he occupied The General Hugh Shelton Chair in Ethics. *Correspondence address*: National Defense University Center for the Study of Weapons of Mass Destruction, Abraham Lincoln Hall, Suite 2400, 260 5th Avenue SW, Bldg 64, Fort Lesley J. McNair, Washington, DC 20319-5066, USA. *Email address*: john.mattox@us.army.mil

Part II

The Ethics of Cyberwarfare

RANDALL R. DIPERT
SUNY (State University of New York) at Buffalo, NY, USA

ABSTRACT *The paper addresses several issues in the morality of cyberwar and cyberwarfare, defined as one nation's attacks on the governmental or civilian information systems of another nation. It sketches the diverse technical ways in which an attack may occur, including denial-of-service attacks and the insertion of various forms of malware. It argues that existing international law and widely discussed principles of Just War Theory do not straightforwardly apply to cyberwarfare, and many forms of cyberwarfare differ from previous forms of warfare in neither injuring nor killing human beings, nor causing lasting physical damage – but can nevertheless cause serious harm to a nation's vital interests. Another dissimilarity with traditional warfare is in the degree of knowledge of the identity of an attacker (the 'attribution problem'). The paper argues that cyberwarfare is not amenable to regulation by international pacts and that we can expect long periods of low-level, multilateral cyberwarfare, a Cyber Cold War, as a game-theoretic equilibrium is sought,. The paper laments the lack of a cyberwarfare policy, and concludes that it is only by applying game-theoretic principles that strategies can be discovered that are both moral and effective in suppressing overall harm to all parties in the long run.*

Introduction

'Cyberwarfare' is a recent term: the Oxford English Dictionary gives its first use as 1994 (OED 2010).[1] The topic of cyberwarfare has received widespread media attention only in 2009 and 2010.

The increasing discussion of cyberwarfare would come as no surprise to anyone who has been following the news in the last decade: cyberattacks on components of US defense cyber-infrastucture, and the probable strategic, military cyberattacks by Russia on Estonia, then on Georgia, (Beaumont 2009) and (USCCU 2009), and probable attacks by China, North Korea, and Iran on US defense and economic targets. Although organized attacks from China on Google and Gmail had a corporate target, the political implications for freedom of speech made them part of a larger cyberconflict and brought a warning from US Secretary of State Hillary Clinton (Clinton 2010) and (Markoff 2009b). In May 2010, the US Department of Defense formed a full command, Cyber Command, headed by a simultaneously promoted four-star general, General Keith Alexander, who also continues to serve as director of the US National Security Agency (NSA). He had earlier noted that the Chinese government has hackers – cybersoldiers – organized into battalions and regiments (Alexander 2007).

Responding to intentional cyberharm ('cyberattacks') of all sorts, including by apolitical and anarchic ('black') hackers, as well as responding to economically motivated industrial cyberespionage, have all been a subject of discussion for decades. What is new is the acknowledgement of attacks that have been coordinated by the central commands of governments (or other political organizations) and that are directed at another country's governmental and military information systems, or at its commercial or infrastructure information systems for political purposes. Most of the public discussion has centered on *defense* against cyberattacks, that is, on governmental, military and economic cybersecurity.[2] What is still more recent is the open discussion of a need for an organized capacity for *offensive* cyberattack on hostile nations and political organizations by technologically advanced democracies.

Cyberwarfare has several unusual features. It is arguably the first major new form of warfare since the development of nuclear weapons and intercontinental missiles. This novelty has also meant that there is at present a virtual policy vacuum: there are no informed, open, public or political discussions of what an ethical and wise policy for the use of such weapons would be.

Second, it is very difficult to determine the source of cyberattacks: this is the 'attribution problem.' This fact would give many cyberattacks credible deniability – especially since in many cases nations can plausibly claim that the attacks may have originated from within their territory but their governments did not initiate them.

Third, many cyberattacks will not be lethal and will not even result in permanent damage to physical objects. This is of course extremely dissimilar from nuclear weapons and from virtually all traditional weapons of war.

Fourth, there are no exotic components to cyberweapons, again very unlike nuclear and other advanced technology weapons, and even unlike chemical or biological weapons. Any computer is a potential cyberweapon and anyone with advanced knowledge of information systems is a potential cybercombatant (Talbot 2010). This makes treaties that would ban cyberweapons virtually impossible from the outset (Libicki 2009).

One feature that cyberwarfare unfortunately shares with warfare using non-cyber weapons is that defense is expensive and liable to failure, while offense is comparably cheap: this parallels the well-known difficulties with nuclear civil defense, with anti-missile technologies, body armor, protection against improvised explosive devices (IEDs), and so on (Talbot 2010). Another feature that cyberwarfare shares with conventional warfare is the low degree of certainty of what will happen with an attack, or in a war, and that there are distant, deleterious side effects that cannot adequately be anticipated – disease, economic consequences, and so on.

My aim in this paper is to focus on the ethical issues surrounding cyberwarfare. I will examine how the widely accepted principles of the Just War Tradition might apply to cyberwarfare, but I will also consider deviations from or additions to this tradition that may be necessary either because of the inadequacy of these traditional principles or because of the sheer novelty of cyberwarfare. I have already signaled some of these hesitations. Traditional discussions of the morality of going to war, or morality in war, have been understandably motivated by the lethality and massive destructiveness of war. But cyberwarfare often will not be like that – although it *could be* both lethal and physically destructive.

I will begin by surveying some issues in the technology and tactics of cyberattacks, since these have ethical implications, and will also discuss issues of strategic policy. Second, I will turn to the separate domains of thinking about the morality of cyberwarfare and of how existing international law applies (or as I will argue, does not apply straightforwardly) to cyberwarfare. Then I will consider in more detail how a theory of morality in cyberwarfare needs to be augmented in two ways: In a *moral* dimension toward a general notion of harm, especially to national vital interests, and away from some of the key concepts in traditional Just War Theory, and in an *ontological* dimension, with a focus away from strictly injury to human beings and physical objects toward a notion of the (mal-)*functioning* of information systems, and the other systems (economic, communication, industrial production) that depend on them. Finally I will discuss what the future of cyberwarfare might be like and some possible remedies or ways we can mitigate its damage.

Cyberweapons: Technical Details and Policy Issues in their Use

Cyberattacks belong to a large genus of all kinds of attacks on information systems. Such attacks include traditional counterespionage and disinformation campaigns, old-fashioned destruction of telephone lines, jamming of radio signals, killing of carrier pigeons (FM 24-5, 1939: ch. 4), and so on. We can restrict ourselves here just to attacks on modern digital information systems, that is, on computers and computer systems: intentional damage to software, hardware, and the operations of information systems. Even the restriction to digital is not especially crucial. (Note that a restriction to 'electronic' information systems is also too narrow since both communication and computation may be performed by utilizing light, other electromagnetic radiation, or any phenomena that permits storage and switching.)

Although it has not been spelled out, most authors use 'cyberattack' to refer only to attacks on digital information systems via the Internet (or maybe, via any sort of network). The possibility exists of attacks on the software of modern information systems via other media (such as CDs of distributed operating systems, applications or data) or secretly incorporated in the BIOS of computers themselves that may be sold by one country to another. The BIOS is the Basic Input/Output System, part of most computers, a form of relatively simple firmware that survives powering down, and that any computer needs in order to know how to load an operating system from some volatile memory device. (Strictly speaking, the BIOS need not be stored in an actual hardware circuit, although most have been, but does need to be relatively non-volatile.)

There has long been discussion of incorporating unique identifying chips into the hardware of each individual computer, in order to reduce the secrecy-of-identity that has engendered much criminal and harmful activity on the Internet. A key cause of this activity is the 'Attribution Problem.' A chip, secretly built into the hardware of a computer or networked device, could also serve as a platform for cyberespionage or a cyberattack, that is, a hardware Trojan horse – see (Markoff 2009b). This approach has so far been widely rejected, but I will return later to the possibility that stronger, universal source identification may be necessary to limit the damage of cyberwarfare. It would be difficult now secretly to insert a chip because even brandname computers are manufactured in diverse countries; furthermore, large numbers of engineers have a hand in the design and testing of every chip and other component, and secrecy would be very difficult.

On the other hand, the worldwide dominance of several 'closed source' operating systems (OS), such as the Microsoft Windows and Apple Mac OS families, would theoretically be susceptible to the hidden incorporation by their creators of sections of the software that identify each copy, track usage, automatically send information to a company or security service, or allow external manipulation via the Internet (a 'back door'). Just such a concern motivated China in 2003 to sign with Microsoft an agreement to grant it access to all of the source code for the Windows operating system. The Chinese were concerned that the US security services, or the Microsoft Corporation, might have incorporated 'back doors' that allow secret access to any system that is running the operating system (Gao 2003). Such access had previously been given to the security agencies of NATO, Russia, and the UK. Although this might seem to be a positive step toward openness, sharing the source code for operating systems and complicated applications like Word, Acrobat, or Excel allows sophisticated analysts in a country like China, Russia, or the US to detect weaknesses in the code that can then be exploited via the Internet. One of the motivations for 'open source' operating systems, such as Linux, and for open source application programs, is to make a secret back door unlikely, since the many thousands of programmers worldwide with access to the source code are more likely to notice weaknesses and suggest fixes than are the hundreds that have access to a proprietary source code. In such large programs it is nevertheless inherently difficult to understand the purpose or exploitable weakness of every line or section of code.

Within the narrowed family of Internet attacks on modern digital information systems via the Internet, there are several very different forms that an attack can take. The classification of cyberattacks is complex, and one can classify cyberattacks along several dimensions. For example in Rowe and Crusty (2010) and Rowe (2006) we see an elaborate classification by the kind of deception that is perpetrated. Another dimension for classification might be the kind and amount of harm. One family of attacks may be termed 'unintrusive' cyberattacks, such as the Denial of Service (DoS) attack. In such an attack, an attacker does not actually gain access to the site. Instead, a site or a server (or even sections of the network) is bombarded by hundreds, or thousands or more, of spurious requests for information per second, so that it cannot successfully respond to them and thus is rendered inaccessible to its intended users. The attacker does not gain access to the internal working of software or hardware, and thus no damage is done to them.

The most successful such attacks in recent years have been Distributed Denial of Service (DDoS) attacks. A programmer designs a malicious piece of software (malware) that embeds itself in hundreds or thousands of computers all around the world. This turns the other computers into remotely controllable slaves, or botnets. Then, either at a pre-established time, or on command from the designer, the embedded malware bombards a targeted site with email or requests for responses. On the Internet, each stream or packet of information is identified by the unique address (the Internet Provider or IP address) from which it originates. In a simple Denial of Service attack, these pieces of information will have the same IP address (although they can be 'spoofed' – their real origin falsified and varied). In direct DoS attacks, the attacking message packets can be blocked, by the server or even at gateways and nodes in the Internet itself. In Distributed Denial of Service attacks, this technique will not work, since the requests to respond will be coming from many different IP addresses, often in different parts of the world.

Many of these techniques of DoS attacks and insertion of malware could be defeated by requiring a confirming dialogue between the source and target ('handshaking'), although the activity of filtering packets with suspicious IP addresses, and the initiation of confirming dialogues, may itself overwhelm the computational power of the target computer. Sites such as Google or sites within the US Department of Defense probably have the computational power to keep up with a huge number of such attacks in real time (for now). A smaller business website operating on a single server might not.

There are nevertheless different ways in which DDoS attacks too can be identified and blocked from disrupting legitimate Internet traffic, such as through telltale common features ('fingerprints' or 'DNA') of their contained messages. One can also track the path of the messages through the Internet, and identify the real source. These and other forensic methods of identifying the sources of DoS attacks and malware can be very expensive in terms of human and computer resources – and thus add to the cost for, and harm to, the attacked party.

The other major form of cyberattack is *intrusive*: malware gains access to sections of a computer's software or stored data through a site. Once there, the malware may alter various pieces of software or data, cause the system to crash or make certain pieces of software inoperable, erase hard-drives, send email messages pretending to be from the user, send information on the software and data back to the malware's author, and so on. If this intrusive malware can then be sent (perhaps in modified form) and infect still other computers, it is a self-replicating. If it is transmitted by attaching itself to another file, program, or email, it is a virus; if it is a free-standing program that can travel through information pathways, it is a worm. In some cases, such a virus does no harm; it is a prank, but its presence is still undesired. The ways a virus spreads are varied: some can, just like biological viruses, alter their own code, so that their outward appearance (equivalent to a biological virus's protein shell) is not immediately recognized as similar to the first versions of the virus. They can also be designed to 'blind' anti-virus software to their existence, or to prevent updates to anti-virus software.

An especially virulent but apparently non-destructive virus is Conficker; early attempts to eradicate it failed (Markoff 2009a). The damage viruses may do and the way they do it are highly varied. Intrusive malware that is not self-replicating and that is at first not easy to detect, or does not at first cause detectable damage (and might even be perceived to perform a useful function), is a 'Trojan (horse).'

Many forms of malware are strictly tools of espionage: they do not directly damage the targeted nation's information systems, or cause damage via these information systems. In this sense, they are not themselves cyberweapons. In such cases, it is only through the use of gathered information by an enemy that a nation's interest may be harmed. Traditionally, mere espionage has not been viewed as a *casus belli* (customary or legitimate reason for going to war), but may bring non-military retaliation, such as the expulsion of diplomats or the limiting of foreign aid or commerce. Consequently, I will not discuss pure cyberespionage further, even cyberespionage that is directed by one nation toward another, since it is not usually an activity that has been considered part of the moral considerations regarding going to war or conduct in war. The ethical considerations in espionage and other intelligence-gathering operations are but one of the several traditionally neglected aspects of the morality of war (although there has been growing interest in this field in the last few years).

Against a nation's military targets, there are essentially three categories for targets of cyberweapons.

First, the cyberweapons may target and impair a main function of military chains of command, namely command and control: communications and information gathering, as well as the communication of precise orders to maneuver, defend, or attack, may be harmed. Command and control data may be blocked, altered, or false reports and commands inserted. Most modern military organizations have hardened communication conduits (such as buried, shielded, or otherwise protected electrical or optical cables), elaborate schemes for encryption, and redundant systems.

Second, weapons or weapon systems can be rendered inoperable for a time or even physically sabotaged by faulty messages or intrusions into their controlling information systems; in the extreme case, they could be directed by an enemy to attack a false target. The most sophisticated computerized weapon systems, such as the Patriot and other anti-aircraft and anti-missile systems, and the Aegis system of the US Navy, could conceivably be rendered inoperable or directed to fire at false targets. These systems have already been proven fragile even in ordinary operation when they are not subject to cyberattacks: there have been friendly-fire and other targeting errors in the case of the Patriot anti-aircraft and anti-missile system in 1991 and 2003, and the disastrous shooting down of an Iranian civilian airliner by the Aegis Combat System (in 1988) aboard the US cruiser *Vincennes* that automates fire control of guns and missiles.

Finally, cyberweapons could target joint-use infrastructure, that is, systems and structures for both civilian and military uses, or even civilian targets with the goal of demoralizing, weakening, or confusing an enemy's military and civilian leadership. Joint-use infrastructure would include a wide variety of computer-guided systems, such as the satellite global positioning system (GPS) network, and energy, communication, water, or sanitation infrastructure, or petroleum and chemical refining systems. Targeting joint-use industries and systems in a lawful military conflict is generally permitted by international law and Just War Theory, but some authors have proposed additional criteria (Walzer 2006: 144–51). Targeting primarily civilian structures and networks is prohibited by international law and by almost all theories of morality in warfare.

The main aspects of chemical processing and energy production and transmission facilities, as well as communications networks, are now largely managed with the help of complex information systems using many modules of software. This software relays data and warnings to human operators, in some cases make decisions itself, and allows remote monitoring and control by engineers. Increasingly, at least in petroleum refining and chemical processing plants, key engineer supervisors are allowed access to the control systems via the Internet. Access by supervising engineers or military officers via the Internet would theoretically place near-instantaneous operational control of these facilities in the hands of the most informed and capable people, while they are at home or traveling. Access to the computer control chemical-processing facilities is typically highly protected, such as by layers of high-quality and frequently changed password protection, as well as accessibility only via a single, identifiable, company-owned computer. Likewise, access to secure military information – and presumably the ability to issue military commands – in the US Department of Defense is controlled by a user's Common Access Card (CAC), whose chip is encrypted and requires a computer to have a special piece of hardware with restricted distribution, the CAC Reader, in addition to the use of high-quality and frequently changed passwords. There have already been several generations of increasingly sophisticated hardware and encryption methods used in the CAC and CAC reader.

Nevertheless, in both industry and military, access via the Internet to this information and control is obviously susceptible to highly sophisticated hacking. The US Department of Defense, and each individual military service, has at least several layers of increasingly secure computer systems. Some of these systems are especially secure networks not immediately accessible via the Internet. The most sensitive such computers are constantly monitored for any unusual activity. This is termed cybersecurity 'defense in depth.' The danger to civilian banking and energy infrastructure has been widely noted in the media and in mass-distribution books, such as (Clarke and Knake 2010).

Public information, including General Keith Alexander's unclassified remarks before the Senate confirmation hearings of the Armed Services Subcommittee (US Senate 2010), indicate that there are hundreds of thousands of probes per hour of Department of Defense and NSA computer systems.

These probes include mostly harmless and unsuccessful attempts at getting past required passwords, or identifying a server's or system's operating system, metadata on its websites (such as the program with which the HTML code was written), etc. The percentage of those that probably originate with foreign governments is classified, as is what has been their level of success. (There are rumors that a majority of these attacks actually originate in the US). This does not include probes or attacks on other US government systems, such as in the Departments of Justice, State, and Homeland Security, or on other Western governments. Among the many officials I have heard or read, all, including General Alexander and the Senators and staff of the Armed Services Committee, are alarmed. (The one exception was a remark by Howard Schmidt (Singel 2009), but this was regarded as a major blunder.) There have been some spectacular recent successes at penetration, such as theft from a contractor of specifications, and thus weaknesses, of the F-35 Joint Strike Fighter.

Yet another peculiarity of cyberweapons is that, once intrusive malware is detected, there will usually be countermeasures and damage repair that can be executed in minutes, hours, or days by a technologically advanced user or country. The malware may be removed or quarantined, the system reloaded (reset, rebooted) with data and software from before the attack, and the incoming data stream can be filtered for this type of malware by looking for suspicious code sequences or IP source addresses. From the point of view of the offensive use of the cyberweapon, this makes many of them 'one-time use' weapons whose effectiveness will likely rapidly diminish. Few traditional weapons have had this characteristic. As an attacker, one will thus *not* want to release one's very best cyberweapon until there are occasions where risking the rapid development of enemy countermeasures might be worth it. An attacker will also probably want to develop offensive cyberweapons 'in depth,' that is, multiple cyberweapons (sometimes called cyber - or logical bombs) that are substantially different in the way they penetrate an enemy information system, and in their distinctive sequence of code, damage they cause, or the means by which they could be detected.

Another aspect of 'cyberweapons in depth' is that sustained, successful development and use of offensive cyberweapons might actually require three or more layers or kinds of attacking software:

1. Reconnaissance or espionage software to determine a system's nature and weaknesses, and to target it uniquely (rather than, say, also damaging civilian information systems).[3]
2. Damage-producing software that harmfully alters the behavior of the enemy's software or produces other undesired effects – possibly including deaths and damage to physical systems. These are the logic- or cyber-bombs themselves.
3. Damage-assessment espionage software that determines the effectiveness of the damage-producing software.

In some cases, reconnaissance and damage assessment might be achieved through traditional intelligence-gathering measures, although nothing would be as useful for future development or assessment of cyberweapons as details about how the damage-producing software actually affected the targeted information system. This is not usually going to be superficially apparent from human, electronic, or satellite surveillance.

Viewed from the point of view of limiting damage in the world, there are several positive and even unique features of cyberweapons. Most cyberweapons will probably fail to achieve their intended or expected degree of penetration and damage (Rowe 2009). In this they will resemble most mutations of biological microorganisms: bioweapons often lack the ability to spread, and harmful organisms are difficult to 'weaponize.'

Second, most cyberweapons will probably not be lethal. Even a massive electrical power failure would not immediately kill many (except those on life support and without backup power, for example).

Third, even if cyberweapons cause intended damage to military or even civilian information systems, they could be designed so that this damage could be reversible (ibid.). One could develop cyberweapons, together with countermeasures or ways of restoring operability or data, that one would eventually share with an enemy or release if an enemy's provocation ceases or if the cyberweapons accidentally cause damage to strictly civilian information systems. The only thing in traditional warfare that is remotely comparable to such reversible damage in cyberwarfare would be the joint production of chemical weapons and antidotes to them.

The Morality of Cyberwarfare: The Easy and Hard Cases

The moral questions of cyberwarfare divide themselves in this way:

1. Is a cyberattack ever morally justified in response to an enemy *conventional* attack?
2. Is a cyberattack ever morally justified in response to an enemy *cyberattack*?
3. Is a *conventional* attack ever morally justified by an enemy *cyberattack*?
4. Is a cyberattack ever morally justified in cases where the enemy has launched neither a cyber- nor a conventional attack? (With United Nations sanction, preemptively, preventively, or for some other reason.)
5. Once a war (cyber- or conventional) has begun what kinds of cyberattacks are morally justified?

Of the five questions, (1) and (2) would seem to be easy questions, subject to the restraints that might be posed by the issues in (5). A cyberattack by nation A on B would seem *prima facie* to justify morally a retaliatory cyberattack by B on A, and one at least as destructive as A's first strike. Likewise, a cyberattack in response to a conventional attack would seem to be at first sight justified, again subject to the issues raised in (5).[4] There may also be other *jus ad bellum* criteria that need to be considered, such as Proportionality and Likelihood of Success, but there do not seem to be any issues here that are peculiar to cyberwarfare.

Cases (3) and (4) are the hard cases. Case (4), such as the moral permissibility of a 'preventive' first-strike, shares many of the same issues and arguments as does a discussion about a non-retaliatory conventional attack. We have just gone through nearly a decade of discussion of the conventional case – the moral issues in the coalition's 'preemptive' attack on Iraq in 2003 – with remarkably little light being shed on the issue. The case of a cyber first-strike is, however, more problematic, because it need not involve widespread death and permanent destruction of physical installations. If a preventative 'attack' is ever justified, then it is likely to be something like a cyberattack, that can be free of the usual death and destruction of traditional forms of warfare.

Cases (2) and (3), being responses to cyberattacks, involve us necessarily in what was earlier called the Attribution Problem. More broadly these are *epistemic* problems that have been ignored by most theorists of the morality of war; namely, how much justification or evidence is necessary concerning the threshold conditions for morally going to war? We can distinguish here between the metaphysical conditions for a moral war, and the epistemic conditions: one is behaving morally if one objectively has extensive evidence that all the necessary conditions for going to war are held, even if it turns out one did not hold. This epistemic uncertainty is one of the peculiarities of responding to a cyberattack that makes it similar to preemptive and preventive war. In the case of a cyberattack, the problem is uncertainty about *who* attacked us, and in the case of preemptive and preventive war, the 'who' is known, but there is uncertainty about *whether* an enemy would indeed, in the future, attack us.

Can we do nothing at all if a cyberattack cripples our industrial, military, and governmental systems when there is some uncertainty about who did it? In another paper (Dipert 2006b) I argued, on game-theoretic grounds, that a preemptive attack can be morally justified if the evidence exceeds a certain threshold of objective likelihood (roughly 90 percent) and if there will be a high level of expected damage to us if we do not preemptively attack. I argued, for consequentialist reasons, that if every state followed such a policy, overall damage to all parties over the long run would be minimized because of a deterrent effect. A similar calculation would seem to hold in the case of a cyberattack when the attacker is not certain: if the expected damage is very large, and it is highly likely that we can identify the attacker, and can likely eliminate the threat or deter others in the future, then we might be morally justified even without certainty in our identification of the attacker. Admittedly, a great deal of moral weight lies on the extent to which we can objectively measure the likelihood of identifying an attacker and in objectively estimating the damage to us if we do not stop this attack, or deter others like it in the future.[5]

I believe Just War Theory, and much – but not all – recent theorizing about the morality of war relies upon rickety meta-ethical foundations. The origins of Just War Theory are to be found, in St Thomas Aquinas (1225–74) and Hugo Grotius (1583–1645), in natural law theory. Yet few contemporary philosophers have a metaphysics that would allow them to maintain such a view. It is also true that Just War Theory, having been endorsed by most industrial democracies and in international law, has also acquired the status of damage-minimizing convention. However, the increasing number of nations, especially non-Western ones, who show no serious effort to endorse or follow this convention – and the unwillingness of other nations to *force* compliance – means that the advantage of a widely accepted convention is lost; it merely handicaps nations with a developed public sense of morality and prevents them from moral intervention.

The only meta-ethical route left to many theorists is one based upon 'intuitions,' legitimized by the 'reflective equilibrium' of John Rawls (1921–2002) and by contemporary 'experimental philosophy'; but it is unclear how these intuitions arose and whether they are reliable – whether, for example, they are not inter-individual ethical intuitions that are vaguely applied to states because of a lack of developed international-political modes of thinking, that is, Michael Walzer's 'domestic analogy.'

The meta-ethical view I am tentatively employing is realistic, diachronic, universalizable, enforceable consequentialism for states. It is 'realistic' not in the sense of dismissing moral thinking (Christopher 1999) but in recognizing that the condition of conflict between societies, as it has been for hundreds (if not thousands) of years, is exacerbated by communication and mobility, and is likely to remain for some time. It is universalizable in the sense that were a policy practiced by all states, then they would be markedly better off than they have been. It is enforceable or 'compellable' in that the practice of this policy would in fact compel other states to follow a similar policy, rather than encouraging free-riders to profit by disregarding it. Universalizability and what I am calling 'enforceability' are related to the desiderata that Robert Axelrod proposed as components of an optimizing strategy (which he proposed was achieved by tit-for-tat).[6] Finally, it is diachronic insofar as it does not look at what is best for this or that state right now, but what is attainable and optimizing in the long run (it is temporally universalizable).

Case (3) is a very hard case indeed: could a cyberattack, perhaps without deaths and permanent destruction, wreak such unjustified harm on a nation that it is morally permissible to launch a conventional attack, which will involve lethality and physical destruction? What if that is the only way of stopping this or future attacks? I will return to these cases.

Cyberwarfare and Traditional Just War Theory

As of the writing of this essay there are two papers mainly about the ethics of cyberwarfare: Arguilla (1999) and Rowe (2009), and the similar Rowe (2010); the writers were not experts in the morality of war, but experts in computer science. A third, book-length monograph was written in consultation with experts on the morality of war (Owens et al. 2009: xiii), but is largely a compendium of existing law. Two characteristics of these approaches are the following: (1) they regard Just War Theory as a settled body of opinion on the morality of war and (2) they accept or argue for the view that Just War

Theory is straightforwardly applicable to cyberwarfare, without the need for modification or addition. I argue that both of these claims need to be questioned.

It is a separate question whether existing national laws and international law are applicable to cyberwarfare, and whether they are sufficient. Laws (statutes, agreements, treaties, etc.) rarely coincide with permitting all of what is morally permissible, and outlawing all of what is morally impermissible. Although it is not my primary focus – nor is it my area of expertise – I hazard some guesses about the legal situation. Roughly, I believe that the legal status of cyberwarfare parallels the situation concerning the morality of cyberwarfare: most legal frameworks do not clearly apply to many instances of cyberwarfare, and cyberwarfare involves aspects of damage or harm that are typically not addressed by law, such as harm to the functioning of information and other systems that might not harm physical objects or persons.

The traditional theory of what is morally permissible in war, Just War Theory, is usually divided into two main questions. First, there is the question of when a country may morally take part in, or begin, a war: *jus ad bellum*. Second, once one finds oneself at war, there is the question about how one may morally fight that war: *jus in bello*. Excellent accounts of traditional theories can be found in Orend (2005), Fotion (2007), and Christopher (1999). The best available account in English of historical thinking on these issues is in Reichberg et al. (2006). The most informed account from the point of view of the recent history of warfare is Walzer (2006). Each of these two aspects of the morality of war have criteria for moral adequacy that are widely (although not universally) accepted by both major past and contemporary thinkers about moral issues of war.

For *jus ad bellum* the criteria are (1) Just Cause, (2) Last Resort, (3) Likelihood of Success, (4) Proportionality, (5) Proper Authority, and (6) Right Intention. Just Cause is regarded as by far the least controversial component of Just War Theory. A Just War Theory that abandoned or extensively modified the traditional understanding of *casus belli* would simply not be Just War Theory as we know it. Consequently, I will focus my remarks on Just Cause. I will make remarks about the other *jus ad bellum* and *jus in bello* criteria in passing.

As Michael Walzer notes in *Just and Unjust Wars* (Walzer 2006: 51–52) the key necessary condition of morally permissible war, Just Cause, is rather blandly and uniformly described as 'aggression' or 'attack' by an enemy. In the United Nations (UN) Charter, what is prohibited in order to maintain peace is broadly characterized as the 'threat or use of force against the territorial integrity or political independence of any state' (UN Charter 1945: Ch. 1, Article 2, Section 4). Later, an 'armed attack' is the only stated condition in the UN Charter under which a nation may justifiably defend itself before the Security Council takes action (UN Charter 1945: Ch. 7, Article 51). Observe that this seems, literally understood, to designate soldiers using 'arms,' roughly, as artifacts for inflicting injury, death, or causing physical destruction of objects.

It is thus a stretch to consider a cyberattack an 'armed' attack, since the artifact doing the damage is a computer (designed for other purposes), or, still more abstractly, an information-theoretic entity. The most prominent and large-scale work on the topic (Owens et al. 2009) considers a cyberattack to be a straightforward instance of an armed attack. This is not obvious. The paradigmatic historical form of aggression or attack is the invasion of the sovereign territory by armed, centrally commanded, enemy soldiers of another state who are prepared to use deadly force. Sometimes this invasion may be momentary, such as at Pearl Harbor in 1941, and the primary harm is extensive death or destruction.

This view of aggression and attack as invasion or destruction is retained by most writers who roughly follow the just war tradition, such as Fotion (2007). The language of 'attack' may be broad and vague enough to include non-paradigm cases of aggression I am about to discuss, although Fotion and most authors give examples only of violent, armed invasions. Yet 'customary' and accepted just causes of war have included a wider array of phenomena that are broadly categorized as *casus belli*: embargos, systematic attacks on and harassment of citizens and businesses abroad, blockades, blocking of necessary supply lines such as a pipeline, attacks on military or civilian ships in international waters, and so on. A cyberattack by one nation on another more resembles one of these. Martin Libicki (Libicki 2009: 179f) notes that NATO explicitly rejected the claim that Russia's cyberattack on Estonia was an act of war that would trigger mutual defense obligations. The National Research Council study (Owens et al. 2009: 257–258) somewhat unsatisfactorily separates the concepts of sanctions and blockade, the latter constituting 'just cause', the former not, regardless of the degree of harm: this makes too much of a legal (not a moral) distinction. Surely sanctions *could* be morally – but not legally – unjust, even if directed by the Security Council or by a regional international organization. Although it was clearly used as a pretext for large-scale war, the obstruction of the movements of goods and people between East Prussia and the main body of Germany in the 1930s (the Danzig Corridor) could reasonably have led to justified acts of war. Likewise the isolation of Berlin from the West by land routes (the Berlin Blockade) could have led to a justified war if the airlift had not been successful. In diplomatic language, and as a warning that these actions might risk war, such events are broadly conceived as intentional damage to *vital* interests of a state. Most discussions of *jus ad bellum* have taken scant notice of them. An interesting exception is Walzer (2006: xiv–xviii), who in the preface to the 4th edition of *Just and Unjust Wars*, calls them 'force-short-of-war,' and notes that it is force and thus morally prohibited nonetheless.

Attacks on another nation that kill or destroy buildings or other artifacts, but that do not include invasion by enemy soldiers, cover a broad range of activity, from small arms fire and artillery fired across a border, to the use of mortars and rockets. Cruise missile attacks or attacks by unmanned aircraft would seem to count as aggression that would, in certain circumstances, provide just cause for a wider counterattack. Even these actions still are traditional in the sense that they involve macro-physical objects that are intentionally directed to intrude into the airspace of another nation and cause physical destruction. One study (Owens et al. 2009: 239f) classifies these as kinetic projectile weapons, placing cyberattack in the category of non-kinetic weapons.

An unprovoked cyberattack by one nation on the civilian or military infrastructure of another nation is thus not very much like traditional, paradigmatic forms of aggression or attack. A cyberattack does not involve intrusions into the territory or airspace by soldiers or even by physical objects. A better analogy would be that a cyberattack is more like Electronic Warfare (EW), such as jamming the radio communications of another nation from outside its borders, or the intentional use of electromagnetic radiation, such as a laser, or an electromagnetic pulse (EMP) weapon to destroy or hinder the functioning of human beings, machinery, or infrastructure from beyond a nation's borders. Exactly this point was noted by General Keith Alexander (2007), who since has become the Commander of US Cybercommand (USCYBERCOM). The field manuals on electronic warfare and information operations (IO) of various military services of several countries have guidelines and policies that consider legal and ethical issues. In the case of radio jamming or a laser or maser attack, there is a nominal sense in which the photons 'invade' the airspace of another nation, but that in itself seems harmless; foreign radio waves constantly pass through nations' airspaces without complaint. In the case of EMP weapons, it is not the electromagnetic radiation, the photons themselves, that violate a nation's sovereignty and cause the ethical problem; it is the secondary production of magnetic fluctuations, and then induced electrical current, that is the problem.

Neither real historical examples nor Just War Theory gives us much direct help in thinking about such matters. Yet we can imagine a cyberattack that caused as much or more damage to a nation's infrastructure and institutions as would more conventional attacks by bombs or artillery. The further physical damage could have been intentional or not, and foreseeable or not. We can even imagine a disruption caused by a cyberattack killing large numbers of human beings – by damaging or maliciously 'invading' the software of a large nuclear reactor, medical information systems, navigation systems, or of passenger aircraft, for example.

The Search for a Moral Aspect of War beyond Violation of Sovereignty: Cyberharm

The broadest useful notion in the discussion of cyber-ethics is *intentional cyberharm*: this is intentional harm caused by an agent, *via* an informatics network such as the Internet, in which the functioning of a system (a person, a machine, software or an economy) is in some way impaired or degraded. An attack is intentional cyberharm to a specific system. A virus playfully set free on the Internet by mischievous hackers is not in this sense an attack, although it might do harm. An attack intentionally causes harm to a specific organization, system, etc. Another target of a cyberattack might be the functioning of an artifact, such as vehicle (say, by disabling it through corrupting the OnStar system), or the healthy functioning of a person (by disabling their pacemaker, and so on). The vehicle or person would be the indirect target of a cyberattack, since by its nature the direct attack is specifically on the functioning of software, etc., and it is the malfunctioning of these systems that causes the intended harm to a person, organism, or artifact.[7]

I focus on cyberattacks (intentional cyberharms) that are instigated or controlled by political organizations (or their military services) on other political organizations or military services. (The usual cases are of states, including nation-states, that have internationally acknowledged sovereignty over a territory.) I use the locution 'political organization' to include insurgent and rebel groups with political goals, as well as political organizations that are not localized to a territory – such as international terrorist organizations. Libicki (2006: 23) puts it this way: '*cyberattack*, for the purposes of this discussion, is the deliberate disruption or corruption by one state of a system of interest to another state.') These are *international cyberattacks*. If the attacks between political entities are sufficiently 'widespread' we might then speak of a cyber*war*. This modifies the useful definition of Brian Orend of 'war' as '*actual, intentional* and *widespread* armed conflict between political communities' (Orend 2005). I would further stipulate that war in its usual sense involves the intentional use of deadly force on human beings. A cyberwar might then not literally be a war in this stricter sense, unless death or severe injury of human beings was the further intended result.

So far as I can see, there are no serious concerns that restrict what a nation may morally take as strictly *defensive* measures to prevent nuisance harm, or to block cyberattacks. However, defensive cybersecurity efforts could violate the privacy or other civil rights of innocent non-state parties, or incidentally cause damage to one's own citizens, economy or computer systems. Defensive actions against cyberattacks could include blocking a system's or nation's Internet from certain foreign or hostile IP addresses and the physical severing of all information conduits (satellite, cable, radio, and so on), when this is possible. As for cyberespionage, there seems to be for now the general understanding that the only possible moral countermeasures are cyber-defenses, including retaliatory cyberespionage. This toleration may change as cyber-espionage and accompanying threats and harm become more serious, and might in the future include diplomatic, economic, and cyber-retaliation.[8]

Another way of categorizing cyberharm uses a (much debated) distinction in computer science, between *data* that an algorithm operates on, and the *algorithm*, a computer or network might be running. There are at least four kinds of entities: the data, and the algorithm (both information entities), the running of the algorithm on the data (an event or process), and the hardware in which this is taking place. *Theft* of data occurs when an unauthorized user gains access to private data – such as schematics of nuclear weapon design or social security numbers, which can then be used to harm the interests of the data's rightful owner. Government-on-government computer espionage falls into this category. These are sometimes casually described as 'attacks'[9]; because they are usually unsuccessful and do no harm; a better word would be that they are hostile probes. I do not consider such espionage a cyberattack, although the further use of information gained to commit cyberharm (or conventional harm) would be.

Martin Libicki (2006: 102f) lumps espionage into a category of mere 'computer network exploitation.' He considers at length whether a cyber-attack might be launched as retaliation for espionage and his answer is, for a variety of interesting reasons, no. Successful attacks require getting past IP-address confirmation exchanges, password protection, and encryption of the data (if any). Countermeasures would include adding further layers of security or, in the worst case, making the data inaccessible altogether from the Internet. *Corruption* of data is perhaps a worse problem, since the changes to the data might be subtle: imagine corruption of a list of the GPS coordinates of all of a hostile nation's fixed targets, or changing bank account data so that it skims pennies from a large number of bank accounts.

Another form of intentional cyberharm involves changing the programs – the algorithms – of a target computer system (or the algorithm of an application program, such as a spreadsheet or inference engine). This may involve appearing to make updates to software (most seriously in the operating system), which in fact causes the software to cease to function, or function in a way undesired by the regular users. This is the realm of the more usual computer viruses. Notice that the effect of changing a program may be to allow easier access to data, which is then stolen or corrupted and then falls into an earlier category.

Returning again briefly to international law, it would appear that these useful ideas of harm (to vital interests) and functioning of systems are poorly addressed. If we imagine that a real war has begun, the only traditionally permissible attacks that damage civilian infrastructure are those whose purpose is to impair an enemy's military operations, and which are militarily necessary to pursue the just goals of a war. This is usually taken to include key components of the electrical grid, for example if these are necessary to support the enemy's military communication and command and control. It may also include components of the transportation infrastructure, such as roads and bridges, airports, and so on, if these have likely military value.

The likely effects of damage to civilian infrastructure with regard to health is especially important, so that in addition to sites such as civilian or military hospitals, which have long been explicitly addressed by international law, this moral principle would seem also to protect water supplies. Indeed, in Protocol I (1977) to the Geneva Conventions, civilian water supply is explicitly protected.[10] The US, in the First Gulf War, and possibly in the Second (Operation 'Iraqi Freedom'), has been accused of intentionally damaging sections of the largely civilian water supply system of Iraq; the matter is complicated, however, since joint military and civilian use is not carefully addressed by the statute and the overall intent of the statute seems to be to prohibit such damage that is used to drive a population out of an area or destruction of objects that are 'indispensable to [human, civilian] survival' rather than merely conducive to good health.

The extensive Protocol I to the Geneva Conventions that covers this general type of damage, seems to be flawed in a way that one can describe as ontological, and thus renders it fundamentally inapplicable to cyberwarfare and certain software- or data- forms of damage to civilian infrastructure.

Namely, Chapter III (Articles 52–56) is concerned with what are there called 'civilian objects.' These are limited, in language and by examples, to material entities such as cultural and religious objects (Art. 53), the environment (Art. 55), the just discussed category (Art. 54) of 'objects indispensable to the survival of the civilian population,' and to '[public] works and installations containing dangerous forces' (Art. 56), such as nuclear (radiation) and dike and dam facilities (flood).

As an introduction to Chapter III there is a general ban on a residual class of non-military objects that are called 'civilian objects.' There are several limitations of this treaty that render it of little application to cyberwarfare, and indeed, to economic warfare. One is that it discusses only (material) *objects*, and not the *functioning* of these objects. This failure is ontological insofar as severe damage to software, data, and operating and control systems does not require the damage or destruction of objects in the usual sense. There are other flaws as well. First, there is an emphasis on prohibiting damage to these entities as military objectives, that is, of intentional attack; this leaves open a wide category of possibly severe incidental damage. Second, the key concept of targetable 'civilian object' is defined very narrowly as those objects that do not 'make an effective contribution to military action and whose total or partial destruction, capture or neutralization, in the circumstances ruling at the time, offers a definite military advantage.'

Special Attributes of Cyberwarfare and Proposed *Jus ad cyber-bellum* Criteria

What I earlier called the 'hard [moral] cases' in cyberwarfare can be answered through analogical reasoning, comparing cyberwarfare with morally similar cases about which we have clearer thinking, and by looking to the ethical foundation that grounds what is wrong with one nation harming another nation in any way. My proposal for the jointly sufficient, and paradigmatic, conditions such that nation B morally may counterattack nation C with a cyber- or conventional attack, after a cyberattack by C, are:

1. The attack of C on B was unjust and substantial.
2. The source of the attack by C was, with overwhelming likelihood, ordered or permitted at the highest levels of a government.
3. Reasonable measures had been taken by nation B to defeat or minimize the cyberharm that a hostile nation or other non-state cyberattacker (black or gray hat hackers) might cause.
4. The expected damage to the enemy (C) is likely to be commensurate to the damage B has suffered, or is the minimum necessary to stop continuing cyberattacks.

Additional conditions would undoubtedly be needed, but they would be shared with whatever criteria one has for morally justified non-cyber wars. This is not to say that minor attacks never permit counterattack, nor that origination with a central governmental or military authority is absolutely necessary.

On (1): if the damage intentionally inflicted on B is primarily to its offensive capability, then this is for ethical purposes insubstantial damage. If damage is done to defensive capability or civilian infrastructure then it should be measured by the amount of increased vulnerability (in the case of harm to defensive installations, such as radar systems) or damage to the civilian population in which measurement takes place in this order: human lives, human well-being, material infrastructure. Thus a cyberattack on the Federal Aviation Authority flight data systems that foreseeably resulted in the crash of several airliners and deaths of hundreds of people, would probably exceed the threshold necessary to launch a conventional attack. Likewise, a massive cyberattack on defenses a nation has against physical attack (such as radar, spy satellites, command and control systems), would risk giving the attacked nation reason to believe that a conventional attack was imminent, and thus possibly trigger the conditions of justified preemptive war (in the technical sense of an attack being imminent and highly likely, see Walzer (2006) and Dipert (2006b)). It is often difficult sharply to distinguish defensive from offensive units, weapons, and installations – such as in command and control systems, including communication.

On (2): in conventional warfare, the identity of the attacker is not usually problematic. In cyberwarfare, identifying the attacker *is* especially problematic – due to problems of identification of IP addresses and the presence of diverse non-state agents with similar capabilities (especially clever black-hat hackers or groups of them). When information technology is concentrated or carefully controlled by a government, as it is in, say, North Korea, then if the IP nationality can be determined with great likelihood, the source is very likely to be the government itself. Likewise the presence of a large, indigenous community of black-hat hackers may give a nation's capacity for offensive military cyberattacks some sort of smokescreen. The enemy cannot determine if the source is governmental or military, or corporate or simply mischievous and anarchic.

On (3): this is an unusual condition, without clear analogy in the case of conventional warfare. We would not normally consider, for moral assessment, whether the attacked nation (B) had built enough bomb shelters or had failed to develop countermeasures to rocket or mortar attacks. Some condition like this nevertheless seems required in the case of cyberwarfare because civilians are in possession of tools that are as destructive as those nations can organize, namely computer programming skills and Internet access, and because one can expect occasional attacks or mischief from them. Civilians do not have tanks.

In my presentations, (3) has proven extremely controversial – and with some justification. My defense of it is analogical, following a mode of ethical reasoning described eloquently as 'Wittgensteinian ethics' in Stroll (1996 and 1998).

The Internet is, to use a quaint analogy, a very polluted stream. It is known to be very polluted with spam, unwanted popups, phishing scams, phony and defaced websites, and so on. We nevertheless continue to use it. We 'drink' from it informatically and 'swim' in it (using it for emails and our own posts), because it offers remarkable access to data and communication, as well as storage and computational power. Yet we are people – and government agencies – who drink from and swim in polluted water. It would be silly to drink large quantities of untreated water from a lake, surrounded by cottages, knowing many have faulty sewage systems that are slowly leaking into the lake, and then become sick, and sue the other cottage owners. A known level of danger requires us to take reasonable precautions – especially since our use of the Internet at all times is not necessary for survival. If our use of the water is especially critical, such as in giving water to infants, then we should take even more precautions: boil the water, etc. A relevant consideration in this analogy is that cyberwarfare is a case of someone deliberately polluting the water (so to speak) to harm us. In a civil suit against such an intentional polluter, I might indeed win a civil suit (or a criminal complaint), but the amount of the settlement would almost certainly be reduced by my failure to take reasonable precautions. This is radically unlike, say, the Katyusha rockets falling on Israeli territory from Gaza – there is not a widespread presumption of a falling rocket danger – and is more like acid rain or other background pollutants.

On (4): we have not yet witnessed (so far as I am aware) a case where sufficient damage was done in order to justify a large-scale conventional attack. It would likely take a large number of deaths or the irreversible crippling of an economy or vital economic sector. Although I hesitate to give terrorists a roadmap, such an attack might be on the control systems of airliners, of nuclear power plants, or of highly automated weapons, damaging large medical information networks, infecting large numbers of individual computer systems that control cars, medical instruments, and so on. Even then, the likelihood of stopping such a destructive cyberattack by a conventional attack may be small, since the source might not physically be located in a small number of sites. Although this is more a matter of the general theory of the justification of wars, I believe that the understanding of 'Likelihood of Success' is not sufficiently cognizant of game-theoretic considerations: goals may be the modification of this and other enemies' behavior in the future, through punishment and deterrence.[11] There is a contradiction between Just War Theory and wise geopolitical thought. The former, endorsed by a majority of the most intelligent moral thinkers, rejects disproportionate attacks (or threats thereof), and wars without a chance of success. Yet a majority of intelligent geopolitical experts, historians, and game theorists, found precisely this gambit – in a nuclear deterrent strategy – to be the best available option. Traditional Just War Theory thus has difficulties both with soft force and with the hardest of force.

A Coming Cyber Cold War?

It now appears that there were concrete proposals for a cyberattack on Iraq in 2003, before the physical attack began (Markoff & Shanker 2009). However, a more likely scenario for the future is that there will be cyberwarfare 'skirmishing' among the major players, perhaps lasting decades. This would take the form of aggressive Internet probing of military and industrial secrets, and perhaps numerous Denial-of-Service attacks, and corruption of data and software, especially if the origins of the attack can be somewhat disguised (China, the US) or if the government is not sensitive to world opinion (N. Korea). Any weakening of industrial or military power of other nations through cyberattacks would likely enhance a nation's own geopolitical power.

In short, what we are likely to see in the next years, perhaps decades, is something like the Cold War between the West and the Soviet Union. The espionage 'cat and mouse games' of the Cold War are well known, and there was also extensive probing of each other's territorial defenses, by the incursion of small numbers of air, sea, and ground forces, never giving sufficient reason to believe that a large-scale attack was imminent. What we are likely to see is the informal development of a similar 'equilibrium' in the accepted quantity and seriousness of cyberattacks.

It is relatively clear what the reasonable (and thus moral) constraints on a Cyber Cold War would be. There should be little targeting of strictly defensive computer control systems. There should be no attacks that disable or panic global financial or economic systems. There should be no interference in the vital economic and security interests of a major power, especially, as a practical matter, one with the power to attack with conventional physical force. (In Libicki (2006: 181) there is, a chart 'Ranking Various Forms of Harm in Cyberspace.' The top most harmful categories include casualties and, still more harmful, 'Interfere with Nuclear Systems' and 'Affect Military Operations.' This is far too unspecific, since disrupting a training operation, even seriously, is not on a par with blinding radar or rendering defensive electronic systems inoperative.)

Obviously, the most desirable ethical principle would be not ever to harm, through cyberattacks, the interests of other peoples. Since it is likely that many global players (industrial and governmental) will not abide by such a principle, then the 'containment' of harm seems to urge the development of defensive and even deterrent offensive cyberwarfare capacity by any nation whose information infrastructure is large and sensitive to harm. The offensive capacity is necessary in case – as is likely – defensive cybersecurity efforts are not sufficiently successful (or too expensive), and retaliatory cyberattacks are thus necessary to achieve some equilibrium.

There are some marked dissimilarities with the Cold War, however. First, a Cyber Cold War would be multilateral rather than bilateral: it would involve many nations, with different interests and not allied by treaty. Furthermore, the parties would include major non-governmental players such as private companies or even individuals or groups of individual hackers, perhaps with political interests. It is unlikely, in the more capitalistic and constitutionally free countries, that national governments can easily rein in these potential corporate and individual cyberattackers. Second, even if a nation's interests are attacked, it will often be difficult to determine immediately which country or organization is the culprit. For example, a harmful cyber-event may be the result of an organized attack by the government of Russia, by rogue elements in the Russian military, by groups of computer attackers tolerated by the government of Russia, by cyberattackers controlled by large criminal syndicates, by organizations supported by Russian oligarchs or corporations, or by individual hackers, political or apolitical. (The example of Russia is not chosen because I believe it to be especially nefarious, but it provides a wide and colorful cast of characters.)

Finally, computer technology is not dependent on any single controllable bit of technological knowledge (such as was possessed by a small group of nuclear physicists c. 1948), or on physical substances (such as U-235). It is remarkable, for example, how cryptological techniques and skill have spread to a wider public. Cyberattack technology is more like an idea than like a physical thing (or person). These facts would seem to make the creation of international treaties governing cyberattacks between governments, and laws within sovereign territories, extraordinarily difficult to develop, verify, and enforce. It is likely for a long time to be a brutish, if highly informed and non-physical, constrained combat between defensive and offensive cyberoperations. It will probably increasingly require greater and greater allocations of money and human resources.

How can the damage from cyberwarfare – and cyberattacks of all sorts – be eliminated or at least mitigated? The Holy Grail would undoubtedly be a solution to the 'Attribution Problem.' However, such a solution would run afoul of the relative anonymity and privacy that Internet access seems to offer, and to which users (in the West) have become accustomed. The alternative is certain identification of every user and every message. Even if users who have inflicted harm within one state could be identified, then if they are in another jurisdiction, and protected by another state, existing legal and diplomatic remedies will be neither quick nor easy. In the case of states with quite different juridical and moral systems, and distinct state interests, the problem is not now even remotely tractable. Progress is extraordinarily slow and impeded by political interests: witness European cybersecurity policy: 'The European Commission in [the] EU [European Union] is planning to establish a European rapid response system for cyber attacks, and may present an EU Internal Security Strategy in October 2010' (CybercrimeLaw 2010) and 'US asks EU to up cyber security' (Euroactive 2010), and this only describes political issues in the North Atlantic world.

Most domestic DDoS and malware attacks (perhaps working as bots for foreign agencies) could be blocked by requiring Internet Service Providers (ISPs) to require that users have recently employed malware-detection software before they are permitted to connect. This information, via an encrypted handshake with the ISP, would block DDoS attacks and severely limit foreign malware attacks. It could be strengthened if detection of foreign DDoS and malware (and messages through the domestic gateways as allowed by the US Foreign Intelligence Surveillance Act court) were reported to a central authority, with the legal proviso that this could not be used in domestic criminal investigations.

I would argue that this may eventually be required by the prerogatives of foreign and military policy within the US Executive Branch, and the foreign policies of other governments. It makes a mess of the distinction in the US between warmaking (Title 10), intelligence-gathering (Title 50), and law enforcment authorities; namely, in order to have an effective military and diplomatic policy, domestic law enforcement may need to identify and rein in domestic attacks on foreign information systems. If these attackers are sophisticated, or organized, then there is a risk that their efforts – whatever their goals – would be confused by foreign powers and corporations with centrally authorized attacks by the US government, and thus be grounds for counterattack and even war. The 'attribution problem' will thus have curious domestic consequences, not just hobbling our knowledge of our foreign attackers, but requiring us to identify and take action against domestic attackers on foreign entities. Even if there is not a suspicion of a government-directed and organized attack, foreign powers could still complain about – and retaliate against – our state toleration of these international cyber-outlaws and polluters.

Conclusions

On close examination, cyberwarfare appears to be almost entirely unaddressed by the traditional morality and laws of war. Because of a traditional emphasis on damage to human lives and material objects, there are not even clear restrictions on 'soft- or cyber-' damage that would leave wholly civilian targets, necessary for the well-being of the population, inoperable for long periods of time but not, in the strictest sense, damage them as objects. In some rare cases, the morally permissible response to a cyberattack may be a harsher retaliatory cyberattack, or even a conventional attack. As noted elsewhere (Owens et al. 1999: 41f) the requirement of proportionality in response to aggression does not require counterattack-in-kind, especially if it would not redress the harm done, diminish the threat, or deter future attack. This seems to open a window in which a morally permissible proper response to a damaging cyberattack is sometimes conventional, physical action.

The relevant entities in cyberwarfare are so unusual in comparison with the ordinary objects of daily life that the only useful way of thinking about them is by analogy. The relevant cyberentities include such things as the functioning of a system, software, and more broadly, information entities. Consequently, moral reasoning about cyberwarfare requires either the consideration of analogies with more traditional moral problems, or broader, less traditionally legalistic, and clearer moral theories capable of application to all possible ethical events and states of affairs.

It is perhaps understandable that the traditional morality of war, and laws of war, would not want to address such vague notions as the welfare or well-being of the civilian population, and harm to them. The most obvious and undebatable damage of war is on human beings as organisms: they die. Nevertheless, a cyberattack may do such extensive damage to the well-being of a populace, and to the functioning of a government, that it would satisfy the *casus belli* (just cause) requirement of reasonable criteria for morally going to war. Likewise, there would seem to be a need for additional moral reasoning, and additions to international law, such that militarily unnecessary damage to non-objects, namely, the functioning of civilian informatic systems, is limited in time of war.

There is widespread agreement (Libicki, Owens et al., and myself) that something like Cyber-arms Control treaties will not be forthcoming and perhaps even would be impossible to enforce. Others have advocated them (Clarke & Knake 2010), seemingly without a full understanding of the difficulties (including in verification).. Cyberarms are very unlike conventional, nuclear, biological, and chemical (NBC) arms (Owens et al. 2009: 293). Part of the problem is epistemic (attribution), but part is also due to the fact that cyberattack tools – unlike NBC weapons – are already held by diverse individuals and non-state actors. Libicki and I tend to believe that the use or threat of cyberattacks as deterrence is morally permissible and even desirable as policy. The National Academies of Science report (Owens et al. 2009: 41 and ch. 9) is less sanguine about the emergence of a deterrent effect.

In writing this paper, three serious worries have occurred to me. First, there is a large array of possible scenarios involving cyberattacks for which there do not exist obvious moral reasoning, or even straightforward analogies, that could guide us. Second, I am disturbed by the extent to which, through easy Internet access, much of our economic and defense informatics infrastructure is vulnerable to attack. I am myself engaged in research in developing systems that require regular access to open information sources (notably a variety of ontology and other web-based resources) that would not be available for long periods if a large-scale cyberwar erupted. I have long been disturbed by the departure from the relatively secure ARPANET (Advanced Research Projects Agency Network) for use in defense applications in the 1980s to a wide-open Internet, that does not have a single or identifiable handful of portals. Third, it is clear especially from General Alexander's comments (notably his 2007 paper) that serious thought is being devoted to the development of policy and strategy. To date it has remained largely shrouded in secrecy. This will be, and maybe even now may be, a serious problem, since the making public of many elements of policy are absolutely required for a deterrent effect.

Finally, it is interesting that, as far as I can see, traditional ethical and political theories – utilitarianism, Kantian theory, natural rights theory, etc. – cast so little light on this new, and difficult domain. Certainly no one of these theories seems at a special advantage in clarifying the issues. Instead, it is only the vaguest of *prima facie* notions that are most useful in elucidating the issues: (human) well-being, harm, intentional or knowing harm, political entities and their roles as agents, and essential function of states in promoting or protecting the well-being of their citizens. A suspicion of the role that highly developed ethical theories might usefully play for issues in the morality of war is one of the curious remarks in the excellent account by Fotion (2007: 1; see also Stroll (1996 and 1998)). It has also been my working assumption that fully understanding moral constraints on warfare requires understanding certain conclusions from game theory and working them into more traditional moral thinking about war. To date, there is virtually no effort in this, more mathematical and logical, direction.[12]

Acknowledgements

I thank Alan W. Dipert, Monika A.E. Dipert, George Lucas, William Mandrick, J. Jeremy Meyers, Richard Schoonhoven, and Catherine Ullman for comments and encouragement on earlier drafts; Ms Ullman, a philosophy graduate student and a cybersecurity expert at my institution provided me with a constant stream of technical articles on cybersecurity and cyberwarfare. At the interface of technical expertise and ethics, I am grateful to Neil Rowe (US Navy Postgraduate School) for having shared his essays with me. At the interface of defense policy and cyberwarfare, I am indebted, *unbekannterweise*, to the work of Martin Libicki. I especially thank Edward Barrett (Stockdale Center, US Naval Academy) and Steven Kershnar (SUNY Fredonia) for discussing with me at length a number of the issues in this paper; Dr Barrett additionally provided me with copies of much of the literature.

Notes

1 I have worked as a contractor for the US Army, working on a Command and Control Ontology with funding from the Army Net-Centric Data Strategy Center of Excellence, Building 1209, Fort Monmouth, NJ 07703. I allude to this research in my discussion of the entities that are necessary to discuss ethical questions in warfare; any views expressed in this essay are strictly those of the author, and not of the US Army or the US Department of Defense.

2 The exclusive concern with defensive cybersecurity operations continues in the May 2010 National Security Strategy of the United States (NSS 2010: 22). There is one oblique reference to offensive cyberoperations in the small section, 'Use of Force': '[Defense of the US requires] credibly underwriting US defense commitments with tailored approaches to deterrence and ensuring the US military continues to have the necessary capabilities across all domains – land, air, sea, space, and cyber.' This remarkable lack of a clear, declared strategy of response to cyberattacks by other nations is extremely problematic, I will argue below. There is considerable confusion in the media about the differences between cybersecurity and cyberwarfare, with the appointment of Cyber Security tsar Howard Schmidt seen as also covering cyberwarfare (it does not). Schmidt did not help things when he declared 'There is no Cyberwar,' when that is not in his portfolio' (Singel 2010).

3 At the writing of this paper, the Stuxnet computer worm was identified. It targeted industrial control systems (marketed by Siemens) and was apparently intended to strike nuclear processing facilities in Iran. Its sophistication and knowledge of industrial control software suggests to many a degree of organization that only a state could develop. Nevertheless, its failure to do what it apparently was intended to do is clearly a targeting problem: it rapidly spread around the world but apparently caused little or no damage in Iran.

4 There is a developed theory of *Prima facie* ethics developed by writers such as W. D. Ross and Louis Pojman. I do not here mean to refer to anything so sophisticated, but rather to what I think most reasonable, and historically and morally informed, people would hazard as an initial guess about moral permissibility. To be sure, its actual moral permissibility will hinge on the details.

5 I am indebted to Steven Kershnar for pointing out the relevance of my earlier work for the Attribution Problem.

6 In Axelrod (1984), anticipated in Schelling (1960) and various works on game-theoretic equilibria in the 1950s and 1960s. This is not to say that any single strategy, such as tit-for-tat, employed by a player against another with any other strategy is always superior, but only that a 'cooperative' optimizing strategy, in the context of conflict, would have these two features. Axelrod's own characterization of desirable features of a strategy when there is conflict are: being nice, retaliatory, forgiving, and clear (Axelrod 1984: 54).

7 These notions of organization, person, artifact, and organism, as well as function and the event of processing have precise, and carefully interrelated, definitions in work in formal ontology for the military that my research group, NCOR, and others are developing. This begins with the Basic Formal

Ontology (Spear *et al.* 2006) including a nascent ontology for all data interchange in the federal government (UCORE-2).

[8] As Martin Libicki notes (2006: xvi and 41f), the threat of cyber-retaliation is less likely to deter than nuclear retaliation, since in the Cold War there would have been no epistemic problem of determining who launched a nuclear attack. In cyberattacks, identification of the attacker may not be immediately evident, and it might be problematic ever to determine the attacker and the attackers' exact affiliation with a state. However, with the possibility of low-level nuclear attacks by non-state organizations or proxy non-state actors, the epistemic parallel becomes much closer.

[9] 'According to the 2009 Annual Report to Congress of the US-China Economic and Security Review Commission, in 2008 the number of reported cyber attacks against the Department of Defense was 54,640. In 2009, from 1 January to 30 June the number was 43,785' (Zifcak 2009: 1).

[10] See Geneva Conventions (1949 and Additional Protocols), Protocol I (1977) to the Geneva Convention: 'Art 54. Protection of objects indispensable to the survival of the civilian population ¶1. Starvation of civilians as a method of warfare is prohibited. ¶2. It is prohibited to attack, destroy, remove or render useless objects indispensable to the survival of the civilian population, such as food-stuffs, agricultural areas for the production of food-stuffs, crops, livestock, drinking water installations and supplies and irrigation works, for the specific purpose of denying them for their sustenance value to the civilian population or to the adverse Party, whatever the motive, whether in order to starve out civilians, to cause them to move away, or for any other motive. ¶3. The prohibitions in paragraph 2 shall not apply to such of the objects covered by it as are used by an adverse Party: (a) as sustenance solely for the members of its armed forces; or (b) if not as sustenance, then in direct support of military action, provided, however, that in no event shall actions against these objects be taken which may be expected to leave the civilian population with such inadequate food or water as to cause its starvation or force its movement.'

[11] A game-theoretic dimension of the morality of war is discussed in the works of Thomas Schelling, especially Schelling 1960, as well as Dipert 2006a, Dipert 2006b, Dipert 2008, and Dipert 2010.

[12] A version of this paper was originally presented at the International Society of Military Ethics (ISME) Conference in San Diego, January 2010, and at the 2010 McCain Conference in April 2010 sponsored by the Stockdale Center at the US Naval Academy.

References

Alexander, Gen. K. B. (2007) Warfighting in Cyberspace, *Joint Forces Quarterly*, 46(3) (July), available at: http://www.military.com/forums/0,15240,143898,00.html; Internet.

Arguilla, J. (1999) Ethics and Information Warfare, in: Z. Khalizad, J. White & A. Marshall (Eds), *Strategic Appraisal: The Changing Role of Information in Warfare*, pp. 379–401 (Santa Monica, CA: RAND Corporation).

Axelrod, R. (1984) *The Evolution of Cooperation* (New York: Basic Books).

Beaumont, C. (2009) Russia 'Helped Co-ordinate' Attacks on Georgian Websites, *Daily Telegraph* (UK), 18 August 2009, accessed 25 October 2010, available at: http://www.telegraph.co.uk/technology/6048978/Russia-helped-co-ordinate-attacks-on-Georgian- websites.html; Internet.

Christopher, P. (1999) *The Ethics of War and Peace: An Introduction to Legal and Moral Issues* 2nd ed. (Upper Saddle River, NJ: Prentice Hall).

Clarke, R. & Knake, R. (2010) *Cyber War: The Next Threat to National Security and What to Do about It* (New York: HarperCollins).

Clinton, H. (2010) Remarks on Internet Freedom, 21 January 2010, accessed 13 September 2010, available at: http://www.state.gov/secretary/rm/2010/01/135519.htm; Internet.

CyberCrimeLaw (2010), available at: http://www.cybercrimelaw.net/Cybercrimelaw.html; Internet.

Dipert, R. R. (2006a) Strategies, Rationality, and Game Theory in the Philosophy of War, Paper, Joint Service Academy Conference on Professional Ethics (JSCOPE), Springfield, VA (January 2006).

Dipert, R. R. (2006b) Preventive War and the Epistemological Dimension of the Morality of War, *Journal of Military Ethics*, 5(1), pp. 32–54.

Dipert, R. R. (2008) Act, Policy, and Mathematical Calculation in the Philosophy of the Morality of War, International Society for Military Ethics Conference (ISME, formerly JSCOPE), University of San Diego, CA (January 2008).

Dipert, R. R. (2010) Game Theory and the Morality of War, Paper, Department of English and Philosophy (West Point, NY: US Military Academy).
Euroactive (2010) US Asks EU to Up Cyber Security, 17 September 2010, available at: http://www.euractiv.com/en/infosociety/us-asks-eu-cyber-security-news-497876; Internet.
FM 24-5 (1939) *Basic Field Manual: Signal Communications* (Washington, DC: War Department, United States).
Fotion, Nicholas (2007) *War and Ethics: A New Just War Theory* (New York: Continuum).
Gao, K.(2003) China to view Windows code, *CNET News*, 28 February 2003, available at: http://news.cnet.com/2100-1007-990526.html; Internet.
Geneva Conventions of 1949 and Additional Protocols, at http://www.icrc.org.
Libicki, M. (2009) *Cyberdeterrence and Cyberwar* (Santa Monica, CA: RAND Project Air Force, RAND Corporation).
Markoff, J. (2009a) Cyberwar: Defying Experts, Rogue Computer Code Still Lurks, *New York Times*, 27 August.
Markoff, J. (2009b) Cyberwar: Old Trick Threatens the Newest Weapons, *New York Times*, 27 October.
Markoff, J. & Shanker, T. (2009) Halted '03 Iraq Plan Illustrates US Fear of Cyberwar Risk, *New York Times*, August 1.
NSS (2010) *National Security Strategy of the United States*, May 2010, accessed 13 September 2010, available at: http://www.whitehouse.gov/sites/default/files/rss_viewer/national_security_strategy.pdf; Internet.
OED (2010) *Oxford English Dictionary*, accessed 13 September 2010, at dictionary.oed.com (word = cyberwarfare); Internet.
Orend, B.(2005) War, in *The Stanford Internet Encyclopedia of Philosophy* (last revisions 2005), accessed 13 September 2001, available at: http://plato.stanford.edu/entries/war/; Internet.
Owens, W., Dam, K. & Lin, H. (Eds) (2009) *Technology, Law, and Ethics Regarding US Acquisition of Cyberattack Capabilities* (Washington, DC: National Research Council of the National Academies of Science).
Reichberg, G., Syse, H. & Begby, E. (2006) *The Ethics of War: Classic and Contemporary Readings* (Oxford: Blackwell).
Rowe, N. (2006) A Taxonomy of Deception in Cyberspace, International Conference on Information Warfare and Security, Princess Anne, MD, USA, March.
Rowe, N. (2009) The Ethics of Cyberweapons in Warfare, *International Journal of Cyberethics*, 1(1), pp. 20–31.
Rowe, N. (2010) Toward Reversible Cyberattacks, The 9th European Conference on Information Warfare and Security, Thessaloniki, Greece, July 2010, in: L. Janczewski & A. Colarik (Eds), *Warfare and Cyber Terrorism* (Hershey, PA: Information Science Reference).
Rowe, J. & Crusty, E. J. (2010) Ch. XII Deception in Cyber Attacks, in: L. Janczewski & A. Colarik (Eds), *Warfare and Cyber Terrorism* (Hershey, PA: Information Science Reference).
Schelling, T. (1960) *The Strategy of Conflict* (Cambridge, MA: Harvard University Press).
Singel, R. (2010) White House Cyber Czar: 'There Is No Cyberwar,' *Wired*, 4 March 2010, accessed 25 October 2010, available at: http://www.wired.com/threatlevel/2010/03/schmidt-cyberwar/; Internet.
Spear, A. (2006) Ontology for the Twenty First Century: An Introduction with Recommendations, accessed 25 October 2010, available at: http://www.ifomis.org/bfo/documents/manual.pdf; Internet.
Stroll, A. (1996) Ethics Without Principles, in: K. S. Johannesen & T. Nordenstam (Eds), *Wittgenstein and the Philosophy of Culture, Proceedings of the 18th International Wittgenstein Symposium*, pp. 310–320 (Vienna: Verlag Holder-Pichler-Tempsky).
Stroll, A. (1998) Ethics without Principles, *Topoi*, 17, pp. 133–147.
Talbot, D. (2010) Moore's Outlaws, *Technology Review* July/August 2010, accessed 25 October 2010, available at: http://www.technologyreview.com/computing/25564/; Internet.
UN Charter (1945) *The Charter of the United Nations*, accessed 25 October 2010, available at: http://www.un.org/en/documents/charter/chapter7.shtml.
USCCU (United States Cyber Consequences Unit) (2009) Overview by the US-CCU of the Cyber Campaign against Georgia in August 2008, accessed 25 October 2010, available at: http://www.registan.net/wp-content/uploads/2009/08/US-CCU-Georgia-Cyber-Campaign-Overview.pdf; Internet.

US Senate (2010) Armed Forces Committee of the US Senate, Hearings on the Confirmation of Lieutenant General Keith B. Alexander, US Army, to be General and Director, National Security Agency/Chief, Central Security Service/Commander, United States Cyber Command, 14 April 2010, accessed 25 October 2010, transcript available at: http://armed-services.senate.gov/testimony.cfm?wit_id=9367&id=4505; Internet.

Walzer, M. (2006) *Just and Unjust Wars* 4th ed. (New York: Basic Books).

Zifcak, N. (2009) New Cyber Chief Faces Dynamic Challenges, *Epoch Times*, 12 October 2009, accessed 25 October 2010, available at: http://www.theepochtimes.com/n2/content/view/26958/; Internet.

Biography

Randall R. Dipert is Charles S. Peirce Professor of American Philosophy at the SUNY University at Buffalo, New York state. From 1995 to 2000 he taught at the United States Military Academy at West Point, NY and he has recently been a contractor (in applied formal ontology) for the US Army. He has published a book on artifacts and action theory, and has co-authored a book on logic. He has published numerous articles on Peirce, logic, artifacts, mathematics, the philosophy of mind, artificial intelligence, and ontology. He has written numerous pieces of software in Prolog and LISP for logic and logic instruction. He has published one paper in the *Journal of Military Ethics* and given presentations on the morality of preventive war and the application of game theory to the morality of war.

'Cyberation' and Just War Doctrine: A Response to Randall Dipert

COLONEL JAMES COOK
US Air Force Academy, Colorado Springs, USA

ABSTRACT *In this essay, I reject the suggestion that the just war tradition (JWT) does not apply to cyberwarfare (CW). That is not to say CW will not include grey areas defying easy analysis in terms of the JWT. But analogously ambiguous cases have long existed in warfare without undercutting the JWT's broad relevance. That some aspects of CW are unique is likewise no threat to the JWT's applicability. The special character of CW remains similar enough to other kinds of warfare; the distinctions are more differences of degree than of kind.*

Does the just-war tradition (JWT) apply to cyberwarfare (CW)? I attempt to answer that question in several ways: in the first section below by rejecting the peculiar logic that says the JWT doesn't apply because CW is not really war; then by offering a positive case for the JWT's relevance to CW; and third, by showing that scenarios dubbed "hard cases" involving responses to conventional and cyber attacks conform nicely to long-established JW principles. In the fourth and final section below I suggest ways in which CW might pose special challenges to JW analysis. I conclude that CW's moral and logical ambiguities, though somewhat new, present differences of degree rather than kind; analogously ambiguous cases have long existed in warfare without undercutting the JWT's broad relevance.

1. Rejecting the stacked rhetorical deck

Consider the claim that cyberspace as a medium and CW as a set of means, tactics, and strategies are so novel as to defy application of the JWT. Another way of expressing the same claim: analogies linking cyberspace and CW to predecessor military media, means, tactics, and strategies are often too weak to be of use to just-war thinkers. We might call this the "claim of disanalogy."

We should evaluate this view in light of two working assumptions:

(1) Much of the JWT—the in bello guidelines as well as the ad bellum just cause, proportionality, last resort, and reasonable hope principles—passes judgment on the *effects* rather than the means or media of aggression. Before the JWT can offer its most useful guidance in any given case, those effects must meet three conditions: (a) they must be known or strongly predicted; (b) they must be linked to known intentional actors; and finally, (c) they must manifest as destruction or imminent destruction—of lives, property, governments, cultures, etc.

(2) Many longstanding strategies, tactics, means, and media of warfare have defied easy application of the JWT because of uncertainty as to their probable effects. Similarly, we can't always identify the agents of violence or their intentions. The Unabomber demonstrated that regrettable fact, as did the corpse on whom the FBI finally pinned the 2001 anthrax attacks. Had either enjoyed a foreign government's or domestic terror group's backing, they might have wreaked greater havoc and might still be operating today. As for foreign-backed terrorists, one need look no further than this year's Times Square Bomber to see how difficult attribution might be. When Faisal Shahzad bought the SUV that was to carry and become part of his bomb, he paid cash yet gave the seller his actual phone number. Even so, he was already seated on a Dubai-bound flight at Kennedy International when he was arrested. One wonders what might have happened if things had worked out even a little differently—if Shahzad had guarded his anonymity when buying the SUV, if the bomb had exploded as planned, and so on.[1] Yet even though the so-called attribution problem can be vexing in cases of non-cyberattack, the JWT continues to offer useful guidance.

Dr. Dipert[2] seems to believe the claim of disanalogy. But he arrives at his opinion after stacking the rhetorical deck. He insists that in contrast to the "lethality and massive destructiveness of war .... cyberwarfare often won't be like that–although it *could be* both lethal and physically destructive" (Dipert 2010: 386). Now if one says up front that CW *often* won't be as destructive as earlier sorts of "war," then it's no surprise if the JWT *often* won't easily apply. How could it if the very definition of CW undercuts the first assumption above? Similarly, in several passages Dr. Dipert indicates that the means, media, tactics, and strategy of CW as he defines it will obscure effects, actors, and intentions. Fair enough. But based on the second assumption above it would then follow—in an uninteresting because nearly tautological way—that the JWT often won't apply in those cases.

I find the position Dr. Dipert attributes to Jeff McMahan much easier to stomach.[3] That position—that many a cyberattack will be a clear-cut casus belli—conforms to my intuitions. I find it commonsensical that the JWT, concerned primarily as it is with effects and intentions rather than means and medium, will often have something interesting to say about CW, a type of warfare than can cause vast, intentional harm.

Dr. Dipert himself undercuts the claim of disanalogy by indicating ways that cyber activities can do grave damage. For instance, he notes that "much of our economic and defense informatics infrastructure is vulnerable to [cyber] attack" (Dipert 2010: 406). There is no reason why such damage could not result from a *military* cyberattack, one we could analyze through the lens of the JWT.

Despite stacking the rhetorical deck, however, Dr. Dipert doesn't much dilute what seems to me his paper's more interesting agenda—offering a lucid primer on CW. Dr. Dipert's discussion provides a taxonomy of cyber activity that suggests (among other things) what should be excluded from the definition of "cyberwarfare" as "war" in the sense meant by the JWT. There are, however, a few parts of his presentation that suffer from the stacked deck's influence. I'll consider those in section 3. below after first considering why and how the JWT applies to CW.

2. The JWT does in fact apply to cyberwarfare: a heuristic case by analogy

If it's true that Dr. Dipert has stacked the rhetorical deck—that he has defined cyberwarfare such that the JWT often cannot apply before he "shows" that it in fact often fails to apply—it still remains to make a positive case that the JWT can inform and evaluate CW.

We grow up accustomed to the idea that information instigates kinetic effects. It's not just that we spur others to action and they in turn actuate us by imparting information, e.g., in the form of words. The baby cries, the mother soothes, the baby falls asleep; the teenager cries, the mother says curfew's relaxed but just for tonight, the teenager disappears 'til dawn; and so on. Our earliest education in the world's ways underscores what experience has already taught us. Fairy tales often feature magic words that overcome otherwise insuperable obstacles: Speak to Rapunzel in the right way and you can get into the tower; say "Rumpelstilzchen" and you get to keep the child that a nasty gnome would otherwise take away. It's no different in military operations: information—transmitted or intercepted information, for example—matters in warfare. We know it mattered to the outcome of the American Revolution that colonials could track British troop and supply movements and communicate their intelligence through command and control systems. Had the old North Church been a modern C2 node, the proper signal—"one if by land, two if by sea"—might never have been transmitted or received via secure BostoNet. British cyberattack troops might have rendered the system unusable through a virus or worm or a denial of service attack. Years later the British might even have co-opted Longfellow's word processor and destroyed "Paul Revere's Ride." We know the allied cryptanalyst effort managed to break the code known as JN-25 thanks in large part to the volume of Japanese message traffic intercepted. Breaking the code in turn allowed the US fleet to engage the Japanese at Midway and, arguably, change the war's course in the space of a few hours. Nowadays cyber activity can send legitimate go codes as well as obstruct, intercept, and fabricate them. Done in a certain way under certain circumstances, those activities could become a means of CW, causing or enabling immense destruction.

Consider an admittedly flawed analogy to cyberspace and the goings-on there: that medium and its cyber-things and –activities are at least a little like the air-space continuum and the aviation that goes on it. In both cyber activity and aviation lots of stuff goes to and fro. Some of that to'ing and fro'ing is patently military; much of it isn't; some of it isn't until it is (think 9/11); and enough of it's so ambiguous that one couldn't craft an accurate Venn diagram, at least not one that wasn't obsolete as soon as it was complete.

Besides aviation, one might draw many similar analogies to *cyberation*[4]—e.g., maritime and even postal ones (recall the 2001 US anthrax attacks and the plethora of letter bombs sent to Israeli targets in the 1970s). All such analogies would share obvious features. First, there's the distinction among realms, cyberspace vs. the rest of the world and the air-space continuum versus the earth and sea. Second, there's movement between and within the realms. Finally, the interfaces are conceptually distinct even if they're somewhat vague in practice, as when a helicopter lowers its payload to earth or a C130 shoves a pallet out the door at low altitude rather than landing.

So does the JWT apply to aviation or navigation? We might justifiably think that anyone who asked such a question had failed to express herself clearly. She must have meant to ask a related question, such as *how* the JWT applies to aviation, or *what* aspects of aviation might be used for war and therefore belong in the JWT's bailiwick. *Of course* the JWT will help us judge many activities under the rubrics of aviation and navigation. That would have been harder to say of aviation before observation balloons and other aircraft had become part of the battlefield's mise en scène. But now it's so obvious as to require no defense. Why shouldn't we consider cyberation in the same way? Why shouldn't we say that just as military aviation is a subset of aviation, so military cyberation is a subset of cyberation?

A weak—and false—answer to those questions is that military aviation is somehow different in kind from anything that can happen in cyberspace. Air- and spacecraft are physical things that can cause physical effects. Not so cyber entities and what they can do, or so someone might claim.

But this alleged difference is misleading. Military aircraft can attack other aircraft, military and civilian, as well as targets on the land and at sea. In more abstract terms, military aviation can directly or indirectly destroy targets in its air-space domain and reach across its interface with the land and sea domains to destroy targets there. If one were to apply the language of CW to aviation, one might say that military aviation falls under the JWT because it can cause great harm across interfaces, and that harm is kinetic: people die and things break.

CW is no different from military aviation in this regard. Its interfaces are not just conceptual; they're real, physical things. However, we should think of this conceptual-physical duality as being less Cartesian than Heideggerian. We should avoid analogies of the sort that suggest mind is to body as cyber is to non-cyber. We should avoid them not only because cyber entities such as processes are physical. More importantly, those who labor across the interface of cyberspace are not working in any important sense differently than aviators or navigators. Heidegger points out that I (as *Dasein*) am inevitably in-the-world, properly conceived, and not somehow less or more so depending on the task at hand; what is "closest" or "next" to me is not first and foremost explicable as a function of Cartesian dualism but rather in terms of what I'm attending to. Further, theoretical vivisections of the world are metaphysically secondary, not primary; when I'm working "in" cyber I'm as much in touch with reality as a plumber. (I'm hesitant to enlist thinkers so cursorily and superficially, but perhaps it's pardonable to indicate if only by allusion that the cyber-versus-non-cyber distinction is often overdrawn.)

Although the analogies are imperfect, the JWT must attend to cyberation for the same reasons it applies to aviation: there are aspects of cyberation that can be used as military means to cause great harm. Consider this quotation from a military cyberation veteran, Jason Healey, speaking about the late 1990s predecessor to the recently-established US Cyber Command: "'It was supposed to be a war fighter unit, not a geek unit.'"[5] Without going into specifics, Healey explains why: cyberation can kill people and destroy infrastructure—within cyberspace and outside it—just as military aviation can. That cyberation is removed from aspects beyond its interfaces in a somewhat unique way doesn't matter all that much. Foreign military aviation can kill a US citizen walking down Main Street by targeting her with a kinetic weapon such as a missile—a direct kill. Or it can kill her by ensuring she has no food, power, medical care, etc., or by kinetically destroying her means of obtaining those things. Obviously cyberation can cause similar deprivation and destruction. Although a bit or a byte can't kinetically hit her on the head, cyberation can cause the missile to launch even though its keepers didn't want that to happen or kill her by diverting a missile intended for another target.

With that in mind, it's not particularly instructive to claim that much of cyberation doesn't fall under the JWT. The same is true of much of aviation and navigation—e.g., what we call "civil aviation." The interesting question is, What cyber activities <u>can</u> fall under JWT? The easy answer is that military cyberation is changing too fast for us to do much more than guess. We know some cyber activities that can kill and destroy; we know how to stop a subset of those things; but we probably don't know but a small fraction of the offensive and defensive possibilities. Breakthroughs such as supercruising might be used in any number of ways in aviation, civilian as well as military. The same is true of advances such as quantum computing: it's too soon to know how they'll apply to cyberation in general or to CW in particular. We do know that some kinds of cyberation can destroy just as some kinds of aviation can. That means CW can meet the test of assumption 1 in the previous section.

What else might motivate the claim that CW isn't real war and that therefore the JWT doesn't apply? One such contention is that so-called cyberweapons are merely the ho-hum spinoff of quotidian cyberstuff such as the office or home PC—at most tools of mischief, that is, rather than weapons of war. We should be skeptical of claims to the effect that "there are no exotic components to cyber weapons" (Dipert 2010: 385). True, cyber weapons need not be exotic. Some are clumsy things that can be conceived and launched by individuals. But consider: homegrown, part-time hackers working alone can do enough damage with worms, viruses, and the like to drive an enormous PC security industry. Norton and McAfee represent an iceberg's tip; the cyber security infrastructure is huge because the threats are extremely varied.[6] Now imagine what large teams with significant budgets and state backing could do. We should expect that some cyber weapons will be very exotic.

Even if some cyber tools are common, we should not assume that therefore they cannot become immensely destructive weapons. Fertilizer components, gasoline, cars, trucks, and cell phones are the stuff of everyday life. They're also useful in building improvised explosive devices (IEDs) and their suicide vehicle-borne variants (SVBIEDs), which have killed more coalition troops in Iraq and Afghanistan than any other single weapon system.

3. Rethinking the "hard cases" and the extent to which received theories apply

With the positive case for the relevance of the JWT to CW in mind, we can consider the ethics of a range of scenarios.

Dr. Dipert poses and answers five "moral questions of CW" (Dipert 2010: 392). While he believes there is prima facie moral justification for cyberattack in response to conventional or cyberattack, he finds two other scenarios problematic. The question of whether a nation may respond conventionally to a cyberattack he considers to be "a very hard case indeed." (Dipert 2010: 394) The other question he finds especially vexing is whether it's ever permissible to launch a first-strike cyberattack "with UN sanction, preemptively, preventively, or for some other reason" (Dipert 2010: 392).

If it's true that the JWT applies to CW, and if it's further true that the JWT concerns itself primarily with effects rather than means or media, these two questions appear to be easily answerable in theory if not in practice. The JWT permits a nation that suffers significant damage from an unjust attack to defend itself and punish the wrongdoer in any number of ways, provided principles such as the reasonable hope of success and proportionality are obeyed. The means of the attack is of secondary importance; the effect—the harm suffered—is what matters most. Similarly, the JWT seeks primarily to regulate the effects rather than the means of any military response to the initial aggression. Military aggression that happens to be a cyberattack is not ipso facto less (or more) damaging than a conventional attack. One simply cannot generalize. No wonder, then, that it's easy to imagine circumstances in which a patently military and vastly destructive cyberattack requires a nation to strike back kinetically. For instance, suppose a hostile nation uses a cyberattack to destroy part of our early-warning capability; we're suddenly blind to missiles and air-breathing threats in one sector. We might then choose to eliminate those threats preemptively using kinetic means.

The parity of CW and other sorts of war with respect to their possible effects should help us evaluate the morality of cyber first strikes as well—one of Dr. Dipert's hard cases. If the JWT endorses some kinds of first strike—to preempt unjust attack, for instance—the effect is again the key. It must conform to relevant JW principles such as proportionality. If it does, we won't pay much attention to the means of attack in most cases. Exceptions prove that rule. Admittedly the perfectly discriminate use of a tactical nuclear weapon would cause alarm because it was a nuclear weapon. That would seem to argue against the notion that the effect is everything, the means nothing. But in fact we'd worry about the nuclear means in such a case because we are ultimately concerned with effects: we'd fear any precedent that might make the use of indiscriminate nuclear weapons more likely. In other words, we'd still focus on effects, not on whether a given attack was carried out with cyber or other means. This is true of any kind of attack, including first strikes. A just-war adherent might find room in the tradition for first strikes that preempt imminent devastating attack (Israel 1967) or obstruct unwarranted harm being done to a third party (Kosovo 1999) but no place for first strikes that prevent vague dangers at an unspecified future date (Iraq 2003). A second just-war thinker might disagree on each point. But their disagreement would not have to do primarily with the *means* of first strike but rather with the effect. In most cases the JWT will care *why* CW is used and what its *effects* are, not that it's CW rather than another kind of warfare.

If the JWT applies to CW, it seems impossible to defend the notion that "traditional ethical and political theories—utilitarianism, Kantian theory, natural rights theory, etc.—cast so little light on this new, and difficult domain [CW]" as opposed to other domains (Dipert 2010: 406). How could that be true of CW and not equally true of other sorts of war? The JWT does not provide axioms; rather, it offers principles derived from meta-theories such as the ones Dr. Dipert mentions. In other words, take away "traditional ethical and political theories" and the JWT has no basis for its pre- and proscriptions. I know of no one who'd disagree with Dr. Dipert that game theory can provide useful insights into the nature and ethics of CW (as it has long done in analysis of non-CW). However, it seems not just counterintuitive but indeed demonstrably false that "traditional ethical and political theories" won't usefully inform our analysis of cyber issues even if those theories by themselves prove insufficient to solve the problems we perceive.

In fact, US directives relevant to CW are already indebted to such theories. Arguably the major post-9/11 documents relevant to CW are the *Federal Information Security Management Act of 2002* (FISMA), the *National Strategy to Secure Cyberspace*, the *National Institute of Standards & Technology Special Publication 800–53*, the *National Military Strategy for Cyberspace Operations*, and the *44th President's National Cyberspace Policy Review*. Each in its own way refers to a theoretical underpinning. The *44th President's National Cyberspace Policy Review*, for example, includes the following language: "But with the broad reach of a loose and lightly regulated digital infrastructure, great risks threaten nations, private enterprises, and individual rights. The government has a responsibility to address these strategic vulnerabilities to ensure that the United States and its citizens, together with the larger community of nations, can realize the full potential of the information technology revolution"[7] One can make too much of a single line in a single document. That granted, however, the statement above is as much related to "traditional ethical and political theories" as anything one would expect to find in documents related to war more generally. Consider this passage from the opening paragraph of the 2010 National Security Strategy: "We live in a time of sweeping change. The success of free nations, open markets, and social progress in recent decades has accelerated globalization on an unprecedented scale. This has opened the door to hundreds of millions of people, and made peace possible among the major powers. Yet globalization has also intensified the dangers we face—from international terrorism and the spread of deadly technologies, to economic upheaval and the changing climate. ... we will be unwavering in our commitment to the security of our people, allies, and partners."[8]

A strong assertion of government's role in guaranteeing security is evident in both statements above, an assertion that one could tie to several political theories. One could argue that such a tether doesn't in itself demonstrate the usefulness of received theory in solving practical contemporary problems. But the similarity in the quotations does suggest that cyber policy is no more distant from "traditional ethical and political theories" than national defense policy is. Perhaps Dr. Dipert is not arguing for the mistaken position; perhaps he simply wants to point out that broad-brush theories are difficult to apply to any specific activity in actual war, cyber or not. If he does mean to say that CW is somehow *uniquely* isolated from ethical and political theory, he might have fallen victim to the stacked rhetorical deck discussed in Section 1 above.

4. Differences

Despite CW's similarities to other forms of war, I agree with Dr. Dipert that CW isn't quite like anything else. There's no perfect analogy between military cyberation and military aviation or military navigation, for instance. CW is not just newer; it's arguably faster-evolving. Here are some potential challenges that the JWT will have to accommodate:

Visibility and accountability. It goes without saying that it will sometimes be hard to defend against CW. But defense aside, what can citizens on the aggressor's side know about their country's CWs? Arguably citizens of democracies will have less visibility and therefore less control of what their governments do in CW than in other kinds of war. This will be a difference of degree rather than kind. Governments can hide, and have long hidden, their uses of other means of war such as military aviation. An executive authority such as a US president can order many kilotons of ordnance to be dropped secretly in places like Laos and Cambodia or on officially off-limits targets. Officers such as General Lavelle can be co-opted and then made the scapegoats if prying journalists or others find out.[9] But if a government sends its military aircraft to attack covertly and lies to its citizens, it's still possible that someone will snap a photo, find some traceable dud ordnance or shrapnel, or otherwise document the activity. CW will take place in the form of physical processes too. Theoretically, those will also be traceable as a nexus of causes and effects pointing to an initiator. But the level of sophistication necessary to perform such a trace, and the relative impenetrability of the areas where those clues might be had, are arguably new in the history of warfare. We've grown used to the notion of embedded journalists, and we can imagine the rogue journalist who goes where she's not supposed to, finds out the covert truth, and then tells the world. But what would the embedded journalist-hacker look like? Could she slip away into the off-limits zones of cyberspace and discover a CW that a government had tried to conceal? I find it implausible. That's not to say journalists don't find it challenging to work in conventional war zones too. And surely it shouldn't be impossible to report on CW and cyber combatants; perhaps it won't even be harder than investigating other kinds of covert operation. Still, CW doesn't have to be enormously easier to cover up than other types of war; a matter of degree can be significant as citizens try to hold their government accountable.

Devolution to blunt instrument. Many weapons technologies are theoretically surgical but become practically indiscriminate. Aviation had traced something like that trajectory by the time strategic bombers attacked Hiroshima and Nagasaki. Many of us hoped "smart" weapons of the last generation had turned that corner. On the other hand, smart weapons killed enough civilians in Afghanistan to prompt General McChrystal to restrict and reduce aerial attacks. Nuclear attacks much more potent and numerous than those of 1945 remain a possibility. Whether or not airborne weapons eventually become surgical in their technical capabilities and unfailingly discriminate in their employment, one can imagine CW methods that a nation might be tempted to use too frequently and broadly. Cyberattacks on industrial control systems (ICSs) may provide a case in point: a nation that can neutralize its enemies' centers of gravity such as power grids through CW rather than kinetic attack will be tempted to do so. That sort of attack is likely to be at least somewhat indiscriminate because it targets dual-use assets. Many other examples come to mind.

Practitioners' evolving identities. In 2005 the U.S. Air Force changed its mission description, putting cyberspace on a par with its longtime bailiwicks of air and space. The new text reads: "Deliver sovereign options for the defense of the United States of America and its global interests – to fly and fight in Air, Space, and Cyberspace."[10] Just a few months ago the Air Force dubbed a number of its former communications-computer officers "cyber operators" and changed their numerical specialty code. In Air Force culture the term "operator" delineates those who are closest to the spear's tip as opposed to those who work in "support" career fields: pilots, navigators, some missile crew members, and now certain cyber specialists are "operators." Certainly cyber operators *might* be targets of enemy attack, and one can imagine scenarios where they are especially endangered. But it seems more likely that many will never face the dangers that, say, infantrymen are used to. The cyber operators' daily grind might resemble that of drone pilots, who suddenly find themselves the subjects of a fast-growing literature. A focal point of that literature is the oddity of "warriors" sitting in air-conditioned control modules in Nevada while piloting armed drones above Afghanistan and Pakistan. If bombs were falling in a one-hundred-mile radius around Las Vegas, one might consider the RPA (remotely piloted aircraft) pilots somewhat analogous to, say, a mortar crew on an active battlefield. As things stand, however, a number of drone pilots are too far from the action for that analogy to hold. They more closely resemble the crew of a missile submarine or perhaps of a B-52 on a mission to fire standoff weapons such as cruise missiles. Yet the Nevada operators are not pursued by other vessels in a hostile medium such as the sea's depths or the troposphere and stratosphere. So what are they? Skilled technicians? Certainly. Officers subject to the UCMJ and beholden to a professional code? Yes. "Soldiers" in other senses? Perhaps not. Cyber operators might be like the RPA pilots in all these senses. The relatively unique status or RPA pilots and cyber operators may well challenge the JWT. For instance, if the cyber operators sleep in Las Vegas and its suburbs, are their homes and neighborhoods military facilities? Has the US adequately separated cyber operators from non-combatants?[11]

Autonomous CW. Arguably defense against some kinds of cyberattack is already largely autonomous. One can imagine advantages if offensive CW followed suit, advantages significant enough to convince molasses-paced carbon units to step aside. (Imagine a cyber equivalent of the US Navy's Phalanx anti-missile system: at some point it *must* function fully autonomously or fail to do its job.) To the extent CW becomes autonomous under the sense-think-act paradigm, it will face some of the same ethical issues as autonomous agents that we don't currently tend to consider cyber weapons. This will raise additional problems, among them:

Under- and overconfidence in cyber technology. In Robert Heinlein's science fiction novel *Friday*,[12] a veteran parabolic ballistic "airline" captain named Ian Tormey admits to the novel's namesake, Friday Baldwin, that a computer could fly the passenger craft more proficiently than he can . . . and often does. But Captain Tormey hastens to add that air travelers require a degree of confidence that most can place only in another human being. To know a computer held their lives in its chips would be unacceptable to many; they'd never be willing to fly. Besides, Captain Tormey explains, the average passenger trusts not just the competence but also the empathy of the protoplasm pilot. The looming worry is that a computer algorithm would lack emotion and so not feel desperate to save the human passengers of the aircraft in case of emergency; maybe it would just give up. Of course a computer might be less likely that a human pilot to disguise a fundamental misanthropy or turn psychopathic under pressure or get drunk a couple of hours before takeoff. Then again, we find it all too easy to suspend our disbelief that computers could act against our interests as we read Arthur Clarke's short story or watch Kubrick's film version.[13] Even in what is arguably Heinlein's most sympathetic portrayal of a computer—Mike in *The Moon is a Harsh Mistress*[14]—the question of confidence is front and center.

To the extent that Mike the computer is self-aware, is his identity that of a capricious child or of a steadfast moral agent? Mike is juxtaposed with his human technician's bionic limb: Mike is a computer that can run a planet and more alone, without any human help at all, while the artificial limb is a computer of sorts that is always and necessarily at the technician's beck and call. This juxtaposition reveals a tension that's presumably here to stay. Reasons to make CW increasingly autonomous (analogous to the reasons why the Navy's Phalanx system must have a fully autonomous mode) will sometimes encounter the doubts of a populace not yet ready to cede important wartime decisions to computers.

The other side of the confidence coin is a sort of techno-hubris. The key principles of the JWT are both well and badly served by technology, depending on one's perspectives and aims. Consumers of news, advertisement, and government policy papers in the U.S. hear that technology such as global positioning satellites and receivers make projection of military power much more efficient. So do stealth technologies that help aircraft penetrate hostile airspaces, and so do smart weapons that can hit their targets with greater accuracy and therefore discrimination than earlier generations of weapons. But consider: during the air campaign over Kosovo, U.S. stealth aircraft, guided at least in part by GPS, bombed the Chinese embassy in Belgrade. The U.S. was quick to call the attack an unfortunate error caused by bad intelligence.[15] Frequently punctuating our idealization, perhaps even idolization of so-called smart weaponry are reports of civilian casualties caused by air attacks in Afghanistan and Pakistan. It's likely the same confidence some now place in air supremacy-cum-smart weapons will transfer to the tools of autonomous CW. It's too easy to forget that even the most sophisticated CW systems will share the vulnerabilities of other weapons. Bad intelligence, for instance, could cause otherwise superb tools of CW to destroy many innocent lives.

An international legal bias favoring cyber-haves versus cyber-have-nots. A troubling literature suggests that the law doesn't just fall silent in times of war, as Cicero suggested, but that powerful countries routinely succeed in manipulating international law in general and LOAC in particular to favor their own ways of war. Dr. Thomas W. Smith notes that "Just war theorists contend that soldiers hold a different moral status than that of civilians and thus assume greater risk to life and limb. The Pentagon has weakened, if not reversed, that assumption. RAND's Project Air Force found that the U.S. military favors a 'liberal interpretation' of legal duties to avoid collateral damage, 'one that permits an extremely high level of force protection so long as an appropriate level of accuracy is ensured."[16] One need not accept all of Dr. Smith's conclusions nor his skepticism of what Dr. Alan Vick et al concluded about the ethics of US air attack on urban targets to understand how his reasoning might apply to CW. To the extent that LOAC will be written by the movers and shakers on the international scene, it's possible that future laws governing the use of cyber weapons could allow something like the "appropriate level of accuracy" that RAND envisioned as US military aviation's goal. Where would that leave groups—national or otherwise—who have no access to sophisticated cyber weapons? Presumably they would employ whatever options the asymmetry of the martial landscape allowed while accusing cyber-capable enemies of war crimes or the like. "Contemporary laws of war *are* humanitarian at the low-tech end, and have been crucial in condemning atrocities, including sexual violence, associated with ethnic and other civil conflicts. But if hi-tech violence is shielded from prosecution, this may sap the moral force of the law and allow low-end offenders to paint themselves as the victims of politicized proceedings."[17]

Conclusion

It's worth repeating that all of these existing and potential problems in the application of the JWT to CW represent differences in degree rather than kind. Although CW presents somewhat new ethical challenges, it is not so unique as to defy analysis by the JWT. On the contrary, the JWT focuses on effects and intentions regardless of the type of warfare waged; those elements will be discernible in CW just as they are in other modes of warfare. Problems such as that of attribution will present complications, but that's hardly new to CW. Similarly, JW thinkers might encounter grey areas where cyber harm falling short of CW complicates analysis. Again, that's nothing new: just as aviation includes not just military but also civilian sectors and a grey area uniting both, what I've called "cyberation" will encompass both broad categories of use and a grey area somewhere at the center. The JWT will not solve all the ethical issues of CW any more than it does for any other kind of warfare. Still, its guidance will be not just relevant but indispensable to the conduct of just CW.

Notes

[1] William K. Rashbaum and Al Baker, "Smoking Car to an Arrest in 53 Hours." The New York Times, 4 May 2010, http://www.nytimes.com/2010/05/05/nyregion/05tictoc.html?scp=14&sq=faisal%20shahzad%20al%20baker&st=cse

[2] This issue of JME.

[3] I cannot find the passage Dr. Dipert refers to in Owens, William et al, eds., *Technology, Law, and Ethics Regarding U.S. Acquisition and Use of Cyberattack Capabilities*. National Academies Press, 2009. Professor Jeff McMahan did participate in an October 2006 meeting of the Committee on Offensive Information Warfare, an effort that supported the overall agenda of Admiral Owens et al. But there is no mention of Prof. McMahan's specific contributions. See Appendix B in http://www.nap.edu/openbook.php?record_id=12651.

[4] In 1866 G.J.G. de La Landelle coined the term *aviation* on the pattern of the word *navigation* (*avis*, bird + *agere*, to drive; *navis*, ship + *agere*). If we take Norbert Wiener's cybernetics (Fr. *cybernétique* <Gk. *kybernetes*, a steersman) and Gibson's derivative, *cyberspace*, would we be tempted to create *cyberation*? Either it's redundant—driving him who steers seems too close to driving him who drives—or else appropriately analogous to *aviation* and *navigation*, since in some sense birds and ships do drive themselves. Close enough: for present purposes it suffices to have a generic term to describe cyber goings-on on in a way that "saves the phenomena" (all three of them—cyberspace, cyber stuff, and cyber activities) in the same way that *aviation* and *navigation* link common parlance and common conception.

[5] *Washington Post*, 23 Sep 2010.

[6] The movie *WarGames* (United Artists, 1983) challenged Hollywood's public to think of "military computers" as those controlling missile sensors and launch facilities. But of course developed nations' defense establishments are tied to an enormous range of industrial cyber infrastructures. So-called industrial control systems (ICSs) are critical to facilities ranging from military bases to private corporations. ICSs supporting Iran's nuclear development provide a recent example of a cyber attack which some might see as "exotic": q.v. John Markoff and David Sanger, "In a Computer Worm, a Possible Biblical Clue." *The New York Times*, 29 September 2010, http://www.nytimes.com/2010/09/30/world/middleeast/30worm.html?pagewanted=1&_r=1&ref=todayspaper.)

[7] http://www.whitehouse.gov/assets/documents/Cyberspace_Policy_Review_final.pdf

[8] http://www.whitehouse.gov/sites/default/files/rss_viewer/national_security_strategy.pdf, p. i.

[9] http://www.nytimes.com/2010/08/08/opinion/08sun3.html?_r=1&ref=richard_milhous_nixon; cf. Charles Stevenson, "General Misconduct." http://www.nytimes.com/2010/08/19/opinion/19stevenson.html?ref=richard_milhous_nixon.

[10] http://www.af.mil/news/story.asp?id=123013440.

[11] Cf. 1977 Protocol Additional ... Part IV. Art. 58. http://www.icrc.org/ihl.nsf/WebART/470-750074?OpenDocument

[12] Heinlein, Robert A. *Friday*. New York: Del Rey, 1983 (1982).

[13] *2001: A Space Odyssey*. MGM 1968. Screenplay Stanley Kubrick and Arthur Clarke. Based on Clarke's short story "The Sentinnel."

[14] Heinlein, Robert A. *The Moon Is a Harsh Mistress*. New York: Orb, 1997 (1966).

[15] *The Guardian* (UK) had a different take. See John Sweeney, Jens Holsoe, Ed Vulliamy, "NATO Bombed the Chinese Deliberately." *The Guardian*. 17 October 1999. http://www.guardian.co.uk/world/1999/oct/17/balkans.

[16] Thomas W. Smith, "The New Law of War: Legitimizing Hi-Tech and Infrastructural Violence." *International Studies Quarterly*. Vol 46, No. 3 (Sep 2002), pp. 355–374, p. 361.

[17] Smith 362.

Reference

Dipert, R. R. (2010) The Ethics of Cyberwarfare, *Journal of Military Ethics*, 9(4), pp. 384–410 (Abingdon, Routledge).

Biography

James Cook is professor and head of the Department of Philosophy at the US Air Force Academy.

WARFARE IN A NEW DOMAIN: THE ETHICS OF MILITARY CYBER-OPERATIONS

Edward T. Barrett

Stockdale Center for Ethical Leadership, United States Naval Academy, Annapolis, MD, United States

While most normative evaluations of military cyber-operations have emerged from the legal community, this article assesses the legitimacy of such operations from a philosophical-ethical perspective. After reviewing the relationship between rights forfeiture and the jus ad bellum *and* jus in bello *criteria of the just war tradition, it applies these criteria to several aspects of cyber-operations, including responses to cyber-activities, the use of cyber-capabilities affecting combatants and civilians, and the use of these capabilities by contractors. Finally, it briefly addresses the legitimacy of limiting rights to privacy and anonymity in service of preventing cyber-harm.*

Introduction

The transformation of computer networks into a vital and vulnerable 'domain' of military operations has generated fresh inquiry into the normative aspects of these activities. Although this inquiry has been predominately legal and published in law reviews, philosophical ethicists have begun to offer their insights (Dipert 2010).[1] Since – unless one is a thoroughgoing legal positivist – domestic and international law and their interpretation by states should be ethically grounded, this article will contribute to an increasingly important conversation by evaluating cyber-operations using philosophical, anthropological, and ethical foundations, and their resulting military ethics frameworks.

The first section briefly outlines these foundations and the resulting normative framework through which cyber-operations should be assessed: the *jus ad bellum* and *jus in bello* criteria of the just war tradition. The second part applies these criteria to identify and evaluate ethical issues germane to responding to and using cyber-force. Finally, I will consider the ethics of preventing cyber-harm by briefly examining the foundations and limitations of rights to privacy and anonymity.[2]

Philosophical Foundations: Dignity, Rights, and Forfeiture

Although readers of this journal are well versed in the just war tradition's *ad bellum* and *in bello* criteria, an outline of their foundations will clarify their application to cyber-operations.[3] Except for right intention, these criteria are grounded in the assertion that a person who is the object of an act normally possesses a dignity that renders the basic preconditions to their flourishing normative and, more pointedly, inviolable.[4] Persons, in other words, are the bearers of pre-political rights that the state exists to secure.

A fully articulated theory of human rights would explore at least three anthropological facets: the ontological ground of this dignity, the nature of our consciousness of it, and precisely which preconditions are necessary to the flourishing of a social being – i.e., liberty versus welfare rights. Because a full-fledged assessment of the various philosophical and theological perspectives (i.e., natural law, Kantian, phenomenological, intuitionist, emotivist, Christian, etc.) on these first two issues transcends this article's purpose, I will merely state my assumptions: while the ontological basis of the dignity upon which human rights are grounded may not be knowable, that dignity is itself knowable. Similarly, this paper will not adjudicate between competing claims about which preconditions to flourishing are due because of human dignity. Instead, I will merely reaffirm the existence of a dignity that is not merely constructed and contractualized, and cannot be overridden by consequence-derived utilitarian calculations; and underscore the fact that both classical and welfare liberals agree that persons possess rights to life and life-sustaining property.

However, if this agreement were the last word, killing obviously would never be permissible. From a philosophical perspective, the concept of forfeiture – applicable to both domestic and international politics – describes the mitigation of these rights.[5] While persons normally bear an inviolable right to life, 'aggressors' possessing the capability and intention to murder have compromised their dignity, and should be killed if necessary by those responsible for defending.[6] Several aspects of this principle need to be unpacked. First, human acts are both transitive and intransitive: they affect the external world and the actor (in this case, the aggressor). Second, persons compromise their dignity to the degree required to become liable to lethal force only when threatening life and life-sustaining property.[7] Third, to compromise their dignity in this way, persons must not only possess these harmful capabilities, but must also intend to use them.[8] While the use of lethal force against unintentional threats is thus unjustified, it may be excusable on epistemic grounds.[9] On the other hand, malevolent intentions must be accompanied by 'active preparation' in order to compromise one's dignity and inviolability. Fourth, since aggressors nevertheless still possess some value, they may be harmed only as necessary, and not cruelly. Finally, those whom a society supports to provide for its defense are morally obligated to use force against aggressors.

Combined, these philosophically derived anthropological and ethical assumptions generate – and allow one to apply – the *ad bellum* and *in bello* criteria of the just war tradition. A just cause for using potentially lethal force exists only when adversaries possess capabilities and active intentions that are murderous. As mentioned above, although such aggressors have forfeited their rights to life at this point, their remaining value requires the use of available non-lethal means *ad bellum*, and the eschewal of unnecessary combatant harm *in bello*. In other words, the killing of a person liable to death must be a last resort and militarily necessary.[10] If necessary uses of force will unintentionally but foreseeably harm non-aggressors whose dignity and rights remain intact (i.e., 'collateral damage'), defenders must also foresee a reasonable chance of attaining the justifying condition in a proportionate manner, both before and during the conflict.[11] The state, given its putative role and capabilities, will normally be the proper body to make these rights-based judgments for all involved. Finally, since – as virtue ethicists highlight – acts affect the actor, these state actors should be motivated only by the just cause.

Cyber-Operations and *Jus ad Bellum* Criteria

These ethical criteria – not merely state practice vis-à-vis laws – must be applied to the three key issue areas concerning military responses to cyber-harms: just cause, attribution, and third-party rights.[12]

Just Cause

Because the principle of forfeiture determines permissible responses to all interpersonal harm, the point at which kinetic and cyber-attacks constitute a just cause or *casus belli* is the same. But unique characteristics of cyber-activities generate at least three pressing questions. First, can an accumulation of events, which individually would not be a sufficient reason to use lethal force, constitute a just cause? Although much cyber-threat discussion has focused on catastrophic damage to critical infrastructure, the ubiquitous theft of defense and economic information that – in the aggregate – significantly impacts national security and prosperity begs the question of how one may respond to this possibility of a 'death by a thousand cuts.'[13] Second, since damage to property may constitute a just cause, can temporary losses of computer functionality also qualify as a *casus belli*?[14] Like kinetic weapons, cyber-weapons can physically destroy or damage computers. But offensive computer operations, because of their potential to be transitory or reversible, can also merely compromise functionality. While permanent losses of functionality create the same effect as physical destruction, temporary functionality losses are unique to cyber-operations and require additional analysis. Third, may anticipatory self-defense be invoked in cases of espionage or logic bombs when potential harms would rise to the level of a just cause?[15] Even if aware of such activities, potential victims will find it impossible to determine when otherwise non-lethal activities are about to be elevated into attacks; evidence of imminent attacks will be undetectable. But even if one accepts the legitimacy of using lethal force without such evidence, and permits anticipatory self-defense based on known capabilities and active intentions, lethal responses to these activities may seem problematic. Legally, espionage is neither a just cause nor a war crime. And except for their locations, logic bombs posing an existential threat resemble nuclear weapons, and the mere possession of nuclear weapons has never been deemed a *casus belli*.

On the first question, the permissibility of responding to accumulated cyber-harms will depend upon situational details. As previously established, persons compromise their dignity, forfeit their right to life, and therefore become liable to lethal force only when intentionally employing, or planning to employ, means known to be capable of murdering or destroying/stealing life-sustaining property. On the one hand, a person attempting or actively intending to murder through a series of harms known to be individually non-lethal but lethal in aggregate is liable to lethal force. Accordingly, domestic laws define homicide via successive doses of poison as murder. On the other hand, successive cyber-intrusions that combine to merely weaken a state militarily and/or economically do not in themselves constitute a just cause for war. While a state using such means as part of a broader military campaign plan would be liable to lethal force, theft that reduces another's relative power and standard of living does not create this liability. Serial thievery is nevertheless unjust and, as in domestic contexts, subject to non-lethal defensive measures and/or punishments. In an international system lacking effective due process and penal mechanisms, states still have the right to respond with non-violent alternatives associated with pre-war last resort measures. The precise nature of past and ongoing thefts of military and economic information is an empirical issue beyond the reach of this paper, but my tentative assessment is that they are neither aggregately existential nor aspects of a campaign, but instead constitute serial thievery that must be dealt with accordingly – not as a *casus belli*.

An ethical analysis of the second issue, concerning whether temporary losses of network functionality would constitute just cause, is similarly situational. Assuming intentionality, the status of an actor who destroys, damages, or temporarily disables network-related equipment will depend upon the indirect or 'knock-on' effects of these acts. A temporary loss of the functionality of life-sustaining property that resulted in loss of life would be a *casus belli*. For example, the temporary disabling of air traffic control networks that caused mass casualties would qualify. In such cases, the ethically important variable is an attack's effect, not its duration. A very brief interruption of air traffic control services could amount to a just cause, while equally brief interruptions of electricity and especially automated banking services would not. The interruption of nuclear warning and/or control capabilities also would not qualify, unless accompanied by certainty of a forthcoming nuclear attack that established the legitimacy of anticipatory self-defense.

One's evaluation of our third *casus belli* issue – responses to cyber-espionage and logic bombs – will hinge on one's position on this contentious topic of anticipatory self-defense. My assumptions are as follows. As previously asserted, aggressors compromise their dignity and become liable to lethal force when possessing the capability and active intent to murder – although their remaining value dictates that circumstances require their death. In cases where imminence (i.e., mobilized forces amassed on a border) – the traditional marker of active intentions – will be undetectable, states have both the right and the duty to defend their citizens against life-threatening attacks that will occur without warning at an unknowable time. However, three important qualifications to the theory and practice of pre-emption are necessary. First, some proponents of what could be called 'non-imminent pre-emption' advocate linking the required degree of certainty about intentions to the magnitude of foreseeable harm, and allowing lower levels of certainty when costs would be high. This linkage and logic are unsound. Forfeiture and subsequent liability to harm are a function of both capabilities and active intentions, and even actors with verified weapons of mass destruction (WMD) capabilities are due the same high standard of certainty regarding their intentions as that applied to less capable actors. Second, in practice, the legitimacy of any such pre-emption will often depend upon overcoming the perennial epistemic challenge of adequately knowing active intentions. Legitimate and effective defense in an era of difficult-to-deter and stealthy threats will thus depend upon sufficient intelligence capabilities and capacities, and senior intelligence and military leaders should be prepared to highlight insufficiencies. Finally, the potential effects of bureaucratic politics must be addressed. Even in cases where intelligence communities possess epistemic clarity about intentions, perceptual biases and self-interest can compromise information available to senior leaders. Senior decision-makers should strive to create institutional structures and climates where intelligence analysts' assessments are correctly interpreted and available.

This understanding of anticipatory self-defense suggests that potentially lethal cyber-espionage activities or logic bombs could constitute a *casus belli*. While espionage through cyber or other means is not a just cause, active intentions to cause destruction and/or damage transform the act from espionage into an attack and perhaps, depending on the intended effect, a just cause. And although the acquisition or deployment of nuclear weapons or logic bombs are not in themselves causes for war, anticipatory self-defense could be warranted in the face of active intentions to use them. However, it must be emphasized that such pre-emptive measures would require absolute clarity about an adversary's capabilities, intentions, preparations, and especially – as we will now discuss – identity.

Attribution

Assuming that the self-defensive use of force is legitimate, the attribution problem adds another complexity to responses. Historically, surprise attacks have occurred, but their perpetrators were known. But technical aspects of computers and the Internet can obviate certainty about a perpetrator's location, equipment, personal identity, and/or institutional affiliation – and can even implicate innocent parties. Circumstantial evidence may be the only link to attackers, requiring difficult response decisions based not on the certainty of one's attacker, but on probability. At least two attribution-related questions arise. First, what degree of certainty about a cyber-attacker's identity must be attained before responding with lethal force? Second, assuming a successful identification, if attribution challenges preclude immediate responses to *casus belli*-level cyber-events, at what point does the just cause shift from self-defense to recovery and/or punishment?

On the first issue: lethal responses must be preceded by certainty about an attacker's identity. Even though a game theory-based analysis suggests that deterrence and long-term aggregate net benefit are maximized by responding when 90% certain, it would be patently unjust to kill innocent parties – whose rights are intact and cannot be sacrificed for the 'greater good' – in these situations (Dipert 2010). It would also be inexcusable to do so. The unintentional killing of an innocent, or the intentional killing of an 'innocent aggressor,' are excusable; but the intentional killing of persons posing no harm is not. Recent DNA-based capital punishment exonerations highlight the validity of these concerns. Since circumstantial evidence – based on motive, means, and opportunity – does not adequately establish certainty, and may be all that is available in cyber-attack cases, denial-based deterrence of future incidents may be the only legitimate response.[16]

An ethical analysis of the second attribution-related issue – responses to serious cyber-attacks after attribution delays, during which no subsequent cyber- or kinetic attacks have occurred – depends upon on the legitimacy of *post facto* just causes, such as punishment and recovery. Since the principle of forfeiture indicates that lethal force may be used only to defend against certain and life-threatening levels of harm, the validity of punishment and recovery as just causes must be evaluated by this standard. Legitimate cases of punishment are thus defensive and akin to pre-emption. Following an attack or war, aggressors who remain grave threats also remain liable to lethal force. But those posing no threat may be liable to non-lethal punishments, but not to lethal force.[17] On the other hand, acts of recovery require a different analysis. In wars of defense or recovery, potential or actual losses must be sufficiently grave. The defensive use of force against a lethal threat only intending insignificant harm can be excused based on uncertainty about these intentions. But wars of recovery require greater analytical care, since harm levels have been established. If the goods at stake would have been insufficient in themselves to defend with deadly force, then a war for their recovery would also be illegitimate.

Therefore, cyber-attack attribution delays may create morally challenging situations for leaders. Even when the recovery of life-sustaining goods is not at stake, decision-makers may face tremendous pressure to lethally punish for past crimes when the criminals are no longer liable to such measures. In these cases, leaders will have to possess the moral courage to resist such pressures.

Third Parties and Sovereignty

Finally, ethical challenges remain even after threshold and attribution problems are solved. Since cyber-attacks may involve – i.e., emerge from or be routed through – innocent third-party states, decisions must be made about the respective rights and duties of attacked and innocent parties. Specifically, at what point does a victim state's duty to protect its citizens' rights trump the sovereignty of a nation-state that is the geographic source, but not the cause, of a lethal attack? As current events and long-standing conventions indicate, this question is not entirely unique to cyber-weapons: ongoing anti-Salafist 'targeted killing' operations respond to threats emerging from non-complicit states, and military aircraft overflight rules address attacks routed through third-party states. But computer network capabilities have increased the likelihood of such situations.

An ethical analysis of these situations begins with an analysis of nation-state sovereignty, the foundation of which is the negative right of individuals to associate. Although human fulfillment is promoted in various ways by the very act of associating, it is also promoted by attaining the goals – such as commercial regulation and self-defense – of associations, including the nation-state. Like all liberty rights, the right to associate implies the duty of others to not interfere with these activities. But, as recent 'responsibility to protect' (R2P) advocates have rightly argued, other nation-states also may have an (to use Kant's formulation) imperfect duty to intervene. State actors who abuse their citizens lose their authority, their citizens have a right to rebel and waive their right to non-interference, and other states capable of providing security have obligations to do so.

However, as the targeted killing debate illustrates, obligations also flow from the inside out. Nation-state obligations transcend borders. Societies are obligated to not only protect their citizens' rights, but also to protect other states' citizens from threats emanating from their jurisdictions. States unwilling to meet this latter obligation, or who are unable to and refuse necessary assistance, lose their right of non-intervention, and threatened states have a corresponding obligation to intervene in order to locate and/or retaliate against attackers. In these situations, the principles of forfeiture and necessity still govern actions toward all rights offenders. If the offense warrants lethal force, and if capture by the host state or US forces is impossible, lethal force may be used. However, extraordinary care must accompany the identification of these targets, and appropriately transparent institutional structures must be created to ensure such care. Of course, retaliation against states themselves is warranted if attribution efforts prove collusion with these citizens.

Cyber-attack victims should be guided by the same criteria. When deemed necessary and likely to succeed in a proportionate manner, lethal kinetic or cyber-force may be used by states to defend against ongoing or anticipated *casus belli*-level cyber-attacks emanating from or traversing through another state. But such responses should be preceded by the opportunity for host state enforcement and/or cooperation, and are impermissible if adequate enforcement and/or cooperation efforts have occurred. When an immediate response is necessary, it should be pre-negotiated with and/or approved by the third-party state, which also should be adequately compensated for damages.

Cyber-Operations and *Jus in Bello* Criteria

In addition to these *ad bellum* issues, the application of cyber-force presents challenges to those attempting to comply with the traditional *in bello* normative constraints of discrimination, military necessity, and due care for civilians. Added to these core *in bello* concerns are ethical debates concerning the use of non-lethal weapons and private security contractors.

Discrimination

Notwithstanding the apparent success of the Stuxnet virus, some have argued that tight linkages between military and civilian networks, and cyber-weapons' inability to distinguish between legitimate and illegitimate targets, will combine to frequently render Internet-delivered cyber-attacks indiscriminate.[18] Kinetic munitions, both ballistic and guided, are subject to well-understood laws of gravity, motion, and aerodynamics, and are therefore capable of striking an intended military target. While their accuracy is imperfect and their post-impact effect on the surrounding space can be difficult to predict, they can at least be aimed. On the other hand, the mechanisms through which cyber-weapons move and affect computers are not widely understood, and one can imagine ubiquitous effects even worse than those of chemical weapons, which cannot be aimed but nevertheless affect a relatively limited area.

However, assessments – critical or positive – of these weapons' capabilities and morality made by academics, and most military and civilian leaders, are likely to be unreliable. Even those possessing an understanding of how these systems might operate are unlikely to possess the requisite high-level security clearances that would allow them to evaluate specific systems.[19] But thus far, from the perspective of concerned ethicists who are cognizant of applicable normative constraints but not system capabilities, the news is good: experience shows that cyber-weapons are not necessarily indiscriminate and thus illegal.[20] The Stuxnet virus, for example, despite penetrating both military and civilian networks, only affected military targets. Unfortunately, because of necessary classification levels, we do not – and will not – have many examples. Therefore, the ethicist's roles will be to ensure that the *in bello* criteria are understood by users of cyber-weapons, and to insist that users are duly diligent in ascertaining technical capabilities and relevant aspects of situations in which they are used.

Military Necessity and Civilian Due Care

Even if cyber-weapons are discriminate, questions of unnecessary harm to combatants and disproportionate harm to civilians remain. Since well-tested, human-launched kinetic weapons operate within natural, stable, and relatively knowable conditions, their effects are predictable. For example, the abilities to thoroughly test aircraft-launched precision-guided munitions under controlled conditions, and then discern actual conditions when using them operationally, allow users to confidently foresee the affected area and damage within it. But cyber-attack effects may be highly unpredictable due to their human-created and thus changing cyber-environment. While mock-ups of targeted systems can be constructed, controlled tests on actual adversary networks are obviously impossible. And a perfect mock-up could not eliminate uncertainty associated with unknown changes that adversaries make to their networks. Short of accurate, real-time intelligence, attacks may 'impact' their target, and then create unexpectedly catastrophic and unjust results.

Nevertheless, as with discrimination, experience again shows that the intended and unintended effects of these weapons are foreseeable, and that cyber-attacks are therefore capable of complying with military necessity and civilian 'due care' constraints. In some cases, their use may be more ethical than kinetic options, and thus obligatory.[21] To these ends, users of cyber-weapons must, again, exercise due diligence in determining their likely effects – which includes adequate weapons testing and adequate intelligence on adversary computer network configurations.

On the other hand, the *in bello* use of rapidly responding autonomous lethal systems lacking a 'human-in-the-loop,' while operationally advantageous, should be cautiously pursued. While the speed required to defend in the cyber-domain may favor autonomously deployed 'active defenses,' ethically adequate counter-measures would have to accurately verify targets and determine which strike options would best satisfy *in bello* criteria. Until these systems are thoroughly tested in credible conditions (which is probably impossible), potentially injurious or deadly counter-measures should be approved by qualified persons. Finally, even if deemed reliable for *in bello* use, autonomous cyber-strike capabilities should not be used prior to an existing conflict. Because autonomous active defenses against kinetic munitions strike their targets over neutral or friendly territory (unless boost-phase systems[22]), their errors are highly unlikely to precipitate a conflict. But a state's use of automated cyber-counter-measures that strike network targets inside another state would be at worst culpably unjust, and at best imprudent. Even more problematic would be private actors' use of such counter-measures, which – illustrating the wisdom of the legitimate authority criterion – could unjustly embroil an entire state in an unjust war. Robots should not be put in a position to start wars.

Targeting of Civilians

Echoing ongoing debates over the use of non-lethal weapons such as the active denial system (ADS), some have suggested that offensive cyber-operations that do not strictly qualify as 'uses of force' – which merely inconvenience – may be intentionally directed against civilians. According to this argument, international law stipulates that civilians and civilian objects should not be 'the object of attack,' but that attacks are defined as violence causing death, injury, damage, and destruction.[23] In other words, operations that do not use 'force' may be directed against civilians and their objects. An additional assertion, unique to the cyber-domain, is that civilian data not transferrable into tangible objects is not a civilian object, and may thus be intentionally targeted.

These conclusions, which follow logically from their premises, demonstrate the danger of purely legal approaches to these issues. In addition to their likely ineffectiveness, such attacks would be unjust. In peacetime or wartime, persons who have not intentionally transgressed others' rights have forfeited none of their own and are not liable to any degree of harm – including inconveniences resulting from data destruction. In a domestic context, such intentional harms would be punishable crimes; in war, they would constitute relatively minor but nevertheless punishable war crimes. When committed by soldiers in war, these acts would undermine the cultivation of warriors disposed to appropriately respect the lives of combatants and especially civilians.

Status and Use of Private Military Contractors

The status of civilians participating in cyber-force application is obviously a different matter. At a certain (albeit difficult-to-define) point, civilians who are affiliated with aggressors and 'directly' participate in aggression become liable to lethal force. While the International Committee of the Red Cross has defended a 'revolving-door' concept whereby these contractors would regain their protected status immediately upon ceasing specific hostile acts, this position is unconvincing.[24] Like their uniformed colleagues, these civilians have diminished their ontological status through ongoing participation in an organization with an aggressive purpose, and arguably can be targeted even when 'off duty.'

Assuming that states are willing to tolerate this blurred line between their belligerents and non-belligerents created by using civilians in combat roles, another question arises. Are civilian citizens appropriate agents of legitimate lethal acts? As the recent debate over the use of private military contractors (PMCs) in combat and combat support roles highlights, the litany of *in bello* concerns includes whether their operational practices and reliability levels unduly endanger US soldiers, and whether they – even in combat support roles such as intelligence analysis and combat training/advising – result in unjust levels of adversary combatant and civilian harm. These latter concerns are attributable not only to adverse incentives (i.e., PMCs are rewarded for producing a healthy principal, not minimal collateral damage), but also to differing institutional cultures (Barrett 2009).

While the use of abundant, computer-savvy civilians as attackers is economically attractive, I believe these culture-based concerns also apply to cyber-force application. As virtue ethicist Alasdair MacIntyre (1984) has argued, just as the human *telos* defines and requires properly human virtues, one's social purpose will also define and require virtues appropriate to that particular function. Since the purpose of the liberal state is to secure a particular subset of nevertheless universal rights, dispositions proper to soldiers will be those conducive to securing the good of all persons – including enemy non-combatants. On the other hand, the purpose of a business firm is to maximize market share by satisfying the desires of consumers able to pay a market price. Accordingly, the virtues corresponding to successful firm activity are not necessarily those of the soldier; selflessness may be a vice when acting within this sphere. To be fair, some soldiers lust for domination, and many security contractors are willing to risk themselves to benefit even unknown civilians. But at least in their professional capacities, soldiers are more likely to be solicitous of human rights than (to borrow Plato's term) producers.

For this reason, the approval authority for the use of potentially injurious or lethal force – including cyber-force – should be reserved for members of the military. Also, threat analyses and training/advising should be supervised by military members, since these activities affect the use of force. The contributions of security and combat support contractors will be essential to military cyber-operations, but character and culture differences, in addition to Title 10 legal responsibilities,[25] require military control of these functions.

Jus ante Bellum: Preventing Cyber-Harm

While this article has focused on *ad bellum* and *in bello* issues, two issue areas accompanying efforts to prevent cyber-harm deserve mention – privacy and anonymity. The basic outlines of the privacy debate are familiar: at the expense of privacy, cyber-security could be enhanced by allowing the government or regulated private sector to inspect Internet-traversing packets for malware signatures. Especially since *Griswold v. Connecticut* in 1965, the US Supreme Court has upheld a constitutional right to privacy. But privacy can also be defended on ethical grounds as a precondition to human flourishing, and as one aspect of the general class of negative rights to be 'let alone.' Knowing or credibly suspecting that one's thoughts, words, and actions were being even monitored, let alone publicized, would inflict undue harm in crucial ways. Private thoughts, words, and actions allow unconstrained deliberation and verification vis-à-vis the good life, and are deeply constitutive of intimate relationships. Privacy also ensures reputational fairness conducive to professional and personal relationships, and prevents data theft.[26] Since one neither waives nor forfeits this right by simply communicating via the postal service or Internet, privacy in the cyber-domain must be taken seriously.

However, one cannot establish *a priori* a conflict between the rights to life and property (i.e., security) and the right to privacy. Assuming that the machine-based inspection of metadata or content would be a cost-effective security measure, the effect on privacy would reside in the technical and procedural details – including how 'suspicious' is defined, whether suspicious packets were simply redirected to 'honeypots' or more closely inspected, and the rules governing closer inspections. While security measures should not transgress privacy, it is quite possible that they would not. The empirical facts must be allowed to speak for themselves. What is certain, however, is the dubiousness of claims for a right to anonymity in the cyber-domain. Anonymity is not a precondition for privacy; and as our discussion of the cyber 'attribution problem' disclosed, anonymity undermines a victim's rights to self-defense and recovery.

Conclusion

Ultimately, this article has been an extended discussion of the applicability and importance of philosophical ethical analyses to military cyber-operations. As mentioned earlier, shorn of their ethical foundations, legal analyses – especially in novel situations – can devolve into strained and/or power-driven interpretations of words and concepts, and ascription to resulting state practices. Attack-response decisions become bogged down in debates over the definition of an 'armed attack' and whether 'self-defense' against 'uses of force' is permissible, civilians are transformed into legitimate targets of cyber-attacks that destroy 'non-transferrable' data and thus merely 'inconvenience,' and debates on the use of Private Military Contractors (PMCs) focus on definitions of 'inherently governmental' instead of the effect on human rights. Legal analyses of military cyber-operations have been brilliant, but their inherent limitations demand that ethicists participate more fully in the conversation.

This issue of the *Journal of Military Ethics* is a solid step in the right direction. While this article has explicated the relevant ethical criteria and applied them to several key issues, articles that follow will delve more deeply into specific facets of military cyber-operations. And although recommendations may vary, what I believe will emerge is a consensus that the venerable just war tradition – rightly understood – is fully capable of rightly guiding struggling lawyers and policy makers.

NOTES

1. Among the outstanding legal thinkers on these issues are Michael Schmitt, Walter Gary Sharp, Sr., Charles Dunlap, Eric Jensen, and Sean Watts.
2. This article is the fruit of a 2011–12 research project conducted at the United States Naval Academy's Stockdale Center for Ethical Leadership by Center staff and the following fellows: Randall Dipert, State University of New York at Buffalo; Chris Eberle, US Naval Academy; Panayotis Yannakogeorgos, Air Force Research Institute; and Bradley Strawser, Naval Postgraduate School. Participating Naval Academy faculty and staff were: Michael Skerker; Major Kyle Phillips, USMC; and CMDR Michael Rak, USN. Additionally, the Center collaborated on this research with representatives from the US Cyber Command, National Security Agency, National Intelligence Council, Central Intelligence Agency, and Verizon.
3. For purposes of this paper, the *ad bellum* criteria are legitimate authority, just cause, right intention, last resort, reasonable chance of success, and proportionality; the *in bello* criteria are discrimination, military necessity, and civilian due care.
4. Contrary to the compelling thought of Perry (2007), dignity alone renders these preconditions inviolable – not-to-be-violated; no other quality, such as 'inviolability,' is necessary.
5. The rich and sophisticated literature on forfeiture includes Wellman (2012), Goldman (1979), Simmons (1991, 1992), Morris (1991), Lippke (2001), and Boonin (2008).
6. The classic formulation of this compromise is from Aquinas:

   > By sinning man departs from the order of reason, and consequently falls away from the dignity of his manhood . . . and falls into the slavish state of the beasts, by being disposed of according as he is useful to others . . . Hence, although it be evil in itself to kill a man so long as he preserve his dignity, yet it may be good to kill a man who has sinned, even as it is to kill a beast. For a bad man is worse than a beast, and is more harmful, as the Philosopher says. (Aquinas 1948: *Summa*, II-II, q. 64, a. 2)

   See also Ross (1955: 60–61).
7. While one could add other harmful acts such as rape, maiming, and enslavement, this article will focus on life and life-supporting property for the sake of simplicity. 'National interest' is an often-cited but obviously insufficient standard.
8. This paper assumes that intentionality includes both a malevolent goal, and knowledge that these capabilities could achieve this goal. I am not certain whether invincible ignorance about capabilities' effects would result in forfeiture and liability, but tentatively assert that it would not. Therefore, a cyber-attacker who caused *casus belli*-level harm, but neither sought nor could have foreseen it, would not incur liability – although a lethal response might be excusable. Assumed, of course, is due diligence in ascertaining the possible effects of a new technology.
9. An alternative way of construing 'unjustified' vis-à-vis unintentional aggressors is to deem them 'not liable' to lethal force. This latter formulation predominates in the outstanding work of Rodin (2003) and McMahan (2009).

10. While Rodin (2003: 40–43) argues that necessity is a constituent aspect of liability, I recommend that the latter be associated with the act itself, not its context. In other words, a potential murderer has forfeited their right to life, and circumstances independent of their act then dictate how one may respond to this liability to harm. In either construal, necessity is a moral requirement. Catholic doctrine of capital punishment has recently integrated this notion of necessity. See John Paul II (1995).
11. Implied in this formulation are two assertions: *ad bellum*, reasonable chance of success and proportionality constraints do not apply when only aggressors will be affected; and when civilians are involved, reasonable chance of success is also normative *in bello*. Note that there are two effects of the unintentionality of this collateral damage: from a virtue ethics perspective, the actor is not adversely affected internally; and relatedly, the actor is not liable to harm.
12. Legal analyses of just cause in the cyber-domain have focused on defining an 'armed attack,' to which UN Charter Article 51 allows self-defensive responses. Death/injury/destruction/damage-causing 'uses of force' of sufficient – using Pictet's taxonomy – 'scope, intensity, or duration' are armed attacks. In order to transcend debates over 'sufficient' effects (assuming parties can agree on the validity of an 'effects-based' approach, rather than 'instrument-based' or 'strict-liability' approaches), this article will prescind from this legal framework. For a helpful discussion on these legal tools, see Carr (2010: 45–75).
13. The level of this theft-based harm is disputed, in part because companies are reticent to report it. A 2011 British Cabinet Office report estimated cybercrime's cost at 1.8% of gross domestic product (GDP), while a subsequent Ministry of Defence study asserted a lower level that was mostly comprised of security measures. See Business and Cyber-security, A Spook Speaks, *The Economist*, 30 June 2012.
14. I am indebted to Randall Dipert for identifying this issue. See also Dipert (2010).
15. Malware creating 'back doors' would also fall under this analysis.
16. Dipert (2013) rightly notes that the 'sum or compound attribution method' would not merely employ technical means. However, non-technical components of this method are likely to be inconclusive (i.e., using electronic and human intelligence sources to verify an attacker's identity) or are morally irrelevant (i.e., inferring from means and motive).
17. I am obviously calling into question the legitimacy of punishing for retributive and utilitarian (i.e., deterrence and public order) purposes, and would argue that public safety and reform should be the guiding purposes.
18. Internal attacks on military facilities via flash drives would be exempt from this critique – unless these facilities were connected to the Internet.
19. This limitation also often applies to kinetic weapons.
20. See, e.g., UN General Assembly (1980).
21. For an extended exploration of this logic, see Strawser (2010: 342–368).
22. Boost-phase ballistic missile defense systems may intercept missiles that are still over a launching state's territory.
23. See UN General Assembly (1977), especially articles 51 and 52, and also articles 35 and 55–57. For an excellent legal perspective on this issue, see Schmitt (2011), especially pp. 114–123. My ethical analysis supports the more restrictive conclusions of Knut Dörmann (2004).
24. See Melzer (2009), especially pp. 70–71.

25. Greatly simplified, US Code Titles 10 and 50 govern military and intelligence activities, respectively.
26. For seminal analyses of the nature and foundations of a right to privacy, see Thomson (1975), Scanlon (1975), Rachels (1975), and McCloskey (1980). See also Alfino and Mayes (2003) and Skerker (2010).

REFERENCES

Alfino, Mark & Mayes, G. Randolph (2003) Reconstructing the Right to Privacy, *Social Theory and Practice*, 29(1), pp. 1–18.

Aquinas, Thomas (1948) *Summa Theologiae* (New York: Benziger Bros).

Barrett, Edward (2009) Market Failure: The Moral Hazards of Private Military Corporations. Unpublished paper delivered to US Naval Academy McCain Conference, Annapolis, MD, 23 April.

Boonin, David (2008) *The Problem of Punishment* (New York: Cambridge University Press).

Carr, Jeffery (2010) *Inside Cyber Warfare* (Cambridge: O'Reilly Media).

Dipert, Randall (2010) The Ethics of Cyberwarfare, *Journal of Military Ethics*, 9(4), pp. 384–410.

Dipert, Randall (2013) Other-than-Internet (OTI) Cyberwarfare: Challenges for Ethics, Law, and Policy, *Journal of Military Ethics*, 12(1), pp. 34–53.

Dörmann, Knut (2004) *Applicability of the Additional Protocols to Computer Network Attacks*, accessed 18 March 2013, available at: http://www.icrc.org/eng/assets/files/other/applicabilityofihltocna.pdf; Internet.

Goldman, Alan H. (1979) The Paradox of Punishment, *Philosophy and Public Affairs*, 9(1), pp. 42–58.

John Paul II (1995) *Evangelium Vitae* (Boston, MA: Pauline Books and Media).

Kant, Immanuel (1964) *Groundwork of a Metaphysic of Morals* (New York: Harper & Row).

Lippke, Richard L. (2001) Criminal Offenders and Rights Forfeiture, *Journal of Social Philosophy*, 32(1), pp. 78–89.

McCloskey, H. J. (1980) Privacy and the Right to Privacy, *Philosophy*, 55(211), pp. 17–38.

MacIntyre, Alasdair (1984) *After Virtue: A Study in Moral Theory* (Notre Dame, IN: University of Notre Dame Press).

McMahan, Jeff (2009) *Killing in War* (Oxford: Oxford University Press).

Melzer, Nilz (2009) *Interpretive Guidance on the Notion of Direct Participation in Hostilities under International Humanitarian Law* (Geneva: International Committee of the Red Cross).

Morris, Christopher W. (1991) Punishment and Loss of Moral Standing, *Canadian Journal of Philosophy*, 21(1), pp. 53–79.

Perry, Michael J. (2007) *Toward a Theory of Human Rights: Religion, Law, Courts* (New York: Cambridge University Press).

Rachels, James (1975) Why Privacy is Important, *Philosophy & Public Affairs*, 4(4), pp. 323–333.

Rodin, David (2003) *War and Self-Defense* (Oxford: Oxford University Press).

Ross, William D. (1955) *The Right and the Good* (Oxford: Oxford University Press).

Scanlon, Thomas (1975) Thomson on Privacy, *Philosophy & Public Affairs*, 4(4), pp. 315–322.

Schmitt, Michael (2011) Cyber Operations and the *Jus in Bello*: Key Issues, in: Raul A. Pedrozo & Daria P. Wollschlaeger (Eds), *International Law and the Changing Character of War*, vol. 87, pp. 89–110. Newport, RI: US Naval War College.

Simmons, A. John (1991) Locke and the Right to Punish, *Philosophy and Public Affairs*, 20(4), pp. 311–349.

Simmons, A. John (1992) *The Lockean Theory of Rights* (Princeton, NJ: Princeton University Press).

Skerker, Michael (2010) *An Ethics of Interrogation* (Chicago, IL: University of Chicago Press).

Strawser, Bradley Jay (2010) Moral Predators: The Duty to Employ Uninhabited Aerial Vehicles, *Journal of Military Ethics*, 9(4), pp. 342–368.

Thomson, Judith Jarvis (1975) The Right to Privacy, *Philosophy & Public Affairs*, 4(4), pp. 295–314.

UN General Assembly (1977) *Protocol Additional to the Geneva Conventions of 12 August 1949: Protection of Victims of International Armed Conflicts (Protocol I)* (New York: United Nations).

UN General Assembly (1980) *Convention on Prohibitions or Restrictions on the Use of Certain Conventional Weapons which May Be Deemed to Be Excessively Injurious or to Have Indiscriminate Effects* (New York: United Nations).

Wellman, Christopher Heath (2012) The Rights Forfeiture Theory of Punishment, *Ethics*, 122(2), pp. 371–393.

Edward T. Barrett is the Director of Research at the US Naval Academy's Stockdale Center for Ethical Leadership. A graduate of the University of Notre Dame, he completed a PhD in political theory at the University of Chicago, and is the author of *Persons and Liberal Democracy: The Ethical and Political Thought of Karol Wojtyla/John Paul II* (Lexington Books, 2010). He joined the US Air Force Reserves after serving nine years as an active duty C-130 instructor pilot, and is currently a Colonel in the Air Staff's Directorate of Strategic Planning. *Correspondence Address*: Stockdale Center for Ethical Leadership, United States Naval Academy, 112 Cooper Road, Luce Hall 201, Annapolis, MD 21402, USA. *Email Address:* ebarrett@usna.edu

THE MORAL PERMISSIBILITY OF AUTOMATED RESPONSES DURING CYBERWARFARE

David Danks* and **Joseph H. Danks****

**Department of Philosophy, Carnegie Mellon University, Pittsburgh, PA, United States Institute for Human & Machine Cognition, Pensacola, FL, United States; **Center for Advanced Study of Language, University of Maryland, College Park, MD, United States*

Automated responses are an inevitable aspect of cyberwarfare, but there has not been a systematic treatment of the conditions in which they are morally permissible. We argue that there are three substantial barriers to the moral permissibility of an automated response: the attribution, chain reaction, and projection bias problems. Moreover, these three challenges together provide a set of operational tests that can be used to assess the moral permissibility of a particular automated response in a specific situation. Defensive automated responses will almost always pass all three challenges, while offensive automated responses typically face a substantial positive burden in order to overcome the chain reaction and projection bias challenges. Perhaps the most interesting cases arise in the middle ground between cyber-offense and cyber-defense, such as automated cyber-exploitation responses. In those situations, much depends on the finer details of the response, the context, and the adversary. Importantly, however, the operationalizations of the three challenges provide a clear guide for decision-makers to assess the moral permissibility of automated responses that could potentially be implemented.

1. Introduction

One striking difference between warfare in the cyber and kinetic domains is the variation in timescales on which they proceed, and so the different types of responses that are suitable, or even possible. Events in cyberwarfare – attacks, defenses, counter-attacks – can occur on the scale of seconds or even faster, whereas most events in kinetic warfare may unfold more slowly over days, weeks, or even months. Since humans cannot respond sufficiently quickly to all relevant cyberwar events, successful cyberwarfare requires the (non-exclusive) use of automated responses: our cyber-network systems must include at least some policies of the form 'If event *E* happens in context *C*, then execute response *R*' (where *E*, *C*, and *R* can all be highly complex), where *R* is performed whenever the triggers – combinations of specific *E*s in particular *C*s – occur without the need, or perhaps even the possibility, of any human intervention. For example, if an intrusion to a computer network is detected with a specific signature (possibly attributable to a specific adversary), then a system might automatically respond in various ways (possibly tailored to that specific intrusion/adversary).

In this article, we examine the ethical permissibility of various types of automated responses. Debates about the morality of acts in warfare have principally focused on actions or responses that proceed immediately from deliberation by relevant human decision-makers. In contrast, we are principally concerned with the conditions for the moral legitimacy of automated responses that occur without any human decision at the moment of initiation.[1] In particular, we present three different barriers – each of which can be made relatively operationally clear – that automated responses must overcome in order to be ethically permissible. Previous arguments about automated responses have focused on only one of these (the *attribution challenge*). We argue that the ethical permissibility of automated responses depends also on the (in)ability to predict rapid sequences of events that can result from the use of automated responses (the *chain reaction challenge*), and on human decision-makers' (in)ability to accurately predict their future preferences and desires (the *projection bias challenge*). These challenges or tests can be used to draw important distinctions between classes of automated responses, as well as help to understand the ethical permissibility of activities in computer network operations (CNO) – in particular, computer network exploitation (CNE or cyber-exploitation) – that occupy an ambiguous middle ground between computer network defense (CND or cyber-defense) and computer network attacks (CNA or cyber-offense).

Any type of warfare requires conditional planning in which decision-makers determine which actions or responses to take if certain events happen in a particular context. In many cases, the context *C*, triggering events *E*, and response *R* can be quite complex. Mission plans in kinetic warfare, for example, often include long lists of contingency plans for the many different possible scenarios that might be encountered. Many of these conditionally triggered actions are publically declared in advance, as in mutually assured destruction plans during the Cold War or mutual assistance clauses in treaties. Although all warfare involves conditional plans, cyberwarfare is distinctive because at least some of those conditional plans will need to be automated in order to function properly on the compressed timescales of cyber-events and are unlikely to be publically declared in advance. That is, some of these conditional plans will need to be implemented so that *R* is automatically triggered whenever the antecedent is satisfied without the need or possibility of human intervention or decision, and without prior notification of the adversary. Few analogues to such automated responses occur in the kinetic domain. Mission plans and war-gaming produce conditional plans, but those are almost always implemented by human decision-makers who can change them 'on the fly' when *C* changes, which is especially important when *C* includes many factors. The closest analogues are doomsday triggers for nuclear weapons that guarantee a nuclear strike whenever certain preconditions obtain, regardless of whether a human is present to launch the (counter-)attack. These are completely different from automated responses in the cyber-domain, however, as doomsday triggers are publicly declared and involve responses *R* constituted by overwhelming (kinetic) force. In contrast, automated cyber-responses are rarely publicly declared, and can involve a much broader range of responses. They thus warrant a closer investigation, particularly into their ethical permissibility.

We focus in this article on the response *R*, although we recognize that there are many interesting questions associated with *C* and *E* (e.g., does the language used to describe *C* and *E* create operational 'blind spots' by grouping together contexts or events

during them should not be subject to the projection bias challenge. The problem is that the relevant realism of these exercises is simply unknown, since it must be realistic in the minds of the participants, rather than in technology, displays, and so forth. That is, the people participating in red team-blue team exercises must be sufficiently engaged in the exercise that they actually manage to place themselves (at least, temporarily) into the relevant context. To the extent that people recognize that it is 'just an exercise' and so maintain cognitive distance, it is unlikely that they will avoid the biases that underlie the challenge. In fact, many studies have asked people to 'imagine' or 'visualize' themselves in the novel context *C* before giving their predictions, and those studies have still found that people have great difficulty predicting their future preferences and desires (e.g., Read & van Leeuwen 1998). Thus, an open empirical question is whether red team-blue team activities are sufficient to overcome the projection bias challenge.

3. The 'Easy' Extremes

The previous section laid out three general challenges to the ethical permissibility of automated responses. We now turn to the problem of how to judge these challenges in some classes of cases, beginning with the extremes of the offense-to-defense spectrum. We take it to be uncontroversial that completely defensive automated responses, such as blocking an attacking intrusion or diverting it to a harmless site, are almost always ethically justified in terms of their impacts on others, including adversaries. In general, actions that *directly influence* only cyber-resources that are under our *legitimate control* will always be permissible in terms of the warfare context (although not necessarily well-advised), regardless of whether they are triggered automatically or implemented by a human. Moreover, this intuition accords with the three challenges outlined in the previous section. The attribution challenge is irrelevant: the response does not directly affect any other systems, so defenders do not (in the extreme case) need to worry about whether their attack attribution was correct.

The chain reaction and projection bias challenges are slightly trickier, but not significantly so. It is possible that an automated response that directly influences the defenders' own computer network systems would trigger a chain reaction involving the adversary (e.g., if the adversary's responses depended on internal features of the defenders' computer network systems), but it is hard to construct a scenario in which the defenders would be responsible for such a chain reaction. The fault would lie with the adversary because (by assumption) the adversary does not have legitimate claim to use (or depend on) the defenders' computer network systems. And although a chain reaction involving only the defenders' own systems might occur and be disastrous in its effects, it would not be ethically problematic in terms of the warfare context. Of course, such a chain reaction could be morally problematic for other reasons (e.g., if it shut down the information technology (IT) systems for the defenders' hospitals), but we leave those considerations aside. Similarly, it is possible – perhaps even likely – that projection bias will be an issue for our decision-making, and that the defenders might well regret the automated response. But that regret would not be over an act of *warfare* for which no one was responsible, assuming the automated response only directly influences systems under the defenders' legitimate control.

Of course, it can sometimes be unclear whether a particular automated cyber-response falls into this class, as both 'directly influence' and 'legitimate control' are notions with fuzzy boundaries. With regards to direct influence, sending back corrupted information to the adversary might appear to not directly influence the adversary, except that the manner of corruption could (unknowingly) lead to adverse events in the adversary's systems. Alternatively, if system *S* is necessary for system *T*'s proper functioning, then actions that directly influence only *S* occupy a vague area between direct and indirect influence with regards to system *T*. Along the dimension of legitimate control, the interdependence of networks can easily lead to situations in which it is unclear whether one can legitimately exert influence on a cyber-system. For example, if there is an official agreement linking systems *A* and *B* with distinct owners, then the legitimate owners of *A* can justifiably claim some authority over *B* by virtue of the legal linkage between the two systems.

An example can show the complexities that can arise when trying to decide whether an automated response is purely defensive. One might think that defenders are always permitted to respond to a presumptive DoS or distributed DoS (DDoS) attack by ignoring requests from a particular range of IP addresses, as ignoring requests seems to directly influence only the defenders' own machines (as they are simply telling them not to acknowledge certain packets). Suppose, however, that this particular server provides (previously negotiated) crucial support for a hospital's critical infrastructure. In that case, the defenders are not necessarily justified in ignoring messages from machines in that hospital, even if it appears that some of those machines are beginning a DoS attack on computer systems for which the defenders are responsible. Instead, they face the additional burden of determining whether the users of those machines are knowing or unwitting participants in the DoS attack. The issue becomes further complicated if a botnet is used to implement a DDoS, where some but not all of the machines in the botnet are part of the critical infrastructure for an organization. Nonetheless, despite these complications, defensive automated responses – at least, the obvious cases – are less problematic from an ethical point of view as they focus on internal cyber-resources over which the defenders have legitimate control.

Offensive automated responses are more ethically complicated. We follow Owens et al. (2009: 9) in understanding offensive cyber-actions (i.e., cyber-attacks) to be 'deliberate actions to alter, disrupt, deceive, degrade, or destroy computer systems or networks or the information and/or programs resident in or transiting these systems or networks.' A key element here is that the attacker does not have legitimate control over the target systems. We further restrict our attention to offensive cyber-*responses* – cyber-attacks prompted by outside actions or events, and typically intended to be retaliatory in nature. The attribution challenge is obviously relevant for such automated responses, as defenders may ethically inflict harm only on those that justly deserve it. However, as we argued in the previous section, this challenge is unlikely to be a substantial barrier in the cyberwarfare context, precisely because there are many more opportunities to learn features of the adversary's attack (e.g., the attack signature) that enable rapid, even automatic, attribution. The chain reaction and projection bias challenges are less easily handled.

In cyberwarfare, adversaries will almost certainly possess the resources to have measures designed to protect their own computer networks. If their measures are all purely defensive, then the chain reaction challenge is unlikely to be a substantial barrier, as actions 'internal' to the adversary are unlikely to trigger a further response from the defenders. Of course, the same resources that enable adversaries to protect their systems also make it substantially more likely that some of those automated counter-responses will be offensive in nature. The defenders may even have positive knowledge (whether through cyber- or other channels) that the adversary does have offensive automated counter-responses in place. That knowledge will, however, rarely translate into knowledge of the precise adversary counter-responses, and so defenders will rarely know anything more than 'a chain reaction is highly possible.'

In general, people have a positive ethical obligation to determine the likely direct and indirect effects of their actions, and so defenders with offensive automated responses have a positive duty to consider possible offensive counter-responses by the adversary, and how those can potentially interact with other automated responses of their own. There are rare situations in which defenders can reasonably conclude that a chain reaction is improbable. For example, suppose the defenders' system *S* has only one offensive automated response *R* (and the rest are defensive), and the adversary system that will be targeted by *R* is known to (almost certainly) have only counter-responses that target *S* (i.e., the source of *R*). In this case, there may be a partial escalation, but no runaway chain reaction is plausible. Such situations are likely to be the exception rather than the rule, however, as the combinatoric possibilities of offensive responses and counter-responses will typically include at least some ethically problematic chain reactions. We thus contend that there is a general presumption against offensive automated responses: they are ethically permissible only if defenders undertake the positive work of determining that the conditions for a runaway chain reaction do not exist.

A similar obligation arises from the projection bias challenge. Defenders have a positive obligation to ensure, or at least try to ensure, that their offensive automated responses will still be endorsed when they are activated in the novel cyberwarfare context. If the response is not endorsed at the later time, then it would be an act of (cyber)warfare for which no individual or group is responsible (since responsibility inheres in intentional rather than accidental acts). Importantly, this positive obligation arises precisely because of the automated nature of the offensive response; if a human decision-maker is 'in the loop,' then this positive obligation largely disappears, as there will be a locus of responsibility. Even when no human intervention is possible, it may be possible to directly determine whether the decision made now (to implement the response) will likely still be endorsed when that response is triggered. First, as noted in Section 2, practices such as red team-blue team exercises might (depending on what empirical data emerge) yield decisions that accurately track the choices that decision-makers would endorse if the actual circumstances arose. Second, much of the psychological research on projection bias has aimed to find predictable patterns in people's prediction errors. As just one example, people systematically underestimate, and essentially never overestimate, the later value (to them) of objects not yet in their possession (the 'endowment effect'; see, e.g., Kahneman et al. 1990, although List 2003 shows that certain experiences can reduce this effect). Just as current value estimates provide a lower bound on later judgments, one might hope that current decisions about automated responses can similarly provide (partial) information about later endorsements. To our knowledge, no empirical data directly address this question, and so the viability of this response strategy is unknown.

Third, and perhaps most plausibly, one could argue that the triggering context for the offensive automated response is sufficiently similar to one's usual decision contexts, and so the decision can be trusted not to change. Projection bias arises when thinking about unfamiliar contexts (e.g., decisions about life-sustaining medical treatments), but not for familiar contexts. For example, people are reasonably good at predicting the clothes they are likely to wear next week, precisely because it is a familiar context. Triggering cyberwarfare contexts are obviously quite unusual for most people, but might be quite common for the relevant decision-makers (e.g., individuals in the US Cyber Command). Such a response to the projection bias challenge would require a detailed analysis of the workflow conditions and decision challenges for such individuals; to our knowledge, no such analysis is publicly available, and so we cannot adjudicate whether this avenue will prove viable. Given that key empirical questions are open, we contend that the positive obligation due to the projection bias challenge remains, and so there is a second presumptive barrier to the adoption of offensive automated responses in cyberwarfare contexts.

4. Cyber-Exploitation: Part of the Tricky Middle Ground

Although the moral permissibility of the clear-cut types of automated responses is relatively straightforward (yes for defensive ones, no for offensive ones), there are other types of cyber-activities whose moral permissibility is much less clear. In this section, we consider the relevance of the three challenges for the ethical permissibility of automated responses that provide, enable, or support cyber-exploitation capabilities and activities. We use 'cyber-exploitation' here to refer to activities in which defenders access an adversary's systems in order to gain an informational advantage, but not to (directly) disable or otherwise adversely affect those systems. Examples of actions that fall under the scope of this term include covert exfiltration of information (without damaging the system), changing entries in a database (without corrupting that database), or installing a back door to facilitate access at later times. Human operators can sometimes perform these actions deliberately, but many of them can also be implemented as part of an automated response in certain contexts. For example, exfiltration routines can be deployed such that information, both open and hidden, is collected and returned automatically on a regular basis.

CNE occupies an unclear middle ground between the offensive and defensive extremes: it is not offensive since defenders are not seeking to damage the adversary's computer network system, infrastructure, or kinetic capabilities, but it is not defensive since defenders are directly influencing computer network systems that are not under their legitimate control. We thus see here the real value of the three challenges, as they provide a practical guide to determining the ethical permissibility of automated responses that fall into this significant middle ground. As before, we contend that the attribution challenge is not a significant one for CNE in the cyberwarfare context, as defenders frequently will have substantial, actionable information about the likely adversary. Considerable resources could of course be required to establish attribution at a high enough confidence level to justify an automated response, but governments under threat or at peril during cyberwarfare would presumably have resources to devote to establish attribution with sufficient confidence. We thus focus principally on the chain reaction and projection bias challenges.

For both of these challenges, we contend that much depends on the details of the particular CNE response, which is perhaps unsurprising given that we are dealing here with 'middle ground' cases. More specifically, we distinguish between automated responses that do not directly lead to changes in the *functioning* of an adversary's systems, and those that do yield such changes (which must be non-adverse, as we are focused on cyber-exploitation, not cyber-attack). The former type includes actions such as exfiltration of information or installation of a back door into the adversary's systems.[9] There is little reason to think that such actions would trigger a chain reaction of automated responses, precisely because they do not influence the functioning of the adversary's systems and so are much harder to detect, and respond to, automatically. Of course, such actions may well prompt a reaction from adversaries when they realize what has happened, but the chain reaction challenge focuses on sequences of actions and reactions that occur without human intervention. Having said that, adverse chain reactions of automated responses are still possible, depending on the adversary's detection capabilities. For example, an adversary may have set up an automated counter-response whenever exfiltration is detected, such as corrupting the information being sent back. Such corrupted information might then trigger an automated action by the defenders; it might even trigger an automated cyber-attack. In such cases, the action-response sequences might spiral out of control, at least in the rapid time frame of automated actions.

The projection bias challenge is not quite as easy to overcome in these cases. If no decision-maker endorses the triggering of this type of automated response *R* (at the moment of triggering), then the issue is to what extent the change (if any) to the adversary's systems constitutes an act of warfare (for which no one would be responsible). If defenders have the ability to make it 'as if' *R* never occurred (e.g., destroying or never using exfiltrated information), then *R* does not constitute an act of warfare, since the adversary has suffered no harm (after the corrective action). For cases such as installing a back door, however, defenders cannot undo the change in the adversary's systems, and so their ethical permissibility is likely to be highly context-dependent. In particular, whether this weakening of the adversary constitutes an act of warfare will depend on many particular details of the adversary's systems, the scope of the cyberwarfare activities, and more. For example, an installed back door could afford the automatic exfiltration of information, either continuously or on a schedule (regular or irregular) to avoid detection by the adversary. A human decision-maker would, however, quickly assume full moral responsibility for the CNE response, including not using the back door or actually disabling it. In summary, CNE-automated responses that do not affect the functioning of the adversary's systems are likely to be ethically permissible, although there are important subtleties to be addressed even for these cases.

Suppose instead that the automated response *R* does induce some (non-adverse) change in the adversary's systems' functioning. For example, *R* might change information in an adversary's database (e.g., of troop strength or locations), or, more interestingly for cyber-cases, change properties of the database (e.g., access privileges). Alternatively, distribution lists could be modified to exclude those adversary actors who need to be 'in the loop,' or to include adversary actors who are not part of a particular decision-making group, but who might have an operational interest in the decisions. Such changes could significantly alter the command-and-control structure of the adversary. Such an automated response would presumably be intended to (mis)lead the adversaries without them recognizing the subterfuge. In this case, it is very hard to know *a priori* whether a chain reaction of automated responses would result, as it would depend on, for example, whether the adversary's systems have programs that scan its databases for particular combinations of factors that prompt automated responses. Such programs certainly exist in other domains; for example, many stock-trading systems automatically make certain trades when specific combinations occur in a central 'database' (e.g., the New York Stock

Exchange). And it is at least possible that the adversary would have programs that could launch automated cyber-attacks under some conditions (e.g., depending on information in a database about the defender's cyber-systems), although no such programs have been publicly disclosed. Unlike many of the other cases we have considered in this article, we cannot say here whether it is likely that a chain reaction could occur or be avoided. However, the chain reaction challenge does provide a (relatively) clear operational test for whether an automated CNE response that changes the functioning of an adversary's systems would be ethically permissible.[10]

The projection bias challenge is a bit more straightforward in this case. We assume that defenders are not willing to disclose their cyber-exploitation activity to the adversary; defenders would not, for example, inform the adversary that they had changed entries in a database. The challenge thus turns on whether the change effected by *R* (e.g., the disinformation) constitutes an act of warfare for which someone should be held morally responsible. This is a particular instance of the more general question of when it is acceptable (in warfare) to deceive an adversary. With the advent of the information age, information operations, especially as they relate to denial and deception, have grown in importance in warfare. Certainly within cyberwarfare, the opportunities for, and the impact of, information operations have exploded exponentially. Full exploration of information operations within the cyber-context is required to determine the boundary conditions of their ethical permissibility.

This more general issue has been the locus of substantial philosophical disagreement whose resolution depends in large part on one's account of the ethics of warfare. The most interesting range of positions holds that disinformation is sometimes, but not always, morally permissible. Accounts in this space typically focus on whether the disinformation is in 'good faith' (Mattox 1998): that is, disinformation is ethically permissible when it falls within the 'rules of warfare.' For example, placing inflatable tanks on Pacific islands during World War II to give the appearance of greater troop numbers is morally acceptable; having a weapons factory masquerade as a hospital is not. We suggest that similar considerations apply in the CNE domain: changing a database entry to make a training facility appear to be a battlefield hospital would be an unethical act of warfare; editing a distribution list so that information is spread (against the adversary's wishes) throughout the command-and-control structure would be acceptable. Thus, the question of whether the projection bias challenge arises for a particular automated cyber-exploitation response will depend on the fine details of the response, and the broader warfare context in which it could occur.

5. Conclusion

Automated responses are an inevitable aspect of cyberwarfare, but there has not been a systematic treatment of the conditions in which they are morally permissible. We have argued that there are three substantial barriers to the moral permissibility of an automated response: the attribution, chain reaction, and projection bias problems. The first has been discussed previously, as it is a characteristic problem for *all* cyber-activities. The latter two seem not to have been discussed previously in the philosophical (or other) literature, but arguably pose a greater barrier in the cyberwarfare context. Moreover, these three challenges together provide a set of operational tests that can be used to assess the moral permissibility of a particular automated response in a specific situation. We have argued that defensive automated responses will almost always pass all three challenges, while offensive automated responses (in the cyberwarfare context) typically face a substantial positive burden in order to overcome the chain reaction and projection bias challenges. Indeed, the Obama administration apparently has arrived at a similar conclusion in its review of the president's powers in the cyber arena: 'the administration has ruled out the use of "automatic" retaliation if a cyberattack on America's infrastructure is detected, even if the virus is traveling at network speeds' (Broad Powers Seen for Obama in Cyberstrikes, *The New York Times*, 4 February 2013, p. A1). Perhaps the most interesting cases arise in the middle ground between cyber-offense and cyber-defense, such as automated cyber-exploitation responses. In those situations, we see that much depends on the finer details of the response, the context, and the adversary. Importantly, however, the operationalizations of the three challenges provide a clear guide for decision-makers to assess the moral permissibility of automated responses that could potentially be implemented.

ACKNOWLEDGEMENTS

Thanks to two anonymous reviewers for their valuable comments on an earlier version of this article. DD was partially supported by a James S. McDonnell Foundation Scholar Award. JHD's work was supported, in whole or in part, with funding from the US government. Any opinions, findings and conclusions, or recommendations expressed in this material are those of the authors and do not necessarily reflect the views of the University of Maryland, College Park, and/or any agency or entity of the US government.

NOTES

1. And because we are interested in automated actions, we focus solely on responses. It is unclear whether it even makes sense to talk about an 'automated' unprovoked action.
2. There are also interesting questions about the combatant status of various individuals involved in the cyberwarfare process. For example, how should we think about someone who writes code but does not actually put it into action? Is such an individual comparable to a worker in a munitions factory, or a military support individual, or some other role? For space reasons, we do not address these issues in this article.
3. We assume throughout that only humans, either individuals or groups of individuals, can (in these settings) be a locus of moral responsibility. In particular, we assume that computers cannot be morally responsible for their actions. We also take as given that these automated responses are at least *potentially* morally justified in particular situations, as the mere passage of time between the decision and the action is not sufficient (on its own) to eliminate the possibility of moral justification. We instead focus on when the potential justification is actual.
4. One might hope to have a function that takes H and R (and undoubtedly other factors) as input and then outputs an appropriate threshold of justification, perhaps expressed as a probability. This possibility seems unlikely for at least two reasons. First, the 'other factors' mentioned in parentheses are likely to be a hopelessly complex set, as almost anything can be relevant under sufficiently unusual circumstances. Second, any proposed threshold would likely be vulnerable to a *sorites*-type objection: if τ is acceptable, then $\tau - \varepsilon$ will almost certainly be acceptable as well, and so we have a wedge to show that no

threshold is defensible. More about the issue of what level of justification is required can be found in, for example, Zimmerman (1997), Rosen (2003), and Guerrero (2007).

5. But see our discussion of defensive automated responses in the next section.
6. In general, a projection bias is the (unjustified) attribution of one's own psychological attributes (beliefs, desires, etc.) to others. Projection bias often involves 'mirroring' one's own perspectives onto an adversary, rather than adopting the adversaries' point of view. In this research literature, one's own future states are regarded as 'other people' relative to one's current state, and so 'future self-projection bias' is a specific case of this more general bias.
7. One particularly striking example that shows the pervasive nature of this phenomenon is that people who buy winter jackets on cold days are more likely to subsequently return those jackets (Conlin et al. 2007). Why? Because they are cold at the moment of purchase and so expect to be cold in the future, and thus overestimate the likelihood that they will use the jacket in the future. (And items that people do not use as frequently as they expect are more likely to be returned.) A second example comes from the stability of older adults' preferences for life-sustaining medical treatment (Ditto et al. 2003). Such preferences were collected from elderly adults three times at yearly intervals. Their preferences on whether to receive a treatment or not were only moderately stable – about 70% were the same each year as the year before. So people cannot anticipate their preferences even for serious life decisions.
8. Of course, a similar possibility exists in the kinetic realm. However, those cases are much more likely to involve the possibility or requirement of action at the moment, and so those actions would presumably be better aligned with actual decision-maker preferences.
9. Installing a back door obviously changes the adversary's systems, but the simple act of installation should not change their *functioning*, assuming that the adversary does not have detection routines for just such back doors. Of course, *using* such a back door could easily change their functioning, but then it would fall into the category of either impactful CNE (if no adverse impacts) or cyber-attack.
10. To be a bit more precise, the challenge only indicates whether the automated response is *im*permissible; meeting the challenge is necessary for permissibility, but not sufficient.

REFERENCES

Campbell, John P. (1968) Individual versus Group Problem Solving in an Industrial Sample, *Journal of Applied Psychology*, 52(3), pp. 205–210.

Conlin, Michael, O'Donoghue, Ted & Vogelsang, Timothy J. (2007) Projection Bias in Catalog Orders, *American Economic Review*, 97(4), pp. 1217–1249.

Dipert, Randall R. (2006) Preventive War and the Epistemological Dimension of the Morality of War, *Journal of Military Ethics*, 5(1), pp. 32–54.

Dipert, Randall R. (2010) The Ethics of Cyberwarfare, *Journal of Military Ethics*, 9(4), pp. 384–410.

Ditto, Peter H., Smucker, William D., Danks, Joseph H., Jacobson, Jill A., Houts, Renate M., Fagerlin, Angela, Coppola, Kristen M. & Gready, R. Mitchell (2003) The Stability of Older Adults' Preferences for Life-Sustaining Medical Treatment, *Health Psychology*, 22(6), pp. 605–615.

Frankfurt, Harry G. (1969) Alternate Possibilities and Moral Responsibilities, *Journal of Philosophy*, 66(23), pp. 829–839.

Guerrero, Alexander A. (2007) Don't Know, Don't Kill: Moral Ignorance, Culpability, and Caution, *Philosophical Studies*, 136(1), pp. 59–97.

Hirokawa, Randy Y. (1980) A Comparative Analysis of Communication Patterns within Effective and Ineffective Decision Making Groups, *Communication Monographs*, 47(4), pp. 312–321.

Kahneman, Daniel, Knetsch, Jack L. & Thaler, Richard H. (1990) Experimental Tests of the Endowment Effect and the Coase Theorem, *Journal of Political Economy*, 98(6), pp. 1325–1348.

List, John A. (2003) Does Market Experience Eliminate Market Anomalies?, *Quarterly Journal of Economics*, 118(1), pp. 41–71.

Loewenstein, George, O'Donoghue, Ted & Rabin, Matthew (2003) Projection Bias in Predicting Future Utility, *Quarterly Journal of Economics*, 118(4), pp. 1209–1248.

Mattox, John M. (1998) The Ethics of Military Deception, unpublished master's thesis, US Army Command and General Staff College, Fort Leavenworth, KS.

Owens, William A., Dam, Kenneth W. & Lin, Herbert S. (Eds) (2009) *Technology, Policy, Law, and Ethics regarding US Acquisition and Use of Cyberattack Capabilities* (Washington, DC: The National Academies Press).

Pereboom, Derk (2000) Alternative Possibilities and Causal Histories, *Philosophical Perspectives*, Supplement 14, pp. 119–137.

Read, Daniel & van Leeuwen, Barbara (1998) Predicting Hunger: The Effects of Appetite and Delay on Choice, *Organizational Behavior and Human Decision Processes*, 76(2), pp. 189–205.

Rosen, Gideon (2003) Culpability and Ignorance, *Proceedings of the Aristotelian Society*, 103(1), pp. 61–84.

Rowe, Neil C. (2010) The Ethics of Cyberweapons in Warfare, *International Journal of Cyberethics*, 1(1), pp. 20–31.

Walzer, Michael (1977) *Just and Unjust Wars: A Moral Argument with Historical Illustrations* (New York: Basic Books).

Wolf, Susan (1987) Sanity and the Metaphysics of Responsibility, in: Ferdinand Schoeman (Ed), *Responsibility, Character and the Emotions*, pp. 46–62 (Cambridge: Cambridge University Press).

Woolley, Anita Williams, Gerbasi, Margaret E., Chabris, Christopher F., Kosslyn, Stephen M. & Hackman, J. Richard (2008) Bringing in the Experts: How Team Composition and Work Strategy Jointly Shape Analytic Effectiveness, *Small Group Research*, 39(3), pp. 352–371.

Zimmerman, Michael J. (1997) Moral Responsibility and Ignorance, *Ethics*, 107(3), pp. 410–426.

David Danks is an Associate Professor of Philosophy & Psychology at Carnegie Mellon University (Pittsburgh, PA), as well as a Research Scientist at the Institute for Human & Machine Cognition (Pensacola, FL). His research focuses on cognitive science, principally applying techniques from machine learning and theoretical frameworks from philosophy to the study of the mind. He has published in a wide range of journals, including *Journal of Philosophy*, *Philosophy of Science*, and *Psychological Review*. He is currently finishing a book manuscript under contract to MIT Press exploring the use of graphical models to capture mental representations.
Correspondence Address: Department of Philosophy, 135 Baker Hall, Carnegie Mellon University, Pittsburgh, PA 15213, USA. *Email Address*: ddanks@cmu.edu

Joseph H. Danks is Research Professor at the Center for Advanced Study of Language at the University of Maryland (College Park, MD). For many years previously, he was Professor of Psychology and Dean of the College of Arts and Sciences at Kent State University (Kent, OH). He has published numerous journal articles, book chapters, and books on a variety of topics in psycholinguistics and cognitive psychology, including how people solve problems; how people read and comprehend written texts; how children learn to read; how skilled translators create texts in the target language; how elderly patients communicate their wishes for end-of-life medical care; how to forecast the plans and intentions of a country's leadership to develop weapons of mass destruction (WMDs); and how to develop psychological profiles of cyber adversaries. *Correspondence Address*: 7005 52nd Avenue, Center for Advanced Study of Language, University of Maryland, College Park, MD 20742, USA. *Email Address*: jdanks@casl.umd.edu

OTHER-THAN-INTERNET (OTI) CYBERWARFARE: CHALLENGES FOR ETHICS, LAW, AND POLICY

Randall R. Dipert

University at Buffalo, Buffalo, NY, United States

Almost all discussions of cyberwarfare, other cyber-attacks, and cyber-espionage have focused entirely on the Internet as the chief means of damage – the Internet as a 'vector,' using a term from the theory of infectious diseases. However there are a variety of means, some of which have already been used, that involve cyber-exploitation using vectors other than the Internet. Malware can be installed in the integrated circuits of computers and servers, but also in any devices attached to them – thumb drives, CDs, printers, scanners, and so on. One can also use various forms of electromagnetic radiation at a distance of meters or more to exfiltrate data or to infiltrate corrupt data. I call this large and diverse family of unwanted interference with the functioning of information processing systems other-than-Internet (OTI) attacks on information systems. Stuxnet, and probably the 2007 Israeli corruption of Syrian air defenses, were OTI attacks. Such OTI techniques are more difficult to develop than creating malware, requiring electronic manufacturing facilities or novel technologies, and are thus accessible only to larger corporations and technologically sophisticated countries. Particularly vulnerable would be countries (like the United States and Europe) whose information processing devices are mainly produced outside of the country. Once exploitations via the Internet become harder to perpetrate, OTI exploitations are certain to grow dramatically, eventually requiring equipment for critical uses to be expensively fabricated in (and transported using) secure facilities; expensive detection measures will have to be instituted. This will create challenges for policy, law, and ethics, as well as greatly increasing the cost of many electronic devices.

Introduction

In this article, I will first summarize what I take to be an emerging consensus among experts about cyberwarfare, understood as largely taking place on the Internet. In the main part of the article I will develop in detail what I think is an approaching and very important mode of cyberwarfare, namely what I call other-than-Internet (OTI) cyberwarfare, which has received only scant attention. This requires some discussion of technical issues that are, as scrupulously as I can determine, based on multiple public sources; they are admittedly mostly speculative. Although the possibility of OTI (and instances of its occurrence) has been mentioned, there has been no open and full discussion of the policy and ethical issues, apart from some remarks by former public officials, former Secretary of the Department of Homeland Security Michael Chertoff, former CIA and National Security

Agency Director Gen. (AF, retired) Michael Hayden, and retired Army General Wesley Clark (Andress & Winterfeld 2011, Chertoff 2011, Shobert 2012).

Summary of Consensus and Background Issues

Let us start with some important definitions. In the public discussion, the term 'cyber-attack' is used very loosely. In this broad sense it can be an act of cyber-espionage, hacking or attempting to hack through passwords and security questions to get to secret governmental or commercial information, such as credit card numbers. It can be the blocking or defacing of a website either directly or by a botnet. Important information can be deleted or altered, or it can be a genuine act of cyberwarfare. Among experts there is now a consensus that an act of cyberwarfare, a cyberwarfare attack, is more narrowly defined as one nation intentionally inflicting harm on the information processing system of another nation, using cyber-means. One delineation of this harm is 'to disrupt, deny, degrade, or destroy information resident in computers and computer networks, or the computers and networks themselves' (Joint Staff 2006).[1] This excludes cyber-espionage; cyber-attacks by individuals and by criminal organizations would also be excluded from counting as cyberwarfare.

The intended harm of a genuine attack may have a further goal, such as to disable or destroy weapon systems, disrupt government communications, or kill or incapacitate combatants or even non-combatants for political effect. These are some of the goals and effects of traditional, kinetic warfare. The reason for excluding cyber-espionage from cyberwarfare is simple. Cyber-espionage or theft of intellectual property has in the past never been accepted as a sufficient moral or legal justification for going to war (Schmitt 2013, Part B Section 8, Rule 66a: 158). The moral reason why cyber-espionage and theft of intellectual property do not justify going to war is that they do not, in themselves, cause harm.[2] It is the subsequent *use* of this intelligence or information that can intentionally bring about harm. There is international law that covers acts of human spies or saboteurs on enemy soil. If the intrusions are just electrons or photons from some part of the electromagnetic spectrum, this may be covered by international telecommunications law, including bilateral agreements. However, this (and electronic warfare in the form of radio jamming) has also never been considered a sufficient justification for going to war. There is an interesting question of whether there could be some quantity of cyber-espionage or intellectual theft, or information about certain secrets, such that it would rise to the level of a moral justification for going to cyber- or kinetic war. This issue is beyond the scope of this article.[3]

Some legal experts, including in the United States Cyber Command, have considered using 'attack' only when the cyber-harm is commensurable with a substantial kinetic attack that would, for example, justify an Article 51 (UN Charter) counter-attack, or trigger a mutual-defense pact such as in the North Atlantic Treaty Organization (NATO). Lesser intentional harmful actions would count as 'actions' in cyberspace, but not cyber-attacks (Brown & Poellet 2012, Schmitt 2013, B.III. Rule 30: 91–93). The occurrence of cyber*war* may further require mutual attacks of a certain level of damage over a period of time (Orend 2005). A 'state of war' between two or more nations or political organizations may require certain declarations and understandings that the parties are at war.

One interesting problem is whether a cyberwar could take place without inevitably becoming kinetic. It is far clearer that if there were a (state of) kinetic war, there would be little reason not to employ some forms of cyber-attack. This and other features sharply separate nuclear from cyberwarfare, since nuclear attacks have come to be considered as a 'resort' even beyond the 'last resort' of a kinetic attack. An attack (i.e., 'armed attack' in the sense of UN Article 51) may morally be countered through a counter-attack by the country first attacked. Thus a cyber-attack may incur a morally justified cyber-counter-attack.[4] Discriminating and proportional acts of cyberwarfare may thus not need to satisfy the 'last resort' condition in just war theory (Lucas 2012). Some such cyber-attacks would lie somewhere on a continuum of 'soft force' or 'force short of war' (Dipert 2010, forthcoming a): this has been widely accepted by geopolitical thinkers but is not yet conceded by many legal and ethical theorists. One critical problem is that even modest cyber-attacks may cause escalating responses; this is carefully addressed in Libicki (2009, 2012). Modest cyber-attacks might be far more discriminating and proportional than international sanctions, or even the unilateral actions permitted by the UN Charter, such as cutting off diplomatic, economic, or communication contact (enumerated in Article 41 as not constituting a 'use of armed force').

The two or more parties in cyberwarfare need not be nation-states, but what is crucial is that the intentional acts are ordered through a command structure and that the goals are ultimately political or military. So if the remnants of al-Qaeda could hire mercenary cyber-warriors or buy cyber-weapons, their use would be an act of cyberwarfare. This would be no more problematic than 'asymmetric' war in other, kinetic cases. There is now a substantial literature on the ethical and legal aspects of asymmetric war. Nicholas Fotion (2008) allows slightly different ethical standards in what he calls 'irregular' war (both for *jus ad bellum* and *jus in bello*). Hacktivist activity, as well as a variety of damage by what may be called countercultural groups, have so far fallen mostly into the category of 'pranks.' However, if the actions are guided by a clear political agenda, there is no reason not to include them as acts of cyberwarfare. This distinction between harmful acts of cyberwarfare and other harmful acts is not without consequences, since non-political acts, including intellectual theft, would be best addressed through a law enforcement paradigm.

Legal and Further Ethical Considerations

I take it to be reasonably well agreed upon that existing international law does not completely cover some important aspects of cyberwarfare. At the very least, past treaties and agreements do not utilize a language that names the phenomena of cyberwarfare. Furthermore, without a tradition of established behavior of nations' use of cyberweapons, legal principles derived from customary practice do not seem to be clearly and obviously applicable (Brown & Poellet 2012). The exception to both of these claims is where the intended or foreseeable effects of cyberwarfare are commensurable with the quality and quantity of damage from traditional kinetic warfare (Schmitt et al. 2004).

A serious problem for the applicability of various theories of the morality of war, or for the application of international law, is what I have elsewhere called the *Ontological Problem* (Dipert 2010). This difficulty is that consequences of traditional warfare involve well-understood entities, and well-understood conditions of damage to them, namely the wounding or killing of human beings (persons and agents) and the permanent destruction of physical entities, notably artifacts (from installations and buildings down to dishes and tables), and mass entities (as opposed to count objects) such as air, soil, and water. In the robust Additional Protocol I to the Geneva Conventions (1977), there are numerous references to protected 'objects', but the examples given are all enduring physical objects (endurants, continuants), not entities that are events, activities, or processes (perdurants) (Smith et al. 2012). The damage or harm in many cyber-attacks will be to the *functioning* of information, and of connected control (e.g., industrial), systems. This may intentionally cause extensive harm that lasts for long periods of time, such as a stock exchange or electrical or communication grid not functioning properly for weeks or months. The only explicit discussions of this ontological issue of which I am aware are in Dipert (2010) and briefly in Schmitt (2013, B.III. Rule 30, Comment 10: 93–94).

Another respect in which traditional law seems to fail is that injury to nations has traditionally been framed as an intrusion on sovereignty. This is usually taken to be the activity of person or object (e.g., a physical weapons system, perhaps unmanned) violating that sovereignty. In cyberwarfare, this sense of the word 'invasion' is stretched to the limit, since the cyber-attacks may consist of an invasion only of photons or electrons, and there may even be a sense in which their entry into a sovereign territory are 'welcomed' through the act of that nation being voluntarily connected to the Internet. For contrary views arguing that there is no substantive, new ethical problem in cyberwarfare, see Cook (2010) and Crisp (2012).

There are some moral analogs of cyberwarfare to be found in weapons that are non-lethal, and whose damage is limited in time or is reversible. (Incapacitating gases such as those used in the Moscow theater hostage crisis of 2002 might require prompt administration of an antidote.) It has long been noted that the main goal of warfare is the disabling and disarming of combatants, not their death or permanent wounding. However, the most effective means of disabling and disarming have, until now, involved killing or wounding human beings and permanently destroying physical objects. Although I do not have any good examples, weapons that would interrupt the functioning of, but not permanently damage, other weapons and dual-use infrastructure would also be morally similar to some cyber-weapons.[5]

It is both the likely kind and quantity of this harm in historical forms of war that has guided ethical thinking about these past, kinetic wars – most obvious in just war theory's conditions of 'last resort' but also of 'reasonable probability of success.' Past wars have tended to be massively destructive to what humans value. Another feature of traditional war that has guided ethical thinking – especially in the era of explosive weapons and weapons of mass destruction – is the likelihood of widespread harm to non-combatants that is unintended and often unforeseen. These are issues of 'discrimination' and 'proportionality.' Cyberwarfare might be different in all of these respects, in being discriminating, proportional, and with few unforeseen effects. In the very limited forms of cyberwarfare that we have seen until now, it appears that the predicted catastrophic forms of cyber-conflict foreseen by e.g., Clarke, Knake, and others (Clarke & Knake 2010) are less likely to occur. The reason is that for the major cyber-powers who have or can control such catastrophic weapons, there would be very good reasons not to use them, because of the likelihood of escalation and massive cyber- or kinetic counter-attack. Instead, the long period of low-level conflict predicted in Dipert (2010) now appears more likely, as well as uses of cyber-weapons by major cyber-powers on far less informationally sophisticated nations. One form of cyber-conflict predicted by Dipert that we have fortunately not yet seen are exchanges between major cyber-powers: United States, China, Russia, and perhaps North Korea, South Korea and other former Soviet republics.

It now appears that some cyber-weapons can be highly discriminating – more precise even than precision bombing. This is contrary to the position taken as recently as Rowe (2009). Discriminating attacks in cyberwarfare have been those on Syria and in the case of Stuxnet. Even the Georgian case was more discriminating and proportional than most kinetic attacks. (These cases of cyberwarfare are discussed below.)

Another item of emerging consensus is that, for now, positive law in the form of non-proliferation and cyber-arms treaties, is both not feasible and indeed not helpful. There are some knowledgeable dissenters from this view, but for now the view expressed in Owens et al. (2009) seems correct, namely that treaties with regard to cyber-arms are not feasible because of their deeply unverifiable nature. Even among the more optimistic commentators (Rowe et al. 2011), there is the admission that a workable treaty will probably not be possible for a decade or more. These treaties could even be harmful because they would mainly restrict the actions of, and thus weaken, the more scrupulous players, such as those democracies with extensive press freedom, where a full range of human rights, even for non-citizens, is endorsed, and who carefully observe some principles of international law and just war theory, such as discrimination.

The Attribution Problem

The attribution problem has been at the heart of many policy and ethical conundrums about cyberwarfare. The attribution would appear to bar many counter-attacks to a cyber-attack, and also hinder any possible deterrent policy for cyberwarfare. It is also at the core of most objections to cyberwarfare treaties. How can we have an effective tit-for-tat or other deterrent strategy if we are not at all sure – or at any rate less sure than usually permitted – of whom we may ethically or effectively counter-attack? Yet Dipert (forthcoming a), Yannakogeorgos (2012a), and Lin (2012) and others have shown a way forward. What is called the attribution problem is really a narrow form of it, what Lin (2012) calls the 'technical' attribution problem: how we can determine, from technical means alone, such as easily-spoofed Internet protocol (IP) source information or untrusted Internet service provider (ISP) logs, where a piece of malware ultimately originated? There seem to be very few historical cases where a state has been attacked but had little information about who has attacked it. So reasoning by intuitions and settled opinions about past analogical cases is unhelpful.

What may be called the sum or compound method of attribution would utilize all forms of intelligence (electronic and human), together with technical information, as well as information about capabilities, means of malware production, and hostile intent. It is then far more likely, even if still more uncertain than we would like, that a malefacting agent can be properly identified. A minority opinion is that there are game-theoretic arguments (Dipert 2006a, 2006b, 2010) that show that solid consequentialist reasons for attacking (such as in pre-emptive or preventive cases) exist, even where there is less certainty of a future or present attacker than we usually require.

Still another – and I think ultimately more promising – way out of the technical attribution problem is to realize that the cyber-resources of a third nation are frequently involved, one that is neither attacker nor attacked. These are the 'pivot' points of attack, bots under the control of a foreign botmaster, and so on. Yet, a close analysis of the circumstances of this third nation's cyber-resources are likely to show that, especially where such means of cyber-attack are well known, as they are now, these nations have been negligent in 'supervising' their ISPs and users' systems, so that they are now being used to attack foreign nations. This would be somewhat similar to a nation A attacking another nation B by using the airfields of a third nation C, which allows or negligently fails to prevent the use of these airfields. So the argument is one that may make an intermediary nation's (C's) cyber-resources an ethically legitimate target if they are culpably negligent in allowing their resources to be utilized in attacks initiated by one nation against another nation. In Jeff McMahan's (2011) language, these intermediate nations (C) have made themselves *liable* to attack by B by this negligence.[6]

But it need not come to that. The attacked nation may immediately inform the intermediary nation of the attack and offer its assistance in stopping such attacks or punishing the real attacker. It may offer technical assistance in, say, rooting out a botnet, or offer free anti-virus software. The shortlist of relatively lawless but technologically sophisticated countries that are likely to be used for such purposes is noted in Yannakogeorgos (2012a, 2012b), so they can be identified, offered technical assistance, and put on notice even before their systems are used for attacks by others. Finally, and only if the intermediate nation is not helpful in stopping such utilization, or if there is an urgent need to stop or block the attack, the attacked nation may attack the intermediate nation's cyber-resources, or apply heavy-handed, invasive cyber-espionage tools in that nation to identify the real attacker. Even in this case, the attacked nation would be under a heavier obligation than in simple two-nation cases of morally justified war to limit its attack only to certain resources of the intermediate nation necessary for the attack, and to avoid collateral damage even to the intermediate nation's own offensive and defensive military capabilities. In the 'no alternative' case, this would appeal to what is in criminal law called a 'doctrine of necessity': attacking the intermediate nation is the only means of stopping or reducing an unjust harm.

One peculiar result of this reasoning is that, seen in this way, the United States may count as being exactly such a relatively lawless source of intermediate attacks on foreign countries. This may require the limiting or filtering of certain data from its information systems into foreign nations' (namely the malware or attacks) and may also require more active surveillance of international internet activity for installed malware within than many nations could permit because of privacy laws. Avoiding this negligence may be necessary as a condition of peaceful and lawful international relations. In the United States, the rarely invoked Neutrality Act (US Code Title 18 I.45 par. 960) is relevant since it forbids persons (individuals or corporations) from attacking a foreign nation (Dipert forthcoming a).

OTI Cyberwarfare: Definitions

The public discussion of cyberwarfare, and even discussion by experts, has almost always assumed that cyberwarfare is, and will remain, a problem specifically engendered by the Internet. That is, it is assumed that the possibility of cyberwarfare emerged, and is bound together with, the emergence of an information system network with worldwide and unrestricted access. Although usually appended to the terminology of cyberwarfare, it is neither necessary for its conceptual characteristics nor for its ethical dimensions that such a system be 'digital' (discrete state) or electronic (contemplating here optical switching). It is nevertheless useful to distinguish cyberwarfare from other forms of information warfare, such as propaganda and disinformation by traditional means. On the other hand, some forms of cyberwarfare, such as denial of service (DoS) and distributed denial of service

(DDoS) attacks, are morally very similar to earlier forms of electronic warfare such as communication radio and radar jamming (Alexander 2007, Dipert 2010).

Of the four (and probably only three) genuine acts of cyberwarfare that can be individuated, and that we have seen to date (autumn 2012), two were cases of OTI cyber-attack (Libicki 2012). These four genuine acts of cyberwarfare are the cyber-attacks on Estonia (2007), Syria (2007), Georgia (2008), and Iran (2010). Some give a longer list of incidents (Wikipedia n.d.), but most of these do not count as nation-on-nation or not as true cyber-attacks. The inclusion of the Estonian case is problematic because the role of Russia's government or military is unclear. The Flame malware (2012) appears to be for cyber-espionage, and may be, with Stuxnet, part of the 'Olympic Games' project and thus not separate from it (Sanger 2012). As I argued elsewhere (Dipert 2010), successful cyber-weapon tactics will involve cyber-reconnaissance before an attack (mapping the systems and identifying vulnerabilities) and cyber-damage assessment tools.

Stuxnet was, according to most sources, intended to be injected into a computer system (a supervisory control and data acquisition (SCADA) local area network (LAN)) by a USB storage device. The Iranian control system was believed to be 'air-gapped,' that is, to communicate in neither direction with the Internet. Stuxnet was designed *not* to spread over the Internet and indeed US officials were supposedly horrified (Sanger 2012) when it appeared on the Internet and was then detected and diagnosed by cyber-security experts (first by Kaspersky Labs). In 2007 Israel launched an air attack on a suspected reactor site (IAEA 2012) as part of 'Operation Orchard.' Before this attack, Syrian air defenses were temporarily disabled by what appears to have been an OTI attack on the air defense networked radar system (IAEA 2012).

There are in fact several stories circulating about the 'vector' of the malware that blinded the Syrian system. One is that the Israelis had inserted a back door in the hardware or software of the Syrian information system to allow activation and blinding of radar on command through the Internet. This is unlikely to be the whole story because the Syrian system was almost certainly air-gapped. (In any case, the malware that was activated through the Internet might not have been injected using the Internet.) A second is that control information was sent into the receiver of the Syrian radar system by an Israeli aircraft. This I continue to think is unlikely because a radar system would not be used for communication. (It is barely possible that the coded signals used to try to insure for the Syrians that the radio echo is not coming from a deceiving enemy transmitter could interact with the Syrian encryption software, and that this then somehow interacts with the antenna control or display system of the computer-controlled system of radar stations.) A third possibility, which I regard as most likely, is that the Syrian system – like many now being built after the 2003 destruction of the buried electrical cables of the Iraqi command and control system – used buried optical cables. A small team of operatives was inserted into Syria; located an optical cable (perhaps discerned by photo-reconnaissance when it was earlier buried) or an electronic amplifier-repeater; managed to 'tap into' the bi-directional flow of information within the cable; and inserted software or raw data that manipulated information processing of the entire air defense network to give a false negative (indicating that there were no intruders) or simply made the system malfunction for a very short period of time. In terms of both a special operation of ground forces, and as a technical feat, this was very sophisticated, if it occurred in this way. To protect future such incursions, and to encourage continued trust in optical cables by its enemies, Israel would have had a strong incentive to disguise these means by disinformation.

So, in two of the more likely scenarios, these cyber-attacks were OTI, although admittedly both were probably attacks on a network (namely, a SCADA and air-defense LAN, respectively). Observe, however, that there was nothing in these two OTI operations that essentially targeted a 'network' operation. The Iranian SCADA system and a single Syrian radar station could – with no important policy or ethical difference – have been made up of one computer and one human operator, who controlled the industrial processes, using information on one screen. In these cases, there is no meaningful sense in which a 'network' is involved.

In the official nomenclature being devised for United States Cyber Command (by an earlier joint task force) and now being modified within Cyber Command, there are descriptions that specifically describe all acts of cyberwarfare as 'network' operations (computer network attack (CNA) and computer network exploitation (CNE)).[7] This is a misleading conceptual mistake, since it focuses our attention entirely on network, and by suggestion, on Internet operations. The target of a cyber-attack is necessarily an information processing system, but not necessarily one that is part of a network. Often there will be an attack on a network, which itself is an information processing system, interpreting, directing, and encoding and decoding data between users' individual information processing systems (computers) using transmission control protocol (TCP)/IP, for example. (Encryption and decryption are special cases of coding where secrecy is a goal; most coding, such as error-correcting codes, is for efficiency or accuracy of transmission, but nevertheless is very much information processing.) It is essential that the 'cyber' part of cyberwarfare emphasizes information *processing*, since information transmission could involve only (old-fashioned) telephone lines, radio communication, or (paper) newspapers and messages, and that is definitely not what we mean by cyber-operations.[8] The original meaning of 'cyber' had this more general meaning not necessarily involving networks, such as in Norbert Wiener's notion of cybernetics, which is the modern source of the word (Weiner 1948). It is not essential for policy, nor is it essential for conceptual or ethical analysis, that a 'network' be involved. In fact, if I am correct that many future cyber-attacks will be OTI, then our developing theories of, and laws and policies for, cyberwarfare use nomenclature that is derived from antiquated past conceptions. I will shortly argue that OTI brings with it special policy and ethical considerations.

Kinds of OTI Cyberwarfare

What then are the forms that this OTI of the future might take? We can start by using the notions of *infiltration* (injection) of data into an information processing system and *exfiltration* of data from an information system, which is the basis for cyber-espionage.[9] There are also cases of the *blocking* of information flow within an information system and the *degrading* of or otherwise modifying that flow. While a cyber-attack might intentionally cease up a system altogether, such as erasing its hard drive or running an algorithm that does not halt, this result can also occur from non-cyber-attacks, such as by a sledgehammer, bomb, or an Electromagnetic Pulse (EMP) attack. Likewise, the degrading of information flow may be the result of heat build-up, a solar storm, or intentional physical damage. More interestingly as possible weapons, such disruption could also come from exposing the electronic information processing system to strong or alternating magnetic fields, which could randomly cause or neutralize current flow in the system.

Infiltration may inject undesired data into a system that does not just alter information in a database, but interferes with the intended ways of processing that data. This is malware. (Looked at in one way, programs and 'apps' are themselves just forms of data about how to manipulate other data – or sometimes even themselves.) Some hostile malware or hardware may exfiltrate data of a target information processing system, modify it in certain ways, and infiltrate these modifications. The erasure or modification of log files (logs of operations) on a computer or a server could be the insertion of 'null' data, or a more subtle variation of the existing data that was there, into the data field of the system that stores records of these operations. Shared with the theory of infectious disease (as in 'virus'), we might speak of the *vector* of this infiltration or exfiltration of data.[10] Vectors are the means by which data infiltration or exfiltration takes place. With respect to the computer (or LAN) considered as a single information system, malware is now usually injected via the Internet, and thus the Internet is the vector. If a targeted system sends data say, to a botmaster, the Internet is the vector for exfiltration.

The foundational technical problem for cyberwarfare and cyber-espionage is how data might be infiltrated or exfiltrated. Consideration of the general problem of OTI cyberwarfare should start with an examination of all of the major kinds of possible non-Internet vectors for the exfiltration and infiltration of data. Cyber-security experts have extensively studied the problem of the Internet as a vector, as well as the use as vectors of various storage devices (USB drives, CDs, DVDs, etc.). Likewise, the lack of security in WiFi networks is well known and would not be used when attached to critical systems. In Symantec's products that I use (together with other anti-malware programs), the program is automatically set to examine USB devices that I attach, CDs or DVDs that I insert, and files that I download or that arrive as email attachments. A large number of instances of malware, as well as nuisance spam, is also detected and blocked at a higher level by ISPs, Google's Gmail, my institution's anti-virus software, and so on. The problem of detecting malware that exploits a previously unknown vulnerability is not impossible. One could have collected information on the signatures of past, benign software; when a piece of software contains sections not on this list, anti-malware software could identify it for more intensive scrutiny.[11] So there is reason for collecting and categorizing benign signatures as well as harmful ones.

We have already seen some progress in solving some of the technical features that gave rise to an earlier generation of one kind of cyber-attack: DoS and especially DDoS attacks. These are now much less likely to bring down or make unusable the websites and servers of cyber-power governments, militaries, or large commercial entities. Consequently, DoS and DDoS attacks have largely been abandoned by sophisticated attackers of information systems. There was also no easy way for criminal enterprises to use these techniques to make money and the 'damage' they caused was short term, reversible, and usually not capable of rising to the effects of kinetic attacks in warfare. The most likely criminal uses of such attacks would be for blackmail (pay up, and we will stop bots from pinging your website) or impairing a competitor's operations.

There is also evidence that there has been successful defense against most malware that 'exploit vulnerabilities' in information processing systems. As noted elsewhere (Rowe 2012), these vulnerabilities were principally the result of the dominance of one, not very secure operating system (Microsoft Windows, with its originally DOS kernels), and of needlessly complicated operating systems and application programs. Microsoft's line-by-line code for an early version of Windows was shared with at least two cyber-powers (the United States and China) and with many commercial developers, speeding the discovery of exploitable vulnerabilities. Apple has been much more secretive, and its operating systems had not achieved much market penetration among major government and commercial systems until recently. The major portions of Linux versions are public, but comparatively compact, and they are inspected and guarded by a legion of programmers. There was also a rapidly changing environment for programming, and with each change in programming environment and languages (1980–2000) came new vulnerabilities, which were patched before a new environment became popular, and the process of patching began again.

Programming environments and languages have now coalesced around the C programming languages and around the Java language. Several recent updates to Java have, in 2012, made it more vulnerable – which is especially problematic since a compiler (or the Java virtual machine) is increasingly installed on most computers.

The US Department of Defense recently reported considerable success in blocking vulnerabilities on its systems, with ever-fewer zero-day vulnerabilities appearing. (Some of its operating systems are proprietary; many are Windows-based.) The major vulnerabilities of Windows are slowly being discovered and patched, and large families of malware can be identified and their untoward effects neutralized by signature-based anti-virus software.

Where, then, will the future cyber-security and cyberwarfare threats come from? I conjecture that efforts and resources will gradually shift to OTI vectors. It is now time to describe some of the families of these.

One large family, already exploited in cyber-criminality and cyberwarfare, is storage devices. These are USB drives, CDs, DVDs, and so on. Malware may also be tucked into downloadable programs and operating systems, since these are now replacing physical storage devices as a means of communicating even large programs and massive items of entertainment data.

The most pernicious malware will be software that is embedded in hardware, specifically in the chips (integrated circuits). These could degrade operations, or alter information flow within the system. It is now virtually impossible to detect such alterations. Modern chips have tens of thousands, or millions, of transistors in a square centimeter. Detecting specific alterations would require microscopic analysis of the chip itself. If altered chips are introduced into only a small number of instances of a chip, then number of chips sampled would necessarily be very large in order to detect them. It is inconceivable that hardware validation would be possible within a reasonable time period. One might suppose that the chips could be tested for their processing properties, but hardware validation is as difficult as software validation ('software verification'): ideally one would test the software for all possible inputs of values. However, for large pieces of software, and thus also for complicated chips or systems of chips, this is virtually impossible. One could imagine a bug planted in software, or a section of a chip, that is only activated when it encounters a very long string of characters that is not likely to be encountered in everyday operations.

The largest target for the insertion of hardware malware would be the basic input/output system (BIOS), the minimal, usually hardwired operating system that is necessary to boot up the main operating system from some storage device. Many computers now have a programmable BIOS: a very durable section of memory (that survives without power). These can be used to make computers more efficient if engineers discover an improvement to a

BIOS once manufacturing has taken place. The source of these changes to the BIOS can be through the Internet or another memory device. Since this allows access to the most secure parts of the system, there is a possibility for mischief.

Any alteration of an information processing system that can be affected by software – any form of malware – can also be injected into the system, on cue, by hardware.

Where do opportunities to install deliberately altered hardware configurations, such as modifications to a chip, occur? They could occur at the chip design phase. If the chip design is not kept secure, anyone with access to the design on its way to the fabrication factory could also make small changes. Once a design reaches a chip fabrication plant, anyone with access to the process that imprints the design onto individual chips could modify it. Every worker and inspector who has access to individual chips, could easily insert an alternative chip, as could anyone with access to the parts as they were being assembled – or who had access to the completed system when it is packed, or shipped, or unloaded. There are likely to be hundreds of people with such access for any given computer or information processing device. True, it would take a conspiracy involving an enemy with access to the best equipment to fabricate the alternative chip. However, it would require only one person along the production chain, who has an ideology or is paid by those who do, to enable a cyber-attack by the vector of hardware. One should keep in mind that the overwhelming majority of computers are assembled in poorly supervised factories abroad, and chips are likewise largely manufactured abroad, by low-paid workers. It is doubtful that hiring practices, supervision, and testing could stymie a determined and sophisticated enemy nation.

This then is the 'supply chain' problem. It does not affect just the endpoint computers in a system, but also every server and switch on a network, and every peripheral that is attached to the computer or network. Every printer, scanner, monitor, and so on, themselves contain many chips and sophisticated circuitry that 'sends signals' to the attached network or computer, and so are also possible vectors for malware insertion or interference with the system's processing capabilities.

Another technological possibility for mischief may establish communication between an enemy and an air-gapped computer without any more usual network connection. It is well known that an electrical signal passing through a conductor creates changing magnetic fields that extend far from it. The principle is electromagnetic induction and is the underlying technology for transformers. A sensitive conducting coil, ideally a cryogenic one parallel to a targeted wire or cable, could theoretically detect and isolate (by interferometry or other techniques) signals that are meters, or hundreds of meters, from the computer or network through which electrical pulses are passing. (I do not know of similar technology that could even theoretically detect signals through optical cables. However, optical cables require various electrical devices along their path to switch and amplify signals, and are thus vulnerable at points.) It is true that these faint magnetic signals can be shielded or counteracted in many ways. However, the possibility also exists of an enemy being able to alter any electronic device that comes reasonably close to the targeted circuitry – such as a watch, a mobile phone, or a low-powered electronic device placed in a non-electrical item of apparel – with a detecting coil, which then 'relays' or transports the detected signals outside of the electromagnetically shielded area using more ordinary technology. Just as signals going through a wire in a computer can be detected outside of it, the same science theoretically allows electrified coils outside of the computer to overpower and inject signals into the computer or electrical cable. (This strikes me as much more difficult, because to be useful, the injected signals would have to have tolerances that are close to the signals normally passing through the wire.)

Perhaps much more workable is the creation of alternative chips placed in information processing systems that have tiny low-power radio-receiving and -transmitting circuits buried within their normal information processing chips. As a result of the development of sensitive radio-detecting devices for uses in magnetic resonance imaging (MRI), and other devices, it is now possible to detect very low-power radio signals, down even to the level of several photons, as with photometry. They could be miniaturized down to a small part of an otherwise normal chip. These devices could use sections of the circuit's wiring as an antenna. They could operate at low frequencies (LF 30–300 kHz) or very low frequencies (VLF < 30 kHz), whose radio waves penetrate solid objects without significant diminution. (Some of this theory and equipment was developed for communication with submerged submarines.) But even at the high-frequency ranges, such as those at which many of our current wireless devices operate (in the gigahertz range), it is possible to traverse fairly long distances. Because of the existence of extremely sensitive radio receivers, the transmitted signal on the target system could be nearly undetectable in the background noise that all electrical devices emit – except to those devices especially designed to detect it. Their signals could be further disguised using coding that simulated noise, and through frequency-hopping spread-spectrum technology that has advanced because of great progress in mobile phone (and previous military communication) technology.

In short, someone with access to the most sophisticated, miniaturized, and low-power radio devices, as well as to any part of a computer, peripheral, or network device anywhere along its supply path, could embed devices that enable them to infiltrate and exfiltrate data (including malware) in any computer or network. Even if the powers of operation are too low, the distances great, or there exists shielding, 'relay' devices could be inserted closer to the source of the signals in the targeted information system.

Even if a nation (or corporation) could prevent data infiltration and exfiltration, by controlling the supply chain, encrypting (*within* devices), shielding, and so on, it would likely only be able to protect a small part of its information processing systems. There would remain the vast majority of systems and computers that could not, because of cost, be so protected. And through these less secure computers and information systems, such as for finance, secretarial, and acquisition operations, would probably flow enough information to allow a sophisticated enemy to assemble the pieces and infer more sensitive information.

There has been a similar problem in the recent past: the infamous history of several buildings housing the US embassy in the former Soviet Union. Even as it fell significantly behind the United States in space, information, and consumer electronics, the former Soviet Union maintained a large number of capable scientists and technicians, and was thus able to penetrate data security in the embassy on numerous occasions and with astonishing ingenuity. The Great Seal (Eagle) now displayed in the National Security Agency's (NSA) National Cryptologic Museum is just a tiny part of the story. As with the difficulty of controlling access along the computer supply chain, one source of the difficulty was the necessity of using some Soviet citizens in mundane building, remodeling, and repairing. A moral of the story is that techniques of detecting data infiltration and exfiltration methods usually fall behind the development of those techniques by a determined enemy with extensive resources.

Ethical and Policy Issues of OTI Cyberwarfare

Given the technologies I have sketched, it is probably likely that cyber-espionage using OTI is going to be more frequent than genuine acts of cyberwarfare, as it is now for Internet-based malefaction. While malware and many of the forms of attack through the internet are performed by individuals and small groups, we can expect that only large, sophisticated players will be capable of OTI cyber-attacks and OTI cyber-espionage. The manufacture of an alternative chip, only slightly different from the intended one, and the placement of an agent or use of an unwitting actor in the supply chain, would be extremely difficult. The costs would probably be large enough to deter criminal actors and vandals.

The emergence of the first damage to an information system from hardware is likely to be mystifying to operators and much harder to localize than are the operations of software malware. As we saw above, sophisticated examples are likely to be undetectable and without any effect until activated. Whatever techniques (internal computer surveillance) there are to detect anomalies of day-to-day operations could also discover software or data anomalies that were injected by the hardware. But it is unlikely that the techniques we are now using could determine the source piece of hardware: there are so many chips and so many circuits on these chips. Some of the malware injected by hardware would undoubtedly be used to destroy evidence of the source, or to plant false leads.

It is likely that only major cyber-powers would be able to utilize this form of cyber-attack using an OTI vector. In any such attack the possible state actors would be very few. Depending on the sophistication of the technology being used, the possible number of actors behind pure cyber-espionage is potentially far larger since it might include allies and even large commercial entities who are seeking intellectual property. There is also less at risk in cyber-espionage, since it is unlikely to escalate to cyber- or kinetic attacks. There may also be some commercial enterprises that are sufficiently capable to engage in OTI attacks and cyber-espionage against other commercial enterprises.

With regard to moral dimensions, OTI cyber-attacks are going to have many of the same characteristics as Internet-based attacks. They can be highly discriminating, such as being placed only in computers destined for a defense or intelligence agency, and only activated when in that environment. One type of chip modification – and probably the easiest to manufacture – would be chips that fail to function or malfunction in particular ways once activated by an internal event (e.g., malfunctioning of a weapon system when it detects that it is in a certain location) or by an outside trigger, via the Internet or a radio signal to an embedded radio receiver. It would be especially undesirable for malfunctioning to occur often, or outside of a narrow set of circumstances, since detection would be easier and the operator could take the entire system offline or replace it when detected. Although it has not been commented upon, in order not to be detected there is automatic pressure in both Internet and non-Internet cyberwarfare on the attacker to discriminate correct targets.

At the sloppiest end of an OTI attack, one could imagine an attacker flooding his enemy's country with defective or altered chips, hoping to get some of them into defense-critical systems. If they were activated by a general signal, location, or timer, this could create chaos throughout the enemy's economy, and invite a serious response. We can also see that they would be a perfect terrorist weapon, wreaking havoc through the entire economy, including on hospitals and flight control systems. However, such means of attack are likely to be available only to large, sophisticated, self-interested countries and companies; risking the blowback from such an indiscriminant activation of a hardware cyber-weapon would rarely be in their interest. (An exception might be Michael Walzer's (2006) justification for using disproportional or indiscriminate weapons in what he calls a supreme emergency (see Christopher 1999).)

One advantage of both Internet and non-Internet cyber-weapons is that they would typically require a defender to spend many times (some say in the order of 10 times) on security (supply-chain scrutiny) compared with what an attacker spends on development (although this is disputed by Martin Libicki (2009), Bruce Scheiner (2012), and others). Rather than achieving full scrutiny of the supply chain, a better approach might be to build three or more redundant systems, each with parts from a different supplier. For every major decision or operation, the same inputs are supplied to these systems and then one takes a 'vote' on the outputs provided by the differently manufactured systems. Such redundancy has been used for decades in 'mission-critical' environments. It is astronomically unlikely that a cyber-attacker could have penetrated two or more different supply chains inserting chips that produced the identical mistaken output. A similar approach for non-hardware-based cyber-defense would be to develop three or more different pieces of software for different platforms (e.g., operating systems), perhaps also coded in different computer languages, but that are intended to perform the same operation.

Because of the needed sophistication in chip design and fabrication, or of other electronic technology, and some knowledge of the supply chain, it is likely that an attacker can be identified with a high degree of certainty. There is thus less of an attribution problem in OTI cyberwarfare. In addition to this rather large cost for fabrication and insertion of chips that must behave almost the same as the intended ones, there is a time delay problem. Namely, in order to facilitate an attack, the hardware malware must be inserted months or years in advance of its use in the enemy's host information processing system. Unless the alterations allowed versatile reprogramming at a distance, they could not be used to respond to a threat, such as in kinetic war, that arose rather suddenly. Even less than for Internet-based malware, they would not allow flexible responses in real time.

The most dramatic effects of rising OTI cyber-attacks and cyber-espionage are likely to have the most impact on world trade and the costs of information systems. To protect against the implanting of pernicious hardware, a nation would (apparently) have to employ stringent quality assurance, supervision, and accounting for many of its information systems. It might require accounting for every chip and part, and searching of key employees arriving at and departing from the facility; it may also require prohibition of all personal electronic devices in the vicinity of sensitive information systems; and may require the expensive redundancy described above. While this may be (barely) possible for comparatively small operations such as the most sensitive data at NSA and Central Intelligence Agency (CIA) headquarters, it may ultimately require massive expansion to a variety of all industries connected to the production of information systems for defense, intelligence, and some commercial operations. It might be enough to require us to go back to using only paper, non-electronic printing devices, and locked filing cabinets. Neil Rowe (2012) has argued for curtailing our need for all the bells and whistles of modern software, and a return to simple computational functionality. Printers should just print American Standard Code for Information Interchange (ASCII) code in a single font, for example. Operating systems (and offered drivers for peripherals) could be much more compact.

There has been some awareness of these problems, notably through the United States' Trusted Foundry program and NSA's Trusted Access Program. There have also been some new reports of the scale of the problem (Old Trick Threatens the Newest Weapons, *The New York Times*, 26 October 2009). However, as noted in (Krepinvich 2012: note 148):

> Only one-fifth of all computer chips are made in the United States. The Defense Department buys only about 2 percent of its computer chips from secure facilities based in the United States. While the Pentagon is expanding the number of plants authorized to manufacture chips for it under the Trusted Foundry Program, production cannot meet the demand for chips for classified military systems.

I am surprised at this investment, because such supply chain scrutiny – long chains involving many individuals – seems destined to fail, namely by the insertion of one corrupted chip by one person.

Ontologically and morally, OTI cyber-attacks using hardware manipulation returns us to a simpler day. Rather ordinary principles of property ownership would apply: factories are owned, as are each component of an information processing system. Trespassing is subject to well-established norms, as opposed to what counts as intrusion or violation of sovereignty by photons and electrons, or by data. It is, at least in theory, easier to trace the history of a physical object than information or a pattern of digits.[12]

Other Normative Considerations

I have assumed that all nations, especially those with an economy, military, government, and infrastructure heavily dependent on information systems, morally may – and prudentially should – block OTI attacks by any normally legal means. This includes erecting trade barriers that would otherwise be considered otiose or even contrary to trade agreements, such as blocking imports of information system parts from certain nations designated as cyber-hostile. A cyber-hostile nation is one that has exhibited a pattern of cyber-attacks or extensive cyber-espionage. (One would ideally separate the two forms of hostility.) I assume that most major trade agreements have exceptions for national security. This is the OTI equivalent of defensive cyberwarfare, namely hardware cyber-security.

Once attacked by an OTI cyber-attack what (morally) may a nation do? What prudentially should it do, beyond strictly defensive measures? The parallel case of what a nation may, or should, do if it is a victim of OTI cyber-espionage is beyond the scope of this article. A first guess in the OTI cyber-espionage case is that one may and should respond in all the ways in which enlightened nations have typically responded to traditional espionage. These include applying soft force such as unilateral changes in diplomatic and economic relations, or launching or increasing one's own cyber-espionage against the 'attacking' nation. These are one's espionage-deterrent tools. In the absence of an international enforcement mechanism, this also amounts to 'punishment' – which probably should be restrained by considerations of discrimination and proportionality.

In the case of responding to a true OTI cyber-attack, one's tools are likely to be restricted. By its nature, there would be great difficulty in responding to OTI cyber-attacks by an in-kind response, an OTI cyber-counter-attack. Inserting the necessary hardware takes a great deal of time: months or years. One's tools then are limited to Internet cyber-counter-attack, other tools of soft force, and only as a last resort – and probably only in response to damage on the order of a kinetic attack/kinetic counter-attack.

The hard question then is the one of the legitimacy of an OTI attack. A still harder question is what hardware tools morally may be installed on a possible future attacker's system, but not activated until needed. As many have noted, the placement of software logic bombs including back doors, and in the case of OTI, hardware devices, is morally and legally analogous to placing explosive charges in possible future attackers' infrastructure and weapons (Schmitt 2013, B.III. Rule 30 Comment 14: 94).[13]

Cyber-powers who are world manufacturers and fabricators of information systems and chips are in a special position with regard to embedding OTI parts in equipment destined for a cyber-hostile nation that can later be used to initiate attacks or cyber-espionage. They are 'playing with fire,' since not only would they be easily identifiable if the hardware alterations were detected, but they also risk losing trust in the products that they market worldwide (possibly in *all* of their products). So the detection of any such hardware alterations could have a devastating economic impact on the initiating nation. There would be a technique that is analogous with spoofing,[14] but difficult to bring off. Spoofing of hardware would be to convince a buyer that a given piece of hardware was fabricated in one secure facility, when it was not, and was perhaps fabricated in one (enemy or insecure) country and inserted in products from another (friendly) country.[15]

If I am correct, OTI cyberwarfare is a natural next step. For the reasons discussed, many forms of OTI cyber-espionage and OTI cyber-attacks are likely to be rare at first. It is nevertheless important that doctrinal, policy, ethical, and legal thinking and writing take account of their possibility.

NOTES

1. The flaws in this and similar definitions are discussed below and in Dipert (forthcoming b). Since the key noun is 'information,' this definition may not literally include damage to the *functioning* of an information processing system.
2. Espionage and theft of intellectual property may, once known, cause the spied-upon country to take expensive countermeasures, such as an enhanced counter-espionage program or abandoning a novel weapon system once its vulnerabilities are made known. However, these costs are chosen by that nation, and thus do not directly and necessarily cause harm. For an excerpt from the rapidly increasing literature on the ethics of intelligence gathering, see Goldman (2009).
3. This consideration was urged on me by Don Howard.
4. We have already entered a time when the appropriateness of certain cyber-responses is being tested in practice to see where non-escalating, game-theoretic equilibria can be found, or established by a convention of practice. The international reaction to Stuxnet has been muted.
5. The most striking instance of weapon-destroying weapons is in the classic science fiction film *The Day the Earth Stood Still* (1951).
6. I think this theory is deeply mistaken on several counts, but it does allow a diminished liability to be attacked, such as what non-combatants may incur if they have supported an unjust war.
7. In a draft document of unknown provenance (possibly from the J-5 of Cyber Command), 'Cyberspace Operations Lexicon,' there are modifications that offer a disjunction, including a target that is a 'system' that could be one computer. However, the attack and attacker are described as essentially network operations: 'actions are taken through the use of computer networks.'

8. Technically, AM and FM broadcasts, both involving 'modulation,' and the conversion of sound into electrical current in old-fashioned telephones or spoken-to-written communication, all involve some element of information processing. In past systems, this processing was typically simple (using very short algorithms), and was between one mode of communication and another, for example electrical to radio signals. Modern information processing systems are much more complex (using many very large algorithms) and can be within one mode, such as electrical signals.
9. This exfiltration need not be destructive of the original data operating in the targeted information system. One typically copies the information content and it is this that is exfiltrated. This unfortunately illustrates the need for an improved and better understood language and ontology of data vs. information content, flow of information, data as contained in arrangements or patterns of charges and magnetism, and so on.
10. Another useful notion borrowed from the biomedical realm is the notion of a 'signature.' There are genetic signatures in DNA or RNA (such as for an inherited disease); there are similarly 'signatures' of various species of malware. In both cases they are sequences of entities (amino-acid pairs or bits) or patterns of such sequences. I am part of a project between my institution's National Center for Ontological Research (NCOR) and Pacific Northwest National Laboratories (PNNL), in its initial phase, to develop the systematic, formal ontology of the general notion of a signature.
11. This is from a personal conversation with Neil Rowe.
12. A difficulty is, however, that many of these OTI techniques come in two phases. The first is their installation or other infiltration of flawed data or malware. This has few ontological problems. The second is its activation. The activation of such malware and its contribution to the dysfunctioning of the information system brings back the ontological problem. Which is the attack (perhaps justifying counter-attack)? How much dysfunction, without permanent damage or deaths, constitutes morally actionable harm?
13. In the subsequent comment (Schmitt 2013, B.III. Rule 30 Comment 15: 94) the authors may be going too far in considering even intercepted attacks without damage as an attack. Cyber-powers may be initiating thousands of minor attacks a day on each other, with the expectation that cyber-defenses will block them.
14. 'Spoofing' is a falsification of the source IP address of a message identified in packets containing it.
15. Many of these ideas were developed while I was Fellow (2011–12) at the Stockdale Center for Ethical Leadership at the US Naval Academy – and in fact some of them may have originated from other members of this cyberwarfare research group. The views are, however, entirely my own, not those of other members or of the US government. A version of this paper was sketched in presentations at the John C. Reilly Center for Science, Technology, Values, Notre Dame University, and later at a workshop on cyberpower at Maxwell Air Force Base, both in August 2012. I benefited much from the discussion and astute comments of participants.

REFERENCES

Alexander, Keith B. (2007) Warfighting in Cyberspace, *Joint Forces Quarterly*, 13 July, accessed 20 March 2013, available at: http://www.military.com/forums/0,15240,143898,00.html; Internet.

Andress, Jason & Winterfeld, Steve (2011) *Cyber Warfare: Techniques, Tactics and Tools for Security Practitioners* (Waltham, MA: Syngress).

Brown, Gary & Poellet, Keira (2012) The Customary International Law of Cyberspace, *Strategic Studies Quarterly*, 6(3), pp. 126–145.

Chertoff, Michael (2011) Statement of Record before the United States House of Representatives Subcommittee on Crime, Terrorism and Homeland Security, 15 November, accessed 20 March 2013, available at: judiciary.house.gov/hearings/pdf/Chertoff%2011152011.pdf; Internet.

Christopher, Paul (1999) *The Ethics of War and Peace: An Introduction to Legal and Moral Issues*, 2nd ed. (Upper Saddle River, NJ: Prentice Hall).

Clarke, Richard & Knake, Robert (2010) *Cyber War: The Next Threat to National Security and What to Do about It* (New York: HarperCollins).

Cook, James (2010) 'Cyberation' and Just War Doctrine: A Response to Randall Dipert, *Journal of Military Ethics*, 9(4), pp. 411–423.

Crisp, Roger (2012) Cyberwarfare: No New Ethics Needed, in: Practical Ethics: Ethics in the News, University of Oxford, accessed 11 April 2013, available at: http://blog.practicalethics.ox.ac.uk/2012/06/cyberwarfare-no-new-ethics-needed/; Internet.

Dipert, Randall R. (2006a) Strategies, Rationality, and Game Theory in the Philosophy of War. Paper presented at the Joint Service Academy Conference on Professional Ethics (JSCOPE), Springfield, VA, January 2006.

Dipert, Randall R. (2006b) Preventive War and the Epistemological Dimension of the Morality of War, *Journal of Military Ethics*, 5(1), pp. 32–54.

Dipert, Randall R. (2010) The Ethics of Cyberwarfare, *Journal of Military Ethics*, 9(4), pp. 384–410.

Dipert, Randall R. (forthcoming a) The Future Impact of a Long Period of Limited Cyberwarfare on the Ethics of Warfare, in: Luciano Floridi (Ed), *Ethics of Information Warfare* (in the series Philosophy of Engineering & Technology).

Dipert, Randall R. (forthcoming b) The Essential Features of an Ontology for Cyberwarfare, in: Panayotis Yannakogeorgos (Ed), *Cyber Power: The Quest for a Common Ground* (Montgomery, AL: Air University Press).

DoD (Department of Defense). (2012) *DOD Dictionary of Military Terms*, accessed 20 March 2013, available at: http://www.dtic.mil/doctrine/dod_dictionary/data/c/10082.html; Internet.

Fotion, Nicholas (2008) *War and Ethics: A New Just War Theory* (London: Continuum Books).

Goldman, Jan (2009) *Ethics of Spying: A Reader for the Intelligence Professional*, Vol. 2 (Lanham, MD: Scarecrow Press).

IAEA (International Atomic Energy Agency) (2012) Report of the Director General, accessed 20 March 2013, available at: http://www.iaea.org/Publications/Documents/Board/2012/gov2012-42.pdf; Internet.

Joint Staff, U.S. Department of Defense (2006) U.S. Joint Publication 3–13 (2006), 'Information Operations', accessed 1 April 2013, available at: http://www.carlisle.army.mil/DIME/documents/jp3_13.pdf; Internet.

Krepinvich, Andrew (2012) *Cyberwarfare: A 'Nuclear Option'?*, Washington, DC: Center for Strategic and Budgetary Assessments (CSBA), accessed 20 March 2012, available at: http://www.csbaonline.org/publications/2012/08/cyber-warfare-a-nuclear-option/; Internet.

Libicki, Martin (2009) *Cyberdeterrence and Cyberwar* (Santa Monica, CA: RAND Corporation).

Libicki, Martin (2012) McCain Conference – Panel on Issues in Cyberwarfare, US Naval Academy, Annapolis MD, 27–28 April; accessed 20 March 2013, available at: http://www.youtube.com/watch?v=bI7TLqTt0H0&feature=plcp; Internet.

Lin, Herbert S. (2012) Escalation, Dynamics, and Conflict in Cyberspace, *Strategic Studies Quarterly*, 6(3), pp. 46–70.

Lucas, Jr., George (2012) Permissible Preventive Cyberwar: Restricting Cyber Conflict to Justified Military Targets, accessed 20 March 2013, available at: http://www.elac.ox.ac.uk/downloads/Permissible%20Preventive%20Cyberwar%20UNESCO%202011.pdf; Internet.

McMahan, Jeff (2011) *Killing in War* (Oxford: Oxford University Press).

Orend, Brian (2005) War, in *The Stanford Internet Encyclopedia of Philosophy*, accessed 20 March 2013, available at: http://plato.stanford.edu/entries/war/; Internet.

Owens, William A., Dam, Kenneth W. & Lin, Herbert S. (Eds) (2009) *Technology, Law, and Ethics Regarding US Acquisition of Cyberattack Capabilities* (Washington, DC: National Research Council of the National Academies of Science).

Rowe, Neil (2009) The Ethics of Cyberweapons, *International Journal of Cyberethics*, 1(1), pp. 20–31.

Rowe, Neil (2012) Expert Testimony on Cybersecurity and Cyberwarfare. Paper presented before the Committee on Ethical and Societal Issues in National Security Applications of Emerging Technologies, National Academies of Science, Washington, DC, 12 January.

Rowe, Neil, Garfinkel, Simson L., Beverly, Robert & Yannakogeorgos, Panayotis (2011) Challenges in Monitoring Cyberarms Compliance, *International Journal of Cyberwarfare*, 1(1), pp. 1–14.

Sanger, David (2012) *Confront and Conceal: Obama's Secret Wars and Surprising Use of American Power* (New York: Crown Publishing).

Schmitt, Michael (Ed) (2013) *The Tallinn Manual on the International Law Applicable to Cyber Warfare* (Tallinn: NATO Cooperative Cyber Defence Centre of Excellence), accessed 20 March 2013, available at: https://www.ccdcoe.org/249.html; Internet.

Schneier, Bruce (2012) *Liars and Outliers: Enabling the Trust that Society Needs to Thrive* (New York: Wiley).

Shobert, Benjamin (2012) China's Capacity for Cyber-war, *Online Asia Times*, 15 March, accessed 20 March 2013, available at: http://www.atimes.com/atimes/China/NC15Ad01.html; Internet.

Smith, Barry (and others) (2012) Basic Formal Ontology (BFO 2.0), accessed 20 March 2013, available at: http://www.ifomis.org/bfo; Internet.

Walzer, Michael (2006) *Just and Unjust Wars*, 4th ed. (New York: Basic Books).

Wiener, Norbert (1948) *Cybernetics: Or Control and Communication in the Animal and the Machine* (Paris: Hermann & Cie and Cambridge Mass.: MIT Press Mass).

Wikipedia n.d. Cyberwarfare, accessed 20 March 2013, available at: http://en.wikipedia.org/wiki/Cyberwarfare; Internet.

Yannakogeorgos, Panayotis (2012a) McCain Conference – Panel on Issues in Cyberwarfare, US Naval Academy, Annapolis, MD, 27–28 April, accessed 20 March 2013, available at: http://www.youtube.com/watch?v=bI7TLqTt0H0&feature=plcp; Internet.

Yannakogeorgos, Panayotis (2012b) Internet Governance and National Security, *Strategic Studies Quarterly*, 6(3), pp. 102–125.

Randall R. Dipert is Charles S. Peirce Professor of Philosophy at the University at Buffalo and a member of the Emerging Technologies of National Security and Intelligence (ETNSI) Group, John J. Reilly Center for Science, Technology, and Values University of Notre

Dame. In 2011–12 he was a Fellow at the Stockdale Center for Ethical Leadership at the US Naval Academy for research on the ethics of cyberwarfare. From 1995 to 2000 he taught at the United States Military Academy at West Point, NY, and he has worked on applied ontology for the US Army. He has published a book on artifacts and action theory, and has co-authored a book on logic. He has published dozens of articles on logic, artifacts, mathematics, the philosophy of mind, artificial intelligence, and ontology. He has published papers and given presentations on the morality of preventive war and the application of game theory to the morality of war. He authored one of the first papers on the subject of the ethics of cyberwafare in *Journal of Military Ethics* (December 2010). *Correspondence Address*: Department of Philosophy, University of Buffalo, Buffalo, NY 14260, USA. *Email Address*: rdipert@buffalo.edu

JUST WAR AND CYBERWAR

Christopher J. Eberle

United States Naval Academy, Annapolis, MD, United States

My aim in this paper is to reflect on a very narrow question: under what conditions might a cyber-attack provide a just cause for war? I begin by articulating what makes for a just cause, briefly address the problem of attribution, and then discuss three broad categories of cyber-attack: those that clearly do not satisfy the just cause requirement, those that clearly do satisfy the just cause requirement, and three ambiguous cases – the destruction of property, the emplacement of logic bombs, and the failure to prevent cyber-attacks. My conclusions are exploratory and suggestive rather than definitive, partly by virtue of the extreme paucity of literature on the moral assessment of cyberwar.

Introduction

My aim in this paper is to reflect on a narrow moral question: under what conditions might a cyber-attack render a recipient morally permitted to respond to that attack with military violence? I raise this question from the vantage provided by the just war tradition (JWT), broadly conceived. A central holding of the JWT is the claim that a polity is morally permitted to wage war only if that polity has a just cause to do so. In what follows, I will reflect on whether or not a cyber-attack can satisfy this 'just cause requirement.'

I want to emphasize from the outset that my discussion leaves unanswered – even unaddressed – a whole range of important normative questions regarding the normative status of cyberwar. So, for example, an important, and morally suggestive, literature addresses my topic from a *legal* perspective – say, whether or not a cyber-attack can constitute an act of aggression that provides a nation-state with a legal right to collective self-defense. I will not address that literature either directly or in detail (Dinstein 2002, Dunlap 2011, Jensen 2002, Schmitt 1999, 2010, Sklerov 2009). I will also refrain from addressing a broad swathe of important *moral* questions regarding the moral status of cyberwar. Most pertinently, my reflections implicate only the just cause requirement and not the JWT's other constituent commitments. So, for example, I will not attempt to specify the bearing of the principle of discrimination on cyber-attacks that inflict non-lethal suffering on non-combatants – one of many important moral questions that cyberwar raises for adherents of the JWT.[1] It seems to me that the narrow focus I strive to maintain throughout is the better part of wisdom: novel questions, with respect to which there exists only a scant literature, are best addressed – initially at least – in piecemeal fashion and without any pretension to systematic treatment.

What Is a Just Cause?

In order to determine whether a cyber-attack can provide a just cause for war, we need to have some clear understanding of the just cause requirement. There are many competing conceptions, so I will begin by the conception I favor. I will offer a number of brief comments in lieu of a more detailed discussion.[2]

First, the just cause requirement is best understood in light of a bedrock understanding of the moral status of individual human beings. That is, each human being has an inherent worth or dignity that grounds a diverse array of natural rights. Most relevant to the JWT is the claim that each human being has a right to life, from which it follows that each of us are bound by a presumptive and defeasible obligation to refrain from killing any human being.

Second, although individual human beings have an elevated moral standing, they enjoy that standing while belonging to groups that matter a great deal to them. We are inveterately tribal creatures. Communal membership is a crucially important component of a human being's moral identity: we define who we are not only by reference to our species membership, but also by reference to the bonds we share with some particular group of human beings. We identify not merely as human beings but also as Americans, or New Yorkers, or Steelers fans, or Buddhists. These communal memberships can powerfully influence our projects, priorities, and aspirations, thereby guiding our practical deliberations, and so shape the way we act.

Third, just as each individual human being has a right to life, so also does each human community have a *prima facie* right not to be attacked with military violence. Correlatively, each community has a *presumptive* obligation not to employ military violence against any and all others.[3] These rights and obligations form the normative core of an irenic moral vision implicit in the JWT: the normal normative condition of the human species is that we are bound together into particular groups, each of which is presumptively prohibited from violently attacking others. Of course, the presumptive prohibition against the use of military violence can be overridden: that is what it *means* for a prohibition against war to be presumptive rather than absolute.

Fourth, the just cause requirement performs a distinctive and morally crucial role in determining when that presumption is overridden. Basic to the JWT is the claim that a *just* war is not absurd, inexplicable reality: it cannot involve organized violence between just any arbitrary pairing of communities. Much to the contrary, if one community wages a just war against another, then there must be some sufficient explanation as to what makes it the case that the use of organized violence is morally permissible between *these* particular communities, here and now. One of the primary functions of the just cause requirement is to mandate that warring communities explain what a community has done to make itself the target of legitimate military violence.

Fifth, put bluntly and dogmatically, this mandated explanation requires warring parties to identify some relevant moral violation – a sufficiently egregious violation of rights. A just war is possible only when military violence is directed against a community that singles *itself* out from among the multitudinous possibilities as a legitimate target of military violence. A moral violation 'marks' a particular target of licit attack so long as it involves the performance of some wrongful act. This is because actions are necessarily correlated with particular agents and so to correlate just causes with wrongful acts is thereby to associate a just cause with particular agents.

Sixth, there is considerable controversy regarding what makes for the kind of violation that can satisfy the just cause requirement. To be sure, certain violations clearly do *not* satisfy the just cause requirement. Suppose, for example, that one polity directs a project of ideological vilification against the members of another: the first wages a propaganda campaign to convince others that the second's members are subhuman, scum of the earth, devoid of rights, and so without moral standing. No matter how persistent, no matter how vitriolic, no matter how large the population wronged, no matter how large the quantity of happiness to be derived from striking back, the vilification of its constituents is clearly not the sort of violation that can provide one nation-state with just cause to attack another with military violence. Again, certain violations clearly *do* satisfy the just cause requirement – one nation-state's use of military violence to invade and subjugate another, minimally decent, nation-state. But between such obvious cases, there is a very large grey area.

Seventh, the philosopher's dream is to be able to specify some principled threshold that marks off the moral violations that do not provide a just cause from those that do. Perhaps there is some such threshold. But even the most cursory familiarity with commonsense moral reflection – even more so philosophical reflection on morality – provides compelling evidence that there is no uncontroversial way to specify it. Once we pass beyond moral platitude, any proposed threshold separating the violations that satisfy the just cause requirement from those that do not will be the object of reasoned disagreement between morally serious and epistemically competent peers.

The central question I address in this essay is which, if any, cyber-attacks can provide the victimized community with a just cause for war. For the most part, my reflections on the moral status of cyber-attacks reflect this broad pattern of agreement and disagreement. That is, some cyber-attacks clearly do not provide for a just cause, some clearly do, and many fall into the nether region between these two extremes. I will explore cyber-attacks that fall into each of these categories.

Just Cause and Attribution

Before I do so, however, let me note – and then set aside – an epistemological and moral problem distinctive to cyberwar. With respect to the use of 'conventional' military violence, the attribution of act to agent is normally unproblematic. When tanks roll over borders, it is normally obvious both *that* they have done so and *whose* tanks they are. But this will not always, or even normally, be the case when aggressors attack with computers, viruses, and logic bombs. In my layman's understanding, a cyber-attack occurs when some party launches an appropriately arranged sequence of 1s and 0s at a vulnerable network of computers in order to interfere with whatever the target is programmed to control. Apparently, these sequences of 1s and 0s can be exceedingly difficult to trace: a cyber-attack can traverse multiple servers in multiple countries, the authorities in those intervening countries might be unable or unwilling to help locate the originating source of the attack, and so even a technologically sophisticated victim might be unable *reliably* to attribute the attack to a perpetrator.[4]

Given the conception of the just cause requirement I have sketched, just what makes for a morally permissible response to an unattributable cyber-attack is perplexing. To be sure, the target of such an attack might *have* a just cause for war and may surely fend it off if possible. But the victims might find that the only feasible way to stop an attack is to strike at its source. This is just what the inability to attribute seems to preclude – although the victims might very well harbor strong suspicions, they cannot reliably identify the perpetrator and so they cannot provide the explanation mandated by the just cause requirement. So the implications of the JWT seem direct: if we have been victimized by an unjust cyber-attack that we cannot reliably attribute to any particular perpetrator, we are nevertheless morally forbidden to respond by attacking the suspected source of that attack.

Now this might not be particularly troubling if attribution is only occasionally problematic – any morally serious human being must grant that we can on occasion be morally forbidden to respond to a genuine wrong. But if cyber-attacks are often or ordinarily resistant to attribution, then the JWT seems to permit us to respond only to cyber-attacks perpetrated by those too clumsy or incompetent to cover their tracks. This restrictive stance is morally troubling – after all, persistent restraint in the face of unattributable attacks surely provides a perverse incentive for hostile parties to develop and deploy the means to exploit such restraint. This is at the very least a highly undesirable – and perhaps practically untenable – moral condition.

I will not even attempt to 'resolve' the problem of attribution in this paper. Indeed, I do not believe that this kind of problem can be 'solved' merely be reflection; far from it: attribution is a 'technical' problem, to be resolved by those with the relevant computer and forensic skills.[5] If technicians cannot determine how to reliably track cyber-attacks, then moral reflection will be stymied by paradox. That is, unless those with the relevant technical skills are able to find some way to reliably determine who launches cyber-attacks, moral reflection on how we may respond to cyber-attacks seems to present us with equally unacceptable alternatives: we ought not to respond to any attack we cannot reliably attribute to a specific course, but we cannot refrain from responding in some way.[6] So it seems to me. On that somewhat deflating note, let me turn to a number of other respects in which cyberwar articulates with the JWT's just cause requirement.

Cyber-Attacks that Cannot Satisfy the Just Cause Requirement

In order for the just cause requirement to be satisfied, we must not only be able to specify a violation that pinpoints some legitimate target, but that violation must be particularly egregious. After all, military violence, in its ordinary uses, kills human beings, human beings have inherent worth or dignity, we may kill human beings only for commensurately weighty reasons, and so we may employ military violence only in response to egregious violations – those that threaten very great goods or weighty rights. Put bluntly, any just cause assessment presupposes some – often difficult and contentious – assessment of the normative weight of a moral violation that putatively provides a just cause for war.

This basic moral point has important implications for our topic. Even cursory familiarity with the literature reveals that the notion of a cyber-attack (and its correlate, 'cyberwar') is used to denote actions that have consequences ranging from the morally trivial to the absolutely devastating.[7] But because only egregious violations can provide a just cause for war, many kinds of cyber-attack cannot. Let me briefly mention two such cases.

Two recent treatments of cyberwar focus on the theft of valuable information. According to Richard Clarke (and his co-author, Robert Knake), the theft of valuable information from American, European, and Japanese facilities by Chinese hackers is:

> ...without precedent in the history of espionage. Exabytes of data have been copied from universities, industrial labs, and governmental facilities. The secrets behind everything from pharmaceutical formulas to bioengineering designs, to nanotechnology, to weapons systems, to everyday industrial products have been taken by the People's Liberation Army and by private hacking groups and given to Chain, Inc. (Clarke & Knake 2010: 59)

Joel Brenner (2011: 51) broadly concurs with this assessment. One of the main theses of his recent *America the Vulnerable* is that 'American and European governments are under constant and withering cyberattacks from China, and these attacks reflect government policy, denials notwithstanding.' He describes a dizzying, and undoubtedly troubling, array of such attacks. So, for example, cyber-attackers have apparently been robbing Uncle Sam blind for years – individual citizens, private corporations, defense contractors, government agencies. They have stolen valuable information – Google's source code, Exxon Mobile's bid data detailing 'the quantities, value, and location of oil discoveries worldwide' (Brenner 2011: 48), and several terabytes of information about the design and performance of the F-35 Joint Strike Fighter. Although the impact of any one of these thefts is rather modest, the accumulation is not. Indeed, Brenner, Clarke, and Knake suspect that the Chinese have orchestrated many of these cyber-attacks as part of a long-term 'unrestricted war,' the aim of which is to weaken the United States economically and technologically and thereby to compromise it militarily (Brenner 2011: 121,125–127, Clarke & Knake 2010: 53). To the uninitiated, what these thieves have been able to get away with is genuinely shocking.

The low-intensity campaign perhaps waged by the Chinese contrasts in important respects with the 'cyberwar' between Estonia and Russia in the spring of 2007 (Davis 2007). Ostensibly in response to an insult to Russian honor – the removal of a statue commemorating the sacrifice of Red Soldiers during the Great Patriotic War – the Republic of Estonia was 'hammered with a distributed denial of service attack from thousands of captive computers organized into botnets. Estonian banks were shut down, remittances from abroad didn't arrive, government departments came to a standstill, big credit card issuers shut down their networks' (Brenner 2011: 128; see also Clarke & Knake 2010: 11). Attribution remains an issue – the Russian government denied responsibility and ascribed the attacks to spontaneous activity on the part of irate cyber-patriots. In any case, the attacks directed at the Estonians had some of the characteristics of a conventional war – a perceived injustice, a coordinated response, planned and predictable harm to the ostensible offender, and so on. Unsurprisingly, some perceived a warlike attack: according to the Speaker of the Estonian Parliament, 'When I look at a nuclear explosion and the explosion that happened in our country . . . I see the same thing . . . . Like nuclear radiation, cyberwar doesn't make you bleed, but it can destroy everything' (Davis 2007 (online)).

Now it is hardly surprising that an Estonian official would react so strongly to a coordinated attack on his country that emanated from the former Soviet Union. After all, violated communities are almost always impelled by circumstance and history to inflate the normative weight of the violations inflicted on them. And this is exactly the correct characterization of the Speaker's response. There is no doubt that the Estonians were wrongly attacked: no 'insult to Russian honor' provided just cause for the cyber-attacks launched against Estonia. Nevertheless, the attacks emanating from Russia killed not a single Estonian citizen. No one was enslaved. No Estonian property was destroyed nor territory seized. The lives of a good many Estonians were disrupted. But human beings are worth far too much to be killed over impeded access to banks, credit cards, and government departments. (This is one of those contentious comparative assessments that, as I mentioned above, are presupposed in any claim to just cause.) The life of no Russian citizen or soldier was rendered forfeit by virtue of the country's role in launching the cyber-attack on Estonia. The attacks on Estonia therefore provided no just cause for war, as would certainly have been the case had Russia attacked Estonia with nuclear weapons.[8]

Similarly with respect to the theft of valuable information: I take it to be obvious that individual cases of cyber-theft cannot provide a just cause for war. You cannot kill human beings to prevent them from stealing Google's source code – nor to prevent performance information about the F-35 falling into the wrong hands. Similarly for an orchestrated campaign of cyber-theft – pile up the purloined information as high as you like, no accumulation is worth the life of a single human being and so cannot provide a just cause for war.[9]

Of course, this is not to say that economic disruptions or information theft are morally negligible. They might very well permit some kind of robust response – diplomatic protest, sanctions, boycotts, and so on. What they do not permit is the use of lethal violence. If this moral assessment of what is worth killing over is correct, then the vast majority of cyber-attacks can provide no just cause for war. Indeed, because most cyber-conflict seems far more akin to intense economic conflict than to conventional war, we should assess most cyber-conflict with moral principles tailored to boycotts not bombs. We should also be exceedingly wary of language that obscures that basic moral reality: the common use of 'cyber-attack' to denote the whole gamut of violations that can be brought about by way of a cyber-attack, from stealing information to bringing down electrical grids, is so morally perilous that we ought simply desist from that practice (Sanger 2012).

Note one, perhaps troubling, implication of this general position: if the moral standing of most cyber-attacks is comparable to the moral standing of ordinary economic conflict, then waging 'cyberwar' will normally be far easier to justify than waging a conventional war. That is, the clear implication of the claim that most cyber-conflict is morally on a par with ordinary economic conflict is that cyber-conflict is far easier to justify than waging war. It also seems to comport with a permissive policy of tit for tat: if A hacks into B's industrial facilities to steal state secrets, then B may return that favor in kind and degree.[10] So, for example, *if* the US government was responsible for using the Stuxnet virus to destroy Iranian centrifuges, and if those centrifuges were in fact to be used only for non-military, economic purposes, then it is plausible to suppose not only that the United States' deployment of Stuxnet wronged the Iranians, but also that the Iranians were thereby morally permitted to respond in kind, say, by deploying malware that inflicts some economically commensurate amount of damage. If the damage inflicted in response is merely economic and so non-lethal, then the retaliatory use of malware that engenders comparable economic costs need not be any more objectionable than the use of more familiar, and clearly permissible, means of responding – say the imposition of economically damaging sanctions on the attacker.[11]

Here it is important to recall an important component of the moral vision that animates the JWT. Although the JWT arguably includes a robust presumption against war, it includes nothing like a presumption against communal conflict generally. As I understand it, the JWT is largely silent about economic competition, ideological conflict, diplomatic dispute, even spying and espionage. It seems to me that there is great moral wisdom in this silence. The besetting sin of the JWT is that its adherents inflate the normative significance of wrongs by which they are victimized – we are strongly inclined to overestimate the moral significance of the wrongs inflicted on the communities with which we have bonded and so to construe comparatively inconsequential wrongs as sufficiently egregious to permit the use of military violence. The reaction of the Estonian Speaker is a rather clear case in point.[12] The JWT's silence regarding the many wrongs that occur in the normal course of non-lethal competition between distinct communities serves as a kind of conceptual prophylactic to this kind of inflationary reaction: wrongs there will always be between communities that engage in economic and diplomatic competition, but so long as the competition remains non-lethal, we ought not think or speak about those wrongs with the language of war.

Cyber-Attacks that Can Satisfy the Just Cause Requirement

If it is clear that many cyber-attacks cannot provide a just cause for war, it is no less clear that some can. For better or worse, technologically advanced countries like the United States have become ever more dependent on the Internet for the provision of essential social, economic, military, and political goods. Dependence on the Internet is directly correlated to vulnerability to cyber-attack: 'If you can do something with software, someone else can undo it with software, and often they can do it remotely' (Brenner 2011: 30). If an air traffic controller uses software to ensure that airplanes do not run into one another, and if the network on which that software runs is connected to the Internet, then some hostile party might be able to access that software and arrange for planes to run into, rather than avoid, one another.[13] Surely that kind of cyber-attack could provide a just cause for war: whether the means employed are explosives or sequences of 1s and 0s, if some hostile party enacts a plan to crash planes full of harmless civilians, it is very plausible to suppose that we have just cause to respond with military violence. A comparable point applies to destructive attacks on electrical grids, nuclear reactors, and other components of a country's 'critical infrastructure': other things being equal, it is morally irrelevant whether a power plant or nuclear reactor is destroyed by a 2000-pound bomb or by the activation of malware by keystroke (Schmitt 2010). (I should note that, although most experts seem to agree that destructive attacks on critical infrastructure are technically possible, they disagree vigorously regarding the likelihood that hostile parties can, or will, actually bring off such attacks.[14])

Let me make four brief comments regarding the narrow range of cyber-attacks that seems clearly able to provide a just cause for war.

First, a cyber-attack on critical infrastructure does not by itself provide a just cause for war. Suppose, for example, that a cyber-soldier from another nation-state hacks into and is able to gain control over some aspect of our air traffic control system. Assume that his 'penetration' of that system provides him with the capacity to kill a large number of human beings. But suppose that he does no such thing. His capacity to wreak havoc lies dormant. Surely his 'penetration' of this component of our critical infrastructure does not provide a just cause for war. After all, he might be trying to determine what is possible rather than enacting any plan to inflict damage.

This first point is closely related to a second: whether a given violation provides a just cause for war normally depends on its role in a wider plan. Some theorists seem to assume that certain triggering events, by themselves, provide a just cause for war – say, that the attack on Pearl Harbor provided the United States with just cause to wage war against

Japan. Although this manner of speaking might be convenient shorthand, it does not adequately specify the violation that provided the United States with a just cause. After all, whether the United States had just cause for war against Japan in 1941 depends in part on whether the attack on Pearl Harbor was an utterly bungled training run or the initiation of a plan to achieve hegemony in the Pacific. What matters – at least as much as the attack itself – is the plan initiated by the attack. The same basic point applies to cyberwar. Whether or not a given cyber-attack provides a just cause for war partly depends on the plan of which it is an enacted component. An attack on the power grid in one small city that is designed to demonstrate mere technical capacity but that accidentally takes out the power grid of a major metropolitan area has a very different moral status than an attack designed to take out the power grid of that major metropolitan area. Despite the fact that the two attacks have exactly the same consequences, the permissible response to the first might very well be a demand for compensation, and to the second, war.

Third, there are severe epistemic barriers to the reliable identification of the plan served by a given cyber-attack. If we cannot reliably attribute an attack to any particular party, then it seems difficult to determine whether that attack serves any plan, much less what that plan is. Again, the effects of cyber-attacks are often unpredictable and so it will be hard to read off a plan from an attack that might very well not have the consequences its perpetrators envisioned. Uncertainty of attribution and unpredictability of effect complicate the assessment of just cause even in the case of destructive cyber-attacks.

Fourth, if destructive cyber-attacks on critical infrastructure can provide a just cause for war, then the range of parties against which it is in principle possible to have a just cause to go to war extends far more broadly than to nation-states. Terrorist organizations are a particular concern. In a familiar dynamic, aggrieved groups that have negligible conventional military power have a strong incentive to develop comparatively inexpensive means of attack that have outsized destructive effects. However great the incentive, there is apparently considerable disagreement as to whether a terrorist group like al-Qaeda has the required expertise (Clarke & Knake 2010: 135).

Contested and Ambiguous Cases

If most cyber-attacks clearly cannot provide a just cause, and if some clearly can, there are many that have a rather ambiguous moral status. I will briefly mention three.

First, what is the moral status of the intentional destruction of valuable property? When a nation-state uses bombs to destroy valuable property – say, instruments used to generate nuclear power – those bombs will very likely also kill human beings. Because the property is valuable, human beings will almost always be located nearby and so the bombs will destroy the people when they destroy the property. But some cyber-attacks destroy property without killing or maiming anyone.[15] The Stuxnet virus seems a case in point and perhaps a harbinger of things to come: my understanding is that the virus sabotaged extremely valuable centrifuges without mussing the hair on the head of a single human being.[16] This naturally raises the question: is the destruction of property ever a just cause for war? If so, when does the destruction of property provide a just cause for war? A number of legal theorists claim that significant property destruction can constitute an act of armed force that triggers a legal right to employ armed force in self-defense (Dinstein 2002: 100). So perhaps, as a matter of law, non-lethal property destruction can provide the target with permission to use military violence.

But then the law of war is more permissive than the morality of war: we ought not organize and carry out a plan to kill human beings merely to prevent the destruction of objects. Of course, all bets are off if the destruction of property also kills and maims human beings: if you destroy the respirator on which my life depends then you are not just destroying property. You are also killing me and others may deploy lethal violence to protect me to prevent you from ruining – or even disabling – my respirator. The claim is that the destruction of property as such – non-lethal destruction – does not provide a just cause for war. We must, however, understand this claim in a sufficiently nuanced manner. The destruction of property – and even the temporary disabling of property, such as an early warning system – that triggers a plan the successful enactment of which kills human beings might provide a just cause for war. If a cyber-attack that disables an air defense system is the triggering act of a planned invasion and conquest, and if the successful enactment of that plan would constitute a sufficiently egregious violation – one sufficient to provide a just cause, then the victim may respond with lethal military violence before any human being is harmed.[17]

Second, what is the moral status of the placement of so-called 'logic bombs?' A logic bomb is a piece of code inserted into a software system that can lie dormant and undetected for extended periods of time but that can be activated to perform some malicious function. As has been recently reported, US intelligence agencies discovered logic bombs implanted in our electrical grid that 'could be used to destroy infrastructure components' (US Electricity Grid Penetrated by Spies, *Wall Street Journal*, 9 April 2009). I take it that a cyber-attack that *actually* took down significant elements of the US electrical grid would constitute a just cause for war. But what about the insertion of code that could be triggered to do so? Does that provide a just cause for war? I doubt it. After all, there are now a number of nuclear weapons directed at our electrical grid (and much else!) that could be activated at a moment's notice. Despite the persistent, dormant, yet easily activated threat posed by those weapons, we have no just cause for war against the countries that have those weapons aimed at us. Neither, therefore, would the insertion of logic bombs in our electrical grid by those same countries provide us with just cause for war. For better or worse, different nation-states will invariably pose all manner of mortal threats to one another and those threats cannot by themselves provide a just cause for war.

Now we might think that these two kinds of threats differ in one morally relevant respect: the fact that hostile nation-states have aimed nuclear weapons at us need not violate our sovereignty – nuclear weapons are located within the confines of the hostile nation-states, whereas logic bombs that threaten our electrical grid must be located within our sovereign territory. Consequently, the latter are a particularly intrusive kind of threat – to which the vulnerable state may respond with proportionate and discriminate military violence.

But I find this response unpersuasive. First, I doubt that violations of sovereignty can bear the moral weight that this response requires – as a number of theorists have recently argued (Norman 1995, Rodin 2003), violations of sovereignty *as such* lack great moral weight. Second, the insertion of logic bombs need not violate sovereignty. No one needs to cross the border that defines the territory of a nation-state to insert a logic bomb, much less force their way in, for it could be the case that our own citizens purchase and use thumb drives that some hostile party has infected with the relevant malware. It could also be the case that they purchase compromised hardware that can be employed to deliver a destructive attack. If the voluntary, albeit unwitting, purchase of infected thumb drives or compromised hardware in the open market by the agents of a nation-state is a violation of its sovereignty, it is a violation too trivial to play any role in providing a just cause for war.[18]

Third, and finally, and ever so briefly, what is the moral status of a state's failure to prevent cyber-attacks that emanate from its sovereign territory? Suppose that the United States is the victim of a destructive cyber-attack on its electrical grid that is carried out by hackers who reside in China but who have not been authorized to launch that attack by the Chinese government. Do we nevertheless have just cause to attack China? There is a large, and I understand unsettled, legal literature on this question and that literature has clear relevance to the moral issues at stake (Sklerov 2009). I will not address this in any detail, other than to say that this question arises in a particularly interesting way for the United States. As I understand it, the United States is ideologically committed to refrain from exercising coercive control over the Internet. Our commitment to a 'free Internet' provides those resident in our country with considerable leeway to attack others. Now suppose that this commitment to a free Internet is morally justifiable – citizens *should* be given considerable freedom. Suppose again that some resident of the United States employs his considerable freedom to launch an attack on another country.[19] Could it be the case that the targeted nation has just cause to attack the United States for an attack that it did not sanction, but that is a predictable result of what is in fact a morally justifiable policy? This is only one of the many moral paradoxes engendered by the prospect of war in cyberspace.

Let me conclude with one final comment. It is not uncommon for advocates of the JWT to claim, as I have, that a just war requires 'a' just cause. But this convenience can mislead if understood to imply that the just cause requirement must be satisfied by one violation, much less some simple and easy-to-specify violation. Wars are complicated affairs, multiple distinct considerations will sometimes count for (and against) a given war, and so it is possible that a number of individually insufficient violations might combine to satisfy the just cause requirement. Perhaps we should understand each of the three kinds of cyber-attack just mentioned in this way: none can by itself satisfy the just cause requirement, but they can play a role in satisfying that requirement – so long as they are corroborated by other relevant considerations.

NOTES

1. This is a question that Bradley Strawser (2012) most ably addressed in a paper recently given at the 2012 McCain Conference on the ethics of cyberwar.
2. I provide a much fuller account in an unpublished manuscript, *God and War: An Exploration*. I should note that implicit in my understanding of the just cause requirement is a conception of justice that has been worked out in considerable detail by Wolterstorff (2010).
3. Many of its most prominent contemporary advocates take such a presumption to be an essential constituent of the JWT. Included in this number are particular theorists – James Childress, Rowan Williams, Robert B. Miller, and C. A. J. Coady – as well as institutional agents – most prominently, the US National Conference of Catholic Bishops. Some adherents of the JWT, notably James Turner Johnson (2009: 250), have decried its inclusion into the JWT as 'grievously wrong,' practically misguided, and unfaithful to tradition.
4. ... The Internet is one big masquerade ball. You can hide behind aliases, you can hide behind proxy servers, and you can surreptitiously enslave other computers to do your dirty work. This is a cruel irony: If you want to shield your identity as you go about your

business, it's extremely difficult. But if you want to hide your identity in order to attack a person or institution, it's unnervingly easy. (Brenner 2011: 32)

On the problem of attribution generally, see Dipert (2010).

5. In a recent address, Secretary of State Leon Panetta (2012) informs us that the US government has 'made significant investments in forensics to address th[e] problem of attribution' and that '[p]otential aggressors should be aware that the United States has the capacity to locate them and to hold them accountable for their actions that may try to harm America'.
6. Of course, there is always the possibility that the perpetrator of an otherwise unattributable cyber-attack admits to having done so, as the US government has admitted to having attacked Iran with the Stuxnet virus (Stunext Was Work of US and Israeli Experts, Officials Say, *Washington Post*, 1 June 2012). So long as that claim of responsibility is credible, the problem of attribution effectively evaporates.
7. Distributed denial of service "attacks," extraction or modification of information, website vandalism, as well as insertion of malicious code designed to damage, or destroy, data and systems, are all referred to as "cyber attack" or even "cyber warfare" regardless of whether these activities result in death, destruction of property, or merely the loss of information. Intrusions and other activity conducted by disgruntled employees, teen-age hackers, and criminals, are typically not distinguished from those by terrorists or foreign intelligence and military personnel. (Huntley 2010: 4)
8. Note that I am not claiming that the DDoS attacks on Estonia provided a just cause but that a military response would have been disproportionate. Rather, I deny that those wrongful attacks were sufficiently weighty as to provide a just cause for war. Bluntly put, the DDoS attacks on Estonia were similar in one morally relevant respect to the 'insult to Russian honor' that putatively provided the Russians with just cause to launch the DDoS attacks: even if the destruction of the commemorative statue dishonored Russia, it was not even remotely a weighty enough violation to provide a just cause for war. Of course, if a cyber-attack does provide a just cause for war, it is possible that war is nevertheless impermissible – should a military response violate the JWT's *ad bellum* proportionality requirement. My thanks to an anonymous referee for raising this point.
9. It might seem that the DDoS attack on Estonia was similar in relevant respects to an unjust blockade; at least it had blockade-like economic consequences. A nation-state that is targeted by a blockade might very well take itself to have a just cause for war. Consequently, a nation-state that is targeted by a DDoS attack, like that directed against Estonia, might very well take itself to have a just cause for war – as did some Estonians. But I doubt that the DDoS attack against Estonia was relevantly similar to the imposition of an unjust blockade. Indeed, it seems to have far more in common with the wrongful imposition of sanctions than with a blockade. Most pertinently, although blockades, sanctions, and DDoS attacks inflict economic harm, only blockades inflict economic harm by threatening lethal military violence. Sanctions and DDoS attacks inflict economic harm absent any lethal violence, threatened or otherwise. Plausibly, unjust DDoS attacks have roughly the same moral status as the imposition of unjust sanctions – so long as they inflict comparable economic harm. Given that the imposition of sanctions does not by itself provide a just cause for war, it is plausible to suppose that DDoS attacks on Estonia

cannot by themselves satisfy the just cause requirement. My thanks to an anonymous referee for raising this point.

10. There are many important moral questions raised by these kinds of 'communal conflicts short of war.' So far as I am concerned they lie beyond the purview of the JWT and – even more so – far beyond the scope of this paper.

11. Iran might have responded to the Stuxnet virus with malware of its own – the so-called 'Sharoom' virus 'virtually destroyed' some 30,000 computers owned by the Saudi Arabian Oil Company Aramco (Panetta 2012). If Sharoom was a response to Stuxnet, then it was almost certainly impermissible, as it failed to target any party to the Stuxnet attack. Nevertheless, *if* the Stuxnet attack was unjust, and *if* the Sharoom virus had been directed at Stuxnet's actual perpetrator(s), it might very well have been a morally permissible response. The destruction of 30,000 computers seems an entirely apt response to the unjust destruction of nuclear centrifuges.

12. The kind of threat inflation committed by the Estonian Speaker is, in my experience, exceedingly common in the ever-proliferating discussions of cyberwar in the media. So, for example, a news commentator responded to Secretary of State Panetta's (2012) recent warning about an impending cyber 'Pearl Harbor' by insisting that the United States had already committed a 'Pearl Harbor' of its own – the deployment of the Stuxnet virus to destroy Iranian centrifuges. But this assessment of the Stuxnet virus is morally obtuse – the two attacks have vastly differential normative weight – with respect both to their immediate destructiveness and to the plans they helped to enact.

13. Richard Clarke: 'As one pilot told me, "Aircraft these days, whether it's the F-22 Raptor or the Boeing 787 . . . all they are is a bunch of software that happens to be flying through the air. Mess with the software and it stops flying through the air"' (Clarke & Knake 2010: 174).

14. My sense is that those most concerned about the dangers of cyberwar have their eye on future attacks made possible by ever-increasing dependence on software, and are more than willing to admit that the cyber-attacks that have actually been carried out have had comparatively mild consequences. See Clarke & Knake (2010: 32).

15. As Randall Dipert (2010) has noted, a cyber-attack can also impede technology from performing its designated function.

16. I rely on the account provided in David Sanger (2012: 188ff).

17. What then of the Stuxnet virus? Because no human being was killed, because there is no indication whatsoever that the damage caused by the Stuxnet virus is in any way a prelude to the future killing of innocents, because the destruction inflicted by the Stuxnet virus was, and looks to be, merely economic, that cyber-attack provides the Iranians with no just cause for war. As I noted above, however, it might provide 'just cause' for retaliatory measures short of war.

18. A comparable point applies to the purchase of compromised hardware, as discussed by Randall Dipert in his essay in this issue of *Journal of Military Ethics*.

19. Apparently, 'patriotic hackers' from the United States launched DDoS attacks on Iranian government websites in order to 'thwart their ability to spread lies and propaganda' (Morozov 2011: 227).

REFERENCES

Brenner, Joel (2011) *America the Vulnerable* (New York: Penguin Press).

Clarke, Richard & Knake, Robert (2010) *Cyber War: The Next Threat to National Security* (New York: HarperCollins).

Davis, Joshua (2007) Hackers Take down the Most Wired Country in Europe, *Wired Magazine*, 21 August, 2007; accessed 10 February, 2013, available at: http://www.wired.com/politics/security/magazine/15-09/ff_estonia?currentPage=all; Internet.

Dinstein, Yoram (2002) Computer Network Attacks and Self-Defense, in: Michael N. Schmitt & Brian T. O'Connell (Eds), *Computer Network Attack and International Law*, pp. 99–119 (Newport, RI: Naval War College Press).

Dipert, Randall (2010) The Ethics of Cyberwarfare, *Journal of Military Ethics*, 9(4), pp. 384–410.

Dunlap, Charles J. (2011) Perspectives for Cyber Strategists on Law for Cyberwar, *Strategic Studies Quarterly*, 5(1), pp. 81–99.

Huntley, Todd (2010) Controlling the Use of Force in Cyberspace, *Naval Law Review* 60, pp. 1–40.

Jensen, Eric (2002) Computer Attacks on Critical National Infrastructure: A Use of Force Invoking the Right of Self-Defense, *Stanford Journal of International Law*, 38(2), pp. 204–240.

Johnson, James Turner (2009) Thinking Historically about Just War, *Journal of Military Ethics*, 8(3), pp. 246–259.

Morozov, Evgeny (2011) *The Net Delusion: The Dark Side of Internet Freedom* (New York: Public Affairs Books).

Norman, Richard (1995) *Ethics, Killing and War* (Cambridge: Cambridge University Press).

Panetta, Leon (2012) Remarks by Secretary Panetta on Cybersecurity to the Business Executives for National Security, New York City, accessed 19 March 2013, available at: http://www.defense.gov/transcripts/transcript.aspx?transcriptid=5136; Internet.

Rodin, David (2003) *War and Self-Defense* (Oxford: Oxford University Press).

Sanger, David (2012) *Confront and Conceal: Obama's Secret Wars and the Surprising Use of American Power* (New York: Crown Publishers).

Schmitt, Michael N. (1999) Computer Network Attack and the Use of Force in International Law: Thoughts on a Normative Framework, *Columbia Journal of Transnational Law*, 37(7), pp. 885–937.

Schmitt, Michael (2010) Cyber Operations in International Law: The Use of Force, Collective Security, Self-Defense, and Armed Conflicts, in: *Proceedings of a Workshop on Deterring Cyberattacks*, pp. 151–179 (Washington, DC: The National Research Council).

Sklerov, Michael (2009) Solving the Dilemma of State Response to Cyberattacks: A Justification for the Use of Active Defenses against States Who Neglect their Duty to Prevent, *Military Law Review*, 201(1), pp. 1–85.

Strawser, Bradley (2012) USNA 2012 McCain Conference: Cyberwarfare: Ethics of Application of Force, accessed 19 March 2013, available at: http://www.youtube.com/watch?v=voJSILP3I2Q; Internet.

Wolterstorff, Nicholas (2010) *Justice: Rights and Wrongs* (Princeton, NJ: Princeton University Press).

Christopher J Eberle (PhD UMass/Amherst) is Associate Professor of Philosophy at the United States Naval Academy where he has taught since 2001. He specializes in the area of religion and politics and publishes regularly in that area including *Religious Conviction in Liberal Politics* (Cambridge: Cambridge University Press, 2002); 'Let a Thousand Flowers Bloom: Religious Reasons in Public Life,' *Journal of Legal Commentary* (2007), and 'Consensus, Convergence, and Religiously Justified Coercion,'

Public Affairs Quarterly (2012). He was a Fellow at the Stockdale Center for Ethical Leadership during 2011 and 2012. He is currently completing a manuscript on religion and war, tentatively titled 'God and War: An Exploration.' *Correspondence Address*: United States Naval Academy, Department of Leadership, Education, and Development, Luce Hall, Annapolis, MD 21402-5022, USA. *Email Address*: eberle@usna.edu

IS STUXNET PHYSICAL? DOES IT MATTER?

Ryan Jenkins

University of Colorado, Boulder, CO, USA

Cyber-weapons are software and software, at least intuitively, is non-physical. Several authors have noted that this potentially renders problematic the application of normative frameworks like UN Charter Article 2(4) to cyber-weapons. If Article 2(4) only proscribes the use of physical force, and if cyber-weapons are non-physical, then cyber-weapons fall outside the purview of Article 2(4). This article explores the physicality of software, examining Stuxnet in particular. First, I show that with a few relatively uncontroversial metaphysical claims we can secure the conclusion that Stuxnet is physical. In particular, there exist instances of Stuxnet that are both located in space and causally efficacious, and this is very strong evidence for their being physical. Second, I argue that the question of physicality is actually irrelevant for the moral evaluation of an attack like Stuxnet because of its undeniably physical effects. Finally, I argue that some features of Stuxnet should make us optimistic about the prospects for discrimination and proportionality in cyberwarfare.

> [Stuxnet] is the first attack of a major nature in which a cyberattack was used to effect physical destruction . . . Somebody crossed the Rubicon.
>
> —Former Director of the Central Intelligence Agency Michael V. Hayden[1]

Introduction

Progress in ethics often depends on settling a prior metaphysical question. Military ethicists may find themselves in such a situation regarding the ethics of cyberwarfare. The just war tradition has been developed around incontrovertibly *physical* means of waging warfare: shooting bullets, dropping bombs, etc. But sensational attacks like Stuxnet seem to usher in a new era of *non*-physical warfare: the virtual genie has been let out of the bottle and there is no putting him back in.

How should we treat efforts to sabotage enemies when those efforts reside *entirely in virtual space*?[2,3] Using Stuxnet as a case study, I will show that this metaphysical question is sometimes irrelevant for the purpose of evaluating cyber-attacks: regardless of whether Stuxnet is a physical thing, it falls under the purview of the traditional just war framework.[4,5]

The advent of cyber-weapon technology has given us pause and we are hesitant to apply our current normative frameworks to cyber-weapons without alteration.[6] Lucas (2010: 294) warns that this 'relentless technological development,' including the development of cyber-weapons, poses a challenge for policy makers and ethicists in the coming years. Among the several technologies that Lucas (2010: 294) has in mind, cyberwarfare remains 'the least thoroughly examined from the standpoint of ethics.' Franzese (2009: 6) laments that 'no consensus has developed in answering these questions. More troubling, various commentators still hold widely divergent views on basic, fundamental questions.' The problem here is that technological development has outpaced the development of our normative theories.[7]

One significant feature of cyber-weapons that makes them different is their apparently non-physical nature. If cyber-weapons are not physical entities, perhaps they are not subject to the same ethical constraints as conventional weapons (i.e. weapons used to wage war in meatspace[8]). In the words of Dipert (2010: 404), 'cyberattack technology is more like an idea than like a physical thing.'[9,10] Lucas observes that the non-physical nature of cyber-weapons seems to place them outside the jurisdiction of the laws of armed conflict (LOAC).[11] And in his seminal article on the ethics of cyberwarfare, Schmitt (1999: 912) also worries that cyber-attacks exist in 'a normative void.' Finally, Taddeo (2012: 212) calls these problems generated by cyber-weapons' non-physical nature 'emblematic of the problems encountered in the attempt to establish ethical guidelines' to govern this new technology.

Much of the scholarship dealing with the legal aspects of cyberwarfare has been focused on Article 2(4) of the UN Charter, which bans signatories from 'the threat or use of force' in their relations with one another.[12] Benatar (2009: 381) explores the difficulties in applying this article to cyberwarfare, the first of which is that 'an overview of some major dictionaries of the English language' yields a narrow understanding of 'force' as designating *physical* force. Jolley (2012: 11) runs into the same problem, finding that the most natural reading of *force* in Article 2(4) is physical force. In Schmitt's (1999: 908) view, Article 2(4) refers to armed force, 'i.e., physical or kinetic force applied by conventional weaponry.' That definition would seem to exclude non-physical means of waging war, for example, the use of cyber-weapons.[13] These observations raise the concern that *non-physical force* might not even be a coherent concept. If *jus ad bellum* and the LOAC concern only the use of physical force, then cyber-weapons seem to generate serious problems for our traditional critical framework, or even demand a new normative framework altogether (see Schmitt 1999: 934). This seems to depend on whether cyber-weapons turn out to be *physical things*.

Stuxnet presents a particularly interesting test case for the ethics of cyberwarfare in this regard since it is one of the first (known) weapons – and the most spectacular to date – to jump the gap from wreaking mere *cyber*-havoc to wreaking *physical* destruction.[14] In the first part of this article, I explore whether Stuxnet is a physical thing. There are two good reasons for answering in the affirmative: (1) that with a few relatively uncontroversial metaphysical premises we can secure the conclusion that instances of Stuxnet exist at a physical location; and (2) that these instances entered into the causal order with other physical objects. In the second part of this paper, I grant that these arguments may be mistaken and I consider the implications both for the assumption that Stuxnet is physical and that it is non-physical. This turns out not to make a difference to the moral status of its use. I will argue that Stuxnet is a special case of a cyber-weapon effecting physical change, which makes it much simpler to evaluate under existing normative frameworks. Finally, I will briefly argue that Stuxnet should make us optimistic about the prospects for proportionality and discrimination in cyberwarfare.

Is Stuxnet Physical?

Stuxnet is a piece of software. There are both physical and non-physical senses of *software*, so let us examine which is more appropriate in this case.

The arguments for Stuxnet's physicality are more compelling and will be examined first. Suppose Stuxnet is just some particular pattern or arrangement of bits. In this case, Stuxnet is an immaterial type whose instances are *realized* whenever some bits are arranged Stuxnet-wise (e.g. whenever an instance of Stuxnet conforms to the environment in which it finds itself). Moreover, it is relatively uncontroversial that instances are located where their realizers are located, so that these individual instances of Stuxnet also have a physical location. (Appealing to our linguistic practices, it would not be unusual for someone to point to an infected machine and say: 'A copy of Stuxnet is *there*, *on* that machine.') Finally, it is also relatively uncontroversial that having a physical location is sufficient for being a physical thing. This is one good reason for believing that there are physical instances of Stuxnet.

Second, these individual instances are incontrovertibly causally efficacious. The argument here need not be very long: it is precisely because some instances of Stuxnet have entered into the causal order in a rather dramatic way that we are having this conversation in the first place. Because these individual instances are both located in space and causally efficacious, we now have two good reasons for concluding that they are physical things.

There remains an ambiguity to clear up. Above I claim that Stuxnet is a non-physical type that has physical tokens. When we ask whether 'Stuxnet is physical,' are we referring to the physical or the non-physical things? The correct answer seems to be this: the 'Stuxnet' that *actually did damage* to Iran's nuclear enrichment facilities was one of the physical tokens of Stuxnet. If need be, we can even strengthen our argument by jettisoning a commitment to the existence of an immaterial Stuxnet-type altogether. Some theorists deny that abstract objects exist at all. These theorists might claim that Stuxnet goes out of existence when all of its physical instances do. We can make do entirely without a Stuxnet-type since we are certain, at least, that there are causally efficacious instances of Stuxnet in the world and we know the thing that *actually* did damage to Iran's centrifuges was one of these physical instances. The Stuxnet we are interested in is (some particular) physical token. In short: it seems we can confidently say that there are instances of Stuxnet that are located in space and are incontrovertibly causally efficacious, and so are physical things.

On the other hand, there persists a strong intuition that software must be non-physical. Rowe (2010: 21) offers that software is 'just bit patterns.' Since patterns themselves are non-physical things, software must be non-physical. Second, software may be understood as a set of instructions given to a computer to execute. In either case, software is a non-physical thing. This coheres well with the folk understanding of software as immaterial, especially when contrasted with hardware. Lucas (2010) and Cook (2010) are helpful in illuminating the source of this common view. According to them, the belief in a sharp metaphysical distinction between cyberspace and the physical realm is a little-examined background assumption, reinforced by the typical uses of cyber-weapons. Lucas (2010: 296) says:

> . . . cyber warfare experts deal almost exclusively in software: their weapons are "virtual" rather than kinetic (although some can be made to do a great deal of physical damage). Theft of data, denial of access, and most of all, deception and psych-ops to sow confusion and demoralization – all these are tools of the intelligence and espionage community, rather than of the combat warrior.

Cook (2010: 414) adverts to the same assumption: that cyberspace's relationship to the outside world is strongly dualist. The one is non-physical, the other physical, and never the twain shall meet. He even goes so far as to compare this view to Cartesian dualism, which holds that there is an essential difference between mental and physical substances.

Ultimately, though, these intuitions are not strong enough to overturn the metaphysical arguments given above. The weight of the reasons supports the view that Stuxnet is a physical thing. In fact, these same arguments seem to show that *all* cyber-weapons are physical things. I am not the first to make a claim like this. Liles et al. (2012: 4) asserts with little argument that cyberspace is a physical terrain and Cook (2010: 414) later plainly states that cyber-weapons are physical things. Regardless, I am inclined to see this as a surprising conclusion. Although I am swayed by the arguments that entail this stronger view, I must leave a fuller exploration for another time. However, luckily for us, whether our current ethical framework is equipped to deal with Stuxnet may not depend on the question of its physicality after all.

Does It Matter?

Either Stuxnet is a physical thing or it is not. Suppose Stuxnet is a physical thing. If this is true, then it is just one more physical attack, and it falls comfortably under *jus ad bellum* as an instance of the use of physical force against an adversary. This much should be straightforward. The more interesting possibility – and the impetus for our discussion – is that Stuxnet is not a physical thing.

Suppose Stuxnet is not a physical thing. Even if this were true, it would still be possible to evaluate the moral status of Stuxnet under our current framework. That is, the bare non-physicality of an attack is not morally relevant. Imagine a hypothetical case of physical aggression that parallels Stuxnet as closely as possible: a commando raid on Iran's uranium enrichment facilities at Natanz that successfully destroys several hundred uranium enrichment centrifuges with no loss of life, harm to civilian infrastructure, or other collateral damage. This analogy is appropriate for examining whether the bare difference between physical and non-physical means for accomplishing the same goal can make a difference to whether an action falls under the purview of *jus ad bellum*. Such a commando raid clearly falls under *jus ad bellum*, and so if there are no relevant differences between it and Stuxnet, we must render the same verdict in the latter case. Let us examine some possible morally relevant differences.

First, we might think that the operation of a worm like Stuxnet is different from a commando raid, which more obviously involves human agency. But upon closer inspection there is no important difference with regard to human agency between these two cases. The same constraints govern launching a cruise missile as do sending in a targeted commando raid. In fact, these cases elucidate the real nature of the difference here: the extent to which human agents are involved in these two cases of aggression is a difference in degree, not in kind. The ways in which humans can wage war vary in their intimacy, from a punch or a stab, to pressing a button to launch a missile, to coding a cyber-weapon that wreaks destruction just as physical as the previous kinds of attacks. The clearest way of understanding this continuum of intimacy is that these actions vary in the number of causal links between some human action and some destructive effect. But the mere number of causal links in a chain does not seem relevant to the wrongness of an action. This first proposed difference is not morally salient.

Second, perhaps it is relevant that a commando raid endangers American lives while Stuxnet does not. In fact, deploying Stuxnet could plausibly risk American lives in at least two ways. First, unleashing Stuxnet may provoke retaliation against American interests.[15] Second, it is conceivable that Stuxnet could break free of its creators' control and wreak havoc on American computer systems, for example, inadvertently corrupting or erasing sensitive military data. Insofar as both of these cases risk some kind of harm to American interests, however, they are morally similar.

What about the fact that a commando raid risks American lives as part of the operation *itself*? That also seems irrelevant. If anything, the lower risk to American lives that Stuxnet represents in comparison to conventional warfare may be taken as a beneficial feature. On the other hand, such a reduced risk may lower the psychological barrier[16] to ordering such operations and this may foreshadow increasing militaristic adventurism or executive overreach. Those trends are indeed worrisome, but others see the reduced risk to American lives as a clear reason to prefer new technologies such as unmanned aerial drones.[17] Those same arguments ought to extend to the use of cyber-weapons. Regardless, it is unclear why a reduced risk to American lives as part of the operation itself should exclude Stuxnet from evaluation under *jus ad bellum* rather than being an advantage of the weapon.

Third, perhaps because Stuxnet is not physical then its deployment cannot count as a violation of Iran's territorial integrity, since it is impossible for something non-physical like software to *invade* a physical space. Since the commando raid more obviously represents such an invasion, there might be an important difference here. But this suggestion is mistaken, as Iran's cyber-territory is properly regarded as *theirs* in the same way that their physical territory is theirs. In this way, Stuxnet constitutes a very real intrusion into Iran's cyber-territory.[18] The notion of cyber-territory may be perplexing at first. But we should be confident in constructing a basic analogy to physical territory.[19] States have a right to the privacy and integrity of their digital resources. Snooping on their electronic communications is analogous to intercepting their mail; sneaking into their hard drives is analogous to rifling through their filing cabinets; and disrupting the operations of their centrifuges may be thought analogous to crossing some physical wires to achieve the same result.

It will be useful to develop this analogy further. In the above example involving a commando raid, it is clear *when* an aggression has taken place: it takes place precisely upon breach of the physical territory of the host country. Borders are easy to locate in the physical realm, but in cyberspace we are vexed by nebulous or even purportedly non-existent boundaries.[20] Pinpointing *when* an aggression has taken place in cyberspace is a more difficult project. Because this project is worthwhile, I will make some suggestive remarks regarding the precise point at which we might *locate* this aggression.

Barney (2001) develops an analogy between cyber-rights of passage and the naval rights of passage outlined in United Nations Convention on the Law of the Sea (UNCLOS III), and so is particularly helpful here. Barney (2001: 63) introduces 'internal cyberspace' as an analogue to a state's internal waters: an area over which a state 'may exercise complete sovereignty.' Internal cyberspace includes Internet infrastructure essential to the proper functioning of information services that bear on national interests, such as finance, transportation, and defense. Importantly, 'Internal Cyberspace is that medium serviced by the State's national telecommunications infrastructure that is normally only accessible to authorized users,' implying that states have already taken measures to institute restrictions on who can access this region of cyberspace (Barney 2001: 63).[21]

Outside states are not allowed just any passage through a state's internal cyberspace. Among the kinds of impermissible passage are passage involving 'any exercise or practice with weapons of any kind' and any passage 'aimed at interfering with any systems of communication or any other facilities or installations of the coastal State' (Barney; 2001: 70, n. 37). These exceptions are analogous to exceptions to the rights of sea passage outlined in UNCLOS III Article 19(2): (b) and (k), respectively. Stuxnet seems to fall afoul of both of these exceptions.[22] It should be no surprise that, on any plausible theory of cyberspace passage, Stuxnet's intrusion into Iranian cyber-territory represented wrongful trespass. This is because Stuxnet traveled on Iran's internal cyberspace without authorization and with malicious intent.

What is left to discern is that precise point at which an 'aggression' took place. By the time Iranian centrifuges were spinning out of control, an aggression had clearly taken place. It probably took place earlier, when Stuxnet found its way onto the computers at Natanz. And it might have taken place even earlier if Stuxnet traveled over other communications infrastructure such as Iran's fiber-optic cables, routers, network bridges, etc. To choose a precise point at which the aggression took place, we would have to push this physical analogy to its limit. We might say that Stuxnet's passage through cyberspace constituted an aggression the instant that it made use of Iran's network infrastructure, and this is probably at the physical territorial boundary of Iran.[23] If we have to choose a precise physical point, this seems as good as any other.[24] To be more specific, the wrongful trespass of Iran's cyber-territory probably took place on either the first length of cable within Iran's physical territory, or, perhaps, at the first tangible element of Iranian Internet infrastructure.

But even if we find the notion of cyber-territory dubious, or are unable to locate the precise point at which a violation of such territory took place, Stuxnet still manifests in physical ways within Iran's *physical* territory. And so we need not be bothered by this purported difference regarding territorial violation.

There seem to be no differences left for us to investigate, so let us confront the question head on: can the bare difference in physicality make a difference to an action's moral status? It seems it cannot, and this is most clearly the case when a purportedly non-physical weapon succeeds in wreaking physical havoc. What seems most morally relevant in these two cases is not whether the means were physical, but whether an action causes damage to a state's tangible assets. That factor is present in both the commando raid and in the case of Stuxnet, and it is hard to see how the non-physical nature of the means can make a moral difference. Cook (2010: 416) agrees, saying 'the means of attack are of secondary importance; the effect – the harm suffered – is what matters most.' Schmitt (1999: 912) sees the intent of Article 2(4) as proscribing actions that are most likely to disrupt peace, security, international cooperation, and other values shared among states and this leads him to conclude that the means of some attack are largely irrelevant. 'What matters, then, are consequences' of some action, rather than the particular instrument, and the locutions 'use of force' and 'armed force' are merely shorthand. In sum, what makes an attack important is its effect, not its means.

Whether Stuxnet is physical is an interesting metaphysical question. However, if my arguments are right, it need not be of interest to military ethicists because it would not make a difference to its moral status. If Stuxnet is a physical thing, it straightforwardly falls under the purview of *jus ad bellum*, especially since it caused physical damage. If Stuxnet were not physical, however, we should still not be deterred. Such a finding is no impediment to evaluating its moral status since a bare difference in the physicality of the means does not make a morally relevant difference to whether an act counts as an aggression.

Stuxnet and the Prospects for a Just Cyberwar

In the end, examination of Stuxnet, I believe, should have us cautiously optimistic regarding the prospects of restraint in cyberwarfare. Here I examine Stuxnet's implications for the *in bello* requirements of discrimination and proportionality.

Stuxnet should lead us to believe that cyber-weapons may excel at minimizing collateral damage. This claim depends in turn on our confidence in the reliability and controllability of cyber-attacks. Rowe (2010) is decidedly pessimistic on this point. After detailing the problems that confront any new military technology, Rowe remarks that it is unlikely that cyber-weapons will *ever* improve in reliability and effectiveness. In support, Rowe (2010: 24) notes that, 'software in general is not getting any more reliable.' While Rowe may be right about the progress of commercial and industrial software, we can plausibly expect cyber-weapons like Stuxnet to be significantly more reliable, since their authors may take special care to control their propagation and destructiveness. Much of this is motivated out of a concern for maintaining stealth in order to ensure success, but the effect is the same: cyber-weapons may demonstrate markedly improved reliability over software in general. In fact, I believe Stuxnet gives us some reason to be hopeful regarding cyber-weapons' capacity for discrimination. Stuxnet contained an impressive array of features that seem designed specifically to limit its spread. For example, each instance of Stuxnet attempts to propagate for only 21 days, only to three other computers, and only if those other computers are running particular software manufactured by Siemens to interface with specific programmable logic controllers.[25,26] These observations underscore cyber-weapons' potential for restraint in wartime.

There are two caveats worth mentioning. The first is standard: that a weapon is only as morally good as the actor who wields it.[27] Cyberwarfare also presents us with frightening possibilities of destruction unleashed by non-state actors, especially because they would likely be much less interested in authoring something as self-restrained as Stuxnet.[28] Second, too much optimism threatens us with the threshold problem: it may be psychologically easier to deploy cyber-weapons, in turn further eroding the hope that war be a last resort. But cyberwarfare also promises the possibility of an age of precise warfare that is truly unprecedented. After all, it is inconceivable that we might design a bomb that we know will *only* damage, for example, computers at a particular range of Internet protocol (IP) addresses that use a particular version of Siemens's supervisory control and data acquisition software to interface with a particular programmable logic controller. This promise of weapons with more robust discrimination capabilities is worth noting.[29]

Examination of Stuxnet also reveals that cyber-weapons offer prospects for improving the proportionality of warfare. Stuxnet included capabilities for communicating with 'command and control servers in Malaysia and Denmark' (Chen 2010: 3), which raises the possibility of continuous two-way communication and oversight. Physical damage done by a cyber-weapon cannot be undone, of course, but damage in progress could be halted, redirected, or otherwise modified in an attempt to reduce its impact. Rowe (2010) gives several recommendations for ethical constraints on cyber-weapons. Among these are that cyber-weapons should be easily controllable and stoppable (e.g. in the case of enemy surrender). Stuxnet's ability to 'phone home' demonstrates that these criteria can be met in practice. Unlike conventional weapons, cyber-weapons can be redirected after they are deployed or aborted at a moment's notice. These capabilities could be harnessed to improve both the discrimination and proportionality of cyberwarfare.

In short, the most optimistic appraisal of cyberwarfare is that it promises a way to vastly improve our abilities to satisfy the requirements of proportionality and discrimination. While that may seem rose-tinted when considering the range of possible cyber-weapons that may be unleashed by the range of possible actors, Stuxnet is proof that cyber-weapons *can* be more morally sound than conventional weapons ever could.

Conclusion

Cyberwarfare presents ethicists and policy makers with a host of unprecedented challenges. Here, I have suggested that not all of these challenges are insurmountable. Worries arising from the incoherence of non-physical force expressed by the authors above are misplaced at least in the case of Stuxnet and perhaps in the case of all cyber-weapons. Even if those arguments are mistaken, and Stuxnet is non-physical, I sided with Cook (2010) and Schmitt (1999) in arguing that Stuxnet's effects – rather than its means – are of primary moral importance. There are no morally relevant differences between the damage that Stuxnet did and the damage done by a hypothetical commando raid on the enrichment facilities at Natanz. If we judge that the latter falls under the purview of *jus ad bellum*, so does the former. Finally, several particular features of Stuxnet should raise our confidence in the prospects for discriminate and proportional cyberwarfare and promise an improvement over conventional warfare in these regards.

ACKNOWLEDGEMENTS

I am indebted to many for their advice on previous drafts. Thanks to Alastair Norcross for his comments while this paper was in its formative stages. Thanks to Robert Rupert, Kathrin Koslicki, Leonard Kahn, and Andrew Chapman for their helpful advice in developing the arguments concerning Stuxnet's physicality. Special thanks are owed to Panayotis A. Yannakogeorgos for his assistance with the technical aspects of the paper. Finally, I am indebted to guest editor Bradley Jay Strawser and two anonymous editors for this journal, as well as Donald Joy and Ben Hale for their helpful remarks on an earlier draft. Responsibility for any remaining mistakes or infelicities lies solely with me.

NOTES

1. We now know who the 'somebody' is to whom General Hayden refers. According to the *New York Times*, Stuxnet was produced in cooperation by American and Israeli cyberwar divisions (Obama Order Sped up Wave of Cyberattacks against Iran, *New York Times*, 1 June 2012).
2. It is worth noting that this move from physical to non-physical warfare is unprecedented. That is, even non-conventional weapons like nuclear, biological, or chemical weapons are deployed in the physical world. On this point, see Dipert (2010: 385).

3. It is true that cyberspace depends on a physical infrastructure for its existence. And it is true that computational work takes energy and must, of course, obey the physical laws of nature (Landauer 1991). See also Yannakogeorgos (2012: 103–104) on the physical characteristics of activity in cyberspace. But there is room for legitimate uncertainty regarding the metaphysical nature of cyber-weapons themselves.
4. It is probably most appropriate to regard Stuxnet and other cyber-attacks as falling under Walzer's (2006: xv) framework of justice in the use of 'force-short-of-war' or *jus ad vim*. There, Walzer urges the development of a *jus ad vim* framework for evaluating methods often used to effect 'regime change' that seem short of war, such as 'smart sanctions,' blockades, and the enforcement of a no-fly zone. It is worth noting that all of these actions fall along a continuum running from non-physical to physical means of aggression. (Stuxnet is interesting in part because it is unclear where we should locate it along this continuum.) Viewed in this light, Stuxnet may represent an extra-judicial means of enforcing UN Resolution 1696 to prevent Iran from enriching uranium: the pursuit of policy by other, *other* means, if you will.
5. I would like to bracket the question of whether Stuxnet itself was justified, i.e., whether Stuxnet passes the requirements of just cause, proper authority, etc. (For what it is worth, Sanger indicates that the United States and Israel seemed to believe Stuxnet was a last resort (Obama Order Sped up Wave of Cyberattacks against Iran, *New York Times*, 1 June 2012).) Moreover, the question of *whether* Stuxnet represents an aggression (or the use of force) seems prior to the question of whether it was justified.
6. On the other hand, the United States officially reserves the right to respond to cyber-attacks with military force (United States 2011: 13). This is an explicit statement that the US regards a sufficiently severe cyber-attack as tantamount to conventional military aggression. I am suggesting we should be careful about such an equivalence.
7. See Franzese (2009: 6): 'the current international legal paradigm predates cyberspace and cannot adequately address the various issues raised by cyberspace.'
8. William Gibson coined both the terms 'cyberspace' and its opposite 'meatspace' in his 1984 novel *Neuromancer*. Suffice to say only the former has found its way into the vernacular.
9. It has been suggested to me that Dipert is perhaps merely being metaphorical here. At any rate, he and the other authors cited are, I contest, latching onto a serious intuitive worry about applying existing frameworks to cyberwar.
10. Schmitt (1999: 888) expresses a similar sentiment: '. . . in the twenty-first century an attack may be nothing more than the transfer of cyber commands from one computer to others.'
11. While the LOAC are not identical to the ethical constraints of just war theory, they serve a similar purpose: to make war, although it is hell, at least a '*rule-bound* hell,' in the words of Orend (2006: 134, emphasis mine). We should therefore expect that these two parallel projects would be beset by similar theoretical problems.
12. Authors have also grappled with the implications of cyberwarfare for UN Charter Article 51, which permits member states to defend themselves only when they have been the subject of an 'armed attack.' With little modification, the worries expressed here regarding the 'use of force' could apply to 'armed attacks' as well.
13. See, for example, Antolin-Jenkins (2005: 134): 'The use of terms of war in cyberspace operations obscures the reality that cyberwar does not fit well into the legal frameworks on war and use of force.'
14. Stuxnet therefore seems to occupy a middle position between the physical and non-physical. Jolley (2012: 24) notes specifically that there exists no consensus about whether Stuxnet's attack on Natanz 'actually constitutes force.'

15. In fact, there is some evidence that Stuxnet did indeed provoke cyber-retaliation against Saudia Arabia and the United States. According to *New York Times* (US Suspects Iran Was Behind a Wave of Cyberattacks, *New York Times*, 13 October 2012):

> American intelligence officials are increasingly convinced that Iran was the origin of a serious wave of network attacks that crippled computers across the Saudi oil industry and breached financial institutions in the United States, episodes that contributed to a warning last week from Defense Secretary Leon E. Panetta that the United States was at risk of a 'cyber-Pearl Harbor.'

The article talks of 'an emerging shadow war of attacks and counterattacks already under way between the United States and Iran in cyberspace,' of which Stuxnet was surely a part, if not the opening salvo.

16. This concern, known as the 'threshold problem,' is explored more below. Lucas (2010: 294) apparently coins the term as it is being used here.
17. See also Strawser (2010).
18. See also Rowe (2010: 22), who asserts that the use of cyber-weapons 'necessarily involves trespassing.'
19. In fact, arguing from analogy is a popular strategy throughout the ethics of cyberwar literature. Cook (2010), Rowe (2010), Franzese (2009), and others have all used this strategy, perhaps as a tentative and temporary answer to the theoretical lacuna that seems to exist in our normative framework.
20. People sometimes colloquially speak as if there are in fact 'no boundaries' in cyberspace. Witness John Perry Barlow's (1996) oft-cited 'A Declaration of the Independence of Cyberspace':

> Governments derive their just powers from the consent of the governed. You have neither solicited nor received ours. We did not invite you. You do not know us, nor do you know our world. Cyberspace does not lie within your borders . . . Ours is a world that is both everywhere and nowhere, but it is not where bodies live.

21. Barney (2001: 65) continues:

> For example, some Internet sites are restricted to authorized users who register, obtain a password, or pay a fee to view materials or buy products or services on the site. Commanders conducting information operations should probably consider these types of owner/operator restrictions, by password or otherwise, as *prima facie* evidence that the site is within the internal Cyberspace of a State.

The fact that the nuclear facilities at Natanz were air-gapped is similarly strong *prima facie* evidence that they should be regarded as within Iran's internal cyberspace.

22. Although, interestingly, under Barney's (2001: 74) framework, Stuxnet's passage through neutral intermediaries does not violate their neutrality.
23. This has led some, following an analogy to the physical world, to propose Internet border inspections. See, for example, Franzese (2009: 39):

> . . . states must also be able to establish a cyberspace border that a state can both monitor and control. Without the capability to perform this function, the concept of sovereignty in cyberspace is meaningless. One approach is to establish Internet

border inspections similar to the physical border crossings that exist today. This approach relies on the limited number of entry and exit points that route Internet traffic in and out of the United States.

24. There is one noteworthy caveat, however: the facilities at Natanz were air-gapped, and so it is unclear whether this talk of Stuxnet transiting cables is applicable in the present case (Farwell & Rohozinski 2011: 24). However, tens of thousands of computers inside Iran were infected with Stuxnet, and so presumably *some* of these instances entered the country the 'old-fashioned way' through fiber-optic cables or other physical Internet infrastructure.
25. For an excellent explication of these and other features of Stuxnet, see Chen (2010) and Farwell and Rohozinski (2011).
26. Stuxnet did infect many computers that it was apparently not meant to, but it is unclear whether it damaged any of them. There was some speculation that Stuxnet was responsible for a loss of power on India's Insat-4 B satellite in July 2010 (Farwell & Rohozinski 2011: 34). The Indian Space Research Organization has since ruled out that possibility (Cyber Threat: Isro Rules out Stuxnet Attack on Insat-4 B, *Economic Times*, 12 October 2010).
27. Lucas (2010: 297) agrees that cyber-weapons, 'if rightly handled,' could render warfare more proportional and discriminate.
28. Admittedly, designing a sophisticated cyber-weapon seems well outside the capabilities of non-state actors at present. However, the technology and expertise are becoming increasingly common. Farwell and Rohizinski (2011: 25) are already convinced that Stuxnet is less impressive than many observers believe, implying that constructing something like it may already be within the reach of cyber-criminals:

    > Some of [Stuxnet's] core technical characteristics, including the use of a DNS-based command-and-control network, make it less stealthy than much of the more advanced malware that criminals use. Stuxnet's core capabilities and tradecraft, including the use of multiple zero-day exploits, render it more of a Frankenstein patchwork of existing tradecraft, code and best practices drawn from the global cyber-crime community than the likely product of a dedicated, autonomous, advanced research programme or "skunk works". Nor is Stuxnet particularly innovative.

    See also Schmitt (1999: 897): 'the knowledge and equipment necessary to mount a computer network attack are widely available; CNA is quite literally "war on the cheap".'
29. As Farwell and Rohozinski (2011: 34) note: 'One important benefit of cyber attack, to be sure, may be its greater opportunity to achieve goals such as retarding the Iranian nuclear programme without causing the loss of life or injury to innocent civilians that air strikes would seem more likely to inflict.'

REFERENCES

Antolin-Jenkins, Vida M. (2005) Defining the Parameters of Cyberwar Operations: Looking for Law in All the Wrong Places, *Naval Law Review*, 51, pp. 132–174.

Barlow, John Perry (1996) A Declaration of the Independence of Cyberspace, 8 February; accessed 20 March 2013, available at: https://projects.eff.org/~barlow/Declaration-Final.html; Internet.

Barney, Steven M. (2001) Innocent Packets: Applying Navigational Regimes from the Law of the Sea Convention by Analogy to the Realm of Cyberspace, *Naval Law Review*, 48, pp. 56–83.
Benatar, Marco (2009) The Use of Cyber Force: Need for Legal Justification?, *Goettingen Journal of International Law*, 1(3), pp. 375–396.
Chen, Thomas M. (2010) Stuxnet, the Real Start of Cyberwarfare?, *IEEE Network*, 24(6), pp. 2–3.
Cook, James (2010) Cyberation' and Just War Doctrine: A Response to Randall Dipert, *Journal of Military Ethics*, 9(4), pp. 411–423.
Dipert, Randall R. (2010) The Ethics of Cyberwarfare, *Journal of Military Ethics*, 9(4), pp. 384–410.
Farwell, James P. & Rohozinski, Rafal (2011) Stuxnet and the Future of Cyber War, *Survival: Global Politics and Strategy*, 53(1), pp. 23–40.
Franzese, Patrick W. (2009) Sovereignty in Cyberspace: Can It Exist? *Air Force Law Review*, 68(1), pp. 1–42.
Gibson, William (1984) *Neuromancer*, 1st ed. (New York: Ace Books).
Jolley, Jason D. (2012) Article 2(4) and Cyber Warfare: How Do Old Rules Control the Brave New World? 12 July; accessed 20 March 2013, available at: http://papers.ssrn.com/sol3/papers.cfm?abstract_id=2128301; Internet.
Landauer, Rolf (1991) Information is Physical, *Physics Today*, 44(5), pp. 23–29.
Liles, Samuel, Rogers, Marcus, Dietz, J. Eric, & Larson, Dean (2012) Applying Traditional Military Principles to Cyber Warfare. Paper presented at 2012 4th International Conference on Cyber Conflict, Tallinn, 5–8 June.
Lucas, Jr, George R. (2010) Postmodern War, *Journal of Military Ethics*, 9(4), pp. 289–298.
Orend, Brian (2006) *The Morality of War* (Peterborough: Broadview Press).
Rowe, Neil C. (2010) The Ethics of Cyberweapons in Warfare, *International Journal of Technoethics*, 1(1), pp. 20–31.
Schmitt, Michael (1999) Computer Network Attack and the Use of Force in International Law: Thoughts on a Normative Framework, *Columbia Journal of Transnational Law*, 37(3), pp. 885–937.
Strawser, Bradley J. (2010) Moral Predators: The Duty to Employ Uninhabited Aerial Vehicles, *Journal of Military Ethics*, 9(4), pp. 342–368.
Taddeo, Mariarosaria (2012) An Analysis for a Just Cyber-Warfare. Paper presented at 2012 4th International Conference on Cyber Conflict, Tallinn, 5–8 June.
United States (2011) International Strategy for Cyberspace: Prosperity, Security, and Openness in a Networked World, accessed 20 March 2013, available at: http://www.whitehouse.gov/sites/default/files/rss_viewer/international_strategy_for_cyberspace.pdf; Internet.
Walzer, Michael (2006) *Just and Unjust Wars*, 4th ed. (New York: Basic Books).
Yannakogeorgos, Panayotis A. (2012) Internet Governance and National Security, *Strategic Studies*, 6(3), pp. 102–125.

Ryan Jenkins is a doctoral candidate in philosophy at the University of Colorado Boulder. His main research interests are in military ethics and consequentialism (especially rule-consequentialism). He has contributed to Springer's *Encyclopedia of Global Justice* and has published on distributive justice and genocide. *Correspondence Address*: Philosophy Department, UCB 232, Boulder, CO 80309-0232, USA. *Email Address*: Ryan.R.Jenkins@Colorado.edu

TWENTY YEARS OF CYBERWAR

John Arquilla

United States Naval Postgraduate School, Monterey, CA, United States

For many years after its introduction in the early 1990s, the cyberwar concept - which outlined a new mode of conflict that would emphasize disruption of information systems and flows - was given little credence. Over the past several years, however, cyberwars of both irregular and somewhat more conventional types have erupted (e.g., see respectively Estonia in 2007 and Georgia in 2008). Global awareness of cyberwar has risen sharply, and many nation-states are preparing their defenses, as well as their capabilities for mounting offensive operations. The American military's declaration in 2011 that cyberspace is a 'warfighting domain' highlights a need to explore the ethical implications and nuances of cyberwar. This article finds that classic jus ad bellum constructs come under great pressure from cyberwar, while jus in bello concerns may prove more manageable. Another key theme is that disproportionate attention is given to the notion of employing cyberwar 'strategically' (i.e., to strike directly at other nations' infrastructures), where its use is less likely to achieve 'victory' and is more problematic ethically. Instead, a focus on the application of cyberwar techniques in battle may lead to shorter, less bloody, 'more ethical' conflicts.

When David Ronfeldt and I introduced the concept of cyberwar in our 1992 RAND paper,[1] we emphasized the notion that the information revolution's effects upon militaries were going to be double-edged. We foresaw that advanced communications, sensing, and weapons guidance systems would tremendously empower armed forces – but also that growing reliance upon these technologies could, if disrupted, gravely imperil them in battle. The idea being that striking, via cyberspace or by other means, at key information-based control functions would have crippling effects on armies, fleets, and air forces – in a manner similar to the way in which classic *blitzkrieg* offensives early in World War II knocked out enemy communications hubs, creating much chaos.

Ronfeldt and I played down the idea of using cyberspace-based attacks strategically; that is, in a manner akin to aerial bombardment of an adversary's homeland infrastructures. But back in the 1990s, the only really serious interest in cyberwar came from those focused on conducting cyberspace-based conflicts in such a fundamentally strategic fashion[2] (Rattray 2001). The normative aspect of our vision as it unfolded, particularly in our *Athena's Camp* (Arquilla & Ronfeldt 1997), was that strategic cyber-attack was both ethically problematic and unlikely to succeed. In other words, cyberwar would fall afoul of the same moral problems and operational limitations that have plagued the use of strategic air power over the past century.[3]

Thus, our preference was to see cyberwar employed in force-on-force battles, where it would enable decisive victories to be won, shortening the duration of conflicts and easing societal suffering. As one of our critics put it, we were trying 'to make war better' (Gray 1997: 3).

Notions of strategic attack aside, the dominant response to our notion that cyberwar was coming soon was 'No, it's not.' Twenty years on, the tide has clearly turned, with only a few hold-outs taking the view that there is less than meets the eye to cyberwar.[4] Given that sustained cyberspace-based attacks were mounted against Estonia in 2007, and that the Russian ground invasion of Georgia in 2008 was accompanied by skillful hacker attacks on key military and governmental command and control sites, it has grown ever more difficult to deny cyberwar's existence.

The physical damage to Iranian centrifuges done by the Stuxnet computer worm in 2010 proved to be another watershed event: now 0s and 1s – i.e., source code – could clearly be used for destructive purposes in the 'real world.' As Ryan Jenkins (2013: 71) notes in his thoughtful article, this action was, ethically speaking, observationally equivalent to a commando raid causing 'no loss of life, harm to civilian infrastructure, or other collateral damage.' A little over a year after the Stuxnet event, Pentagon officials formally (and finally) declared cyberspace to be a 'warfighting domain.'[5]

Many assume that the United States worked with Israel to develop and deploy Stuxnet; it is also widely believed that Russia committed or directly incited the earlier acts of strategic cyberwar in Estonia and the battlefield use of cyber-attacks in close support of ground operations in Georgia. Whoever was behind these attacks, it is clear that cyberwar *is* here. So, in a world where cyberwar is on the rise, a comprehensive enquiry into its ethical implications – such as that undertaken by the authors who have contributed to this issue of *Journal of Military Ethics* – is already long overdue. They have done much good service in raising important questions about which aspects of traditional just war theory may or should make their way into the strategic discourse about information-age conflict.

In particular, I am drawn to the discussion of the long-standing problem of 'perpetrator anonymity,' which has, for the past two decades, plagued those of us who have worked in this field. Contrary to Randall Dipert's (2013: 38) assertion that '[t]here seem to be very few historical cases where a state was attacked but had little information about who attacked it', the list of disruptive strikes and serious intrusions that have never been adequately 'back hacked' is a very long one. It stretches from the Moonlight Maze penetrations into sensitive Pentagon information systems back in the late 1990s, to the costly disruptions caused by the Code Red and Nimda viruses that were launched around the time of the 9/11 attacks, and on to more recent events not to be mentioned by name.[6] The United States is just one of many countries to have come under such attacks without being able to pierce the veil of anonymity that masks most perpetrators.

Thus, there are times when we do not have a clue as to who the perpetrator may be. Then there are times when identification can be described as 'possible,' occasionally even 'probable.' I can think of only a few events in which perpetrators were identified beyond doubt – and these were rather benign instances of relatively low-threat non-state actors. And so it is the 'possible' and 'probable' cases that tempt us to consider retaliatory action – and here is where ethical dilemmas come quickly to the fore.

What sort of action can be taken? A cyber-attack in kind and of like scale? A more traditionally military response against physical targets? Can one retaliate justly without knowing the identity of the guilty party? These are some of the key ethical questions prompted by the looming age of cyberwar. The answers to them are not at all clear. For example, does the principle of proportionality require an in-kind cyber-response to a cyber-attack? Or can a more physical response be allowed, even a declaration of war followed by military operations?

Christopher Eberle has usefully explored this question and come to some quite sensible conclusions. Leaving aside for a moment the perpetrator ambiguity problem, the 2007 cyber-attacks on Estonia, in his view (Eberle 2013: 59), did not rise to a level that justified physical retaliation. For Eberle, the key item of information needed has to do with parsing the intent of the attacker. If the goal is to cause serious disruption, then some kind of military response could be justifiable (ibid.: 61–2). This seems a very good 'ethical metric.' Yet, using his logic one sees that Estonia *was* indeed a case where disruption was intentional, and so could have been responded to by physical action.

Estonia, of course, cannot really threaten Russia militarily by itself; but it is a North Atlantic Treaty Organization (NATO) member and it was attacked. Tallinn could have demanded its allies to come to the aid of a member state. And while there was no (openly) definitive evidence of a Russian role in the series of cyber-attacks, there was very high confidence that they arose with Moscow's knowledge and approval – and that they fit a known pattern of 'agitprop' activities quite consistent with old-style Soviet disruption strategies (Blank 2008: 227-247). So, given malign intent and the probable identity of the perpetrator, would it have been appropriate, and proportionate, to respond with physical force?

For me, the relatively modest, contained economic damage from the attacks was low enough, and the risk of escalation to a greater war high enough to combine to argue against physical retaliation. To use a concept from Thomas Aquinas, initiating retaliatory action in this case would likely have done more harm than good. In his words, such action would not have led to 'the advancement of good or the avoidance of evil.'[7]

Escalatory risk will not always be present, but in my view crossing the threshold from virtual attack to physical retaliation – either by the destruction of property alone, or by the use of lethal force – is generally a bad idea. Such action should be approached with extreme caution. There may be times when this can be done – e.g., when the initial damage inflicted is costly, escalatory risk is low, and the perpetrator's identity has been established beyond doubt – but I suspect that these conditions will all be met only on very rare occasions.

My caution on this point grows from what I see as a very applicable analogy: the use of physically destructive, possibly even lethal force in response to cyber-attack is not unlike the 1950s concept of 'massive retaliation,' the doctrine of threatening to employ nuclear weapons in response to *any* form of aggression against some innocent party. There were two problems that plagued this policy: (1) sponsors of aggression were likely to be nuclear-armed themselves, so they had a credible escalatory response at the ready; and (2) high-level American policy makers quickly realized that this doctrine was predicated on a highly disproportionate use of retaliatory violence. Thus, the policy of massive retaliation became, as Thomas Schelling (1966: 190) described it, a doctrine 'in decline almost from its enunciation.'

The theme of escalatory risk that I have introduced here touches upon one of the principal concerns that David and Joseph Danks have with automated responses to cyber-attack. They call this escalatory spiral a 'chain reaction' and, while they assert (rightly) that it will be hard to know 'whether it is likely that a chain reaction could occur' (Danks & Danks 2013: 29), their view should be assessed in profoundly cautionary terms. Their resolution of the problem, a recommendation to embrace some purely defensive automated responses while generally eschewing artificial intelligence-driven retaliation, seems the right answer; however, I would suggest that it could be ethically acceptable to use automated systems to do all the preparation for a retaliatory cyber-strike – as long as the disruption-causing phase of the actual attack remained under human control.

In this essay my intention has been to move from some of the specific problems that have bedeviled notions of fighting just cyberwars – especially attribution, proportionate response, and escalation – to contemplate larger, more general ethical issues that bear upon whether classical just war theory remains useful for thinking about the dilemmas posed by conflict in, and from, the virtual domain. Much of this broader strand of enquiry is addressed in Edward Barrett's thoughtful article, and has formed a central interest of my own over the years as well.[8]

Perhaps the biggest ethical questions about cyberwar are to be found by exploring the classical *jus ad bellum* notion of 'right purpose.' Traditionally, there has been no doubt that self-defense is the gold standard of right purpose. One moves to shakier ground when the issue is whether a pre-emptive strike – i.e., an attack launched when one believes the enemy is readying a strike – can also be considered 'just.'

For example, Israel's leaders were of the opinion that an Arab coalition was planning an attack in the summer of 1967, and that pre-emptive action was justified. But some historians have since argued that an attack was not imminent – clouding the ethics of the situation significantly. And even at the time there was a widespread belief that the Israelis struck first because they 'could not afford to wait while the diplomats negotiated. Three months of mobilization [without actually going to war] could wreck the economy' (Associated Press 1967: 16). Clearly, pre-emption can be problematic.

Even worse is the notion of preventive war, the idea of using force against a *potential* threat. The most troubling example of this kind of action, from an ethical perspective, was the American-led invasion of Iraq in 2003. The rationale for the war was the belief that Iraq was developing weapons of mass destruction, and had to be stopped before 'the smoking gun turned into a mushroom cloud.' No such mushrooms were found, casting an ethical pall over this act of preventive war.

Do these cautionary tales of pre-emptive and preventive wars apply in cyberspace? Given that attacking with bits and bytes rather than with bombs and bullets might disrupt but would not destroy, even a mistaken act of pre-emption would seem to have a much lower ethical downside than would a physical act of pre-emption. This would hold as long as an act of cyber-pre-emption was not followed immediately by a physical attack or invasion. The same sort of logic might be applied to the concept of preventive cyberwar, again because the physical damage would almost certainly be non-lethal (as in the Stuxnet attack). Given the minimization of physical destruction, cyber-pre-emption and/or prevention are likely to prove irresistibly attractive.

Whatever the ultimate impact of ethical concerns on taking pre-emptive or preventive action via cyber-means – and such means should be defined in the broad terms that Randall Dipert rightly suggests go beyond just the Internet - both these strategies have practical problems as well. If a pre-emptive or preventive strike is not followed by actual physical military operations, the benefit of either type of action is likely completely lost. Pre-emption, even correctly applied, will only inform the potential aggressor that he has lost the element of surprise and must improve his information security and determine to strike later, at a more advantageous moment. A preventive action, such as was mounted in the Stuxnet attack, did not end the proliferation threat – and surely induced the Iranians to improve their cyber-security. No, without the complementary use of physical force as a follow-up, cyber-pre-emption and prevention will likely be of practical use on only rare occasions.

Moving on from right purpose, other *jus ad bellum* concerns seem also to come under some stress from cyberwar. For example, in an age of networks of all sorts, that includes hacktivists, terrorists, insurgents, and a heaping helping of lone wolves increasingly able to mount serious cyber-attacks, how is the principle of 'duly constituted authority' to be viewed? In the virtual domain, the nation-state has nothing like the monopoly on conflict that it has enjoyed in the physical world – although it must be noted that, in the age of al-Qaeda, and in a world replete with armed factions conducting civil wars, the nation-state's monopoly is already breaking. If almost anybody with the right skill set can start a cyberwar, what becomes of duly constituted authority? And, turning this question on its head, if a hacker network is encouraged by a nation-state to wage cyberwar, is the network, all other things being equal, acting on the basis of duly constituted authority? These are troubling questions.

The just-war notion of 'last resort' also comes under pressure when the virtual domain is added into the conflict mix. Cyberwar is a means that can be used early, easily, and highly effectively. It is stealthy, and will undoubtedly feed the military Moloch of 'surprise attack,' the *sine qua non* of success in war from time immemorial. And in addition to the sheer attractiveness of increasing one's chance of winning a war by starting with a cyber-surprise, there is the prospect of achieving that victory while inflicting less loss of life and other damage on the enemy society, and suffering less oneself as well. What is not to like? So here, too, I see another of the foundations of *jus ad bellum* being undermined.

Where cyberwar may feed temptation to skirt classical guidelines about going to war justly, I think it is quite possible that cyber-strategy and tactics may make *waging* war justly a bit easier. A cyber-attack is hardly ever going to rise, in my view, to the level of a disproportionate act, as there is a big difference between cyber-disruption and physical destruction. Indeed, as I noted earlier, the only really significant worry about taking action that might be viewed as disproportionate will arise when a party that has suffered costly disruption from a cyber-attack is contemplating responding with a physical, possibly lethal, counter-blow.

How about non-combatant immunity? For Michael Walzer (1977: 151), whose deep study of ethics in conflict settings led to his visualization of a 'war convention' spanning times and cultures, the simple rule 'is that noncombatants cannot be attacked at any time.' Walzer was of course contemplating this convention prior to the advent of cyberwar; but the onset of conflict in the virtual domain should lead us to ask whether the notion of non-combatant immunity needs to be reconsidered.

One key issue in play here has to do with the use of cyber-attacks on commercial and financial sites as a form of virtual economic warfare. Clearly, the costly and disruptive effects of such attacks make it obvious that non-combatants have been 'targeted,' but not for lethal violence. Another area where non-combatant immunity might be violated would be in cyber-attacks on communications infrastructures. In the United States, for example, the military relies largely on privately owned and operated information transmission systems. Thus, any attempt to take down military command and control would necessarily have to target civilian infrastructure, creating significant 'collateral damage' (again, non-lethal vis-à-vis civilians) in the form of costly disruptions to both personal and commercial communications.

Does the foregoing violate the notion of non-combatant immunity? Teams of lawyers scrutinizing the relevant international law, namely the Protocol Additional to the Geneva Conventions of 12 August 1949, could probably argue both sides of this question compellingly. This is so because of the operative language used to define an attack: 'violence against the adversary.'

While the Geneva definition seems to imply a focus on bloodshed, attacks on property, including infrastructure like oil pipelines, have long been viewed as violent – even as acts of terrorism. On the other hand, the fundamentally non-lethal nature of cyber-attack on civilian targets[9] may suggest some room for operating, particularly since existing legal notions of non-combatant immunity contemplate some allowable amounts of collateral damage in wartime operations. As legal scholar David Fidler (2012: 80) has put the matter, this view 'might make assessment of many types of cyber-operations easier because their impacts will not match or exceed what *jus in bello* permits in terms of conventional military violence.'

At this point, it seems that a kind of ethical 'bottom line' assessment might be discernible about cyberwar, in two parts. First, *jus ad bellum* comes under great pressure in the key areas of right purpose, duly constituted authority, and last resort. However, the apparent benefits of waging preventive or pre-emptive war, concepts with a lineage dating from Thucydides and Francis Bacon,[10] are largely illusory. Second, it seems that *jus in bello* considerations come off rather better in the areas of proportionality and non-combatant immunity – although there is a bit of complexity in the parsing of notions of acceptable collateral damage.

In sum, the virtual wilderness, artificially created but vast and continuously growing, largely ungoverned, has already seen conflict migrate to it. Given cyberspace's complexity and uncontrollability, it seems that Hobbes's view of wars of 'all against all' is more likely to obtain than Rousseau's notions about the possibility of harmony, even nobility prevailing on the electronic frontier. With conflict inevitable, the need for deeper, fresher understanding of war ethics only grows.

NOTES

1. Our RAND think-piece was then published the following year as a journal article: Arquilla and Ronfeldt (1993).
2. This strand of thought about cyberwar – including a deep analysis of the air power analogy – is discussed thoroughly by Rattray (2001).
3. On this point, see Pape (1996).
4. The most articulate exposition of the anti-cyberwar perspective can be found in Rid (2012).
5. See Pentagon to Unveil Cybersecurity Strategy, *The Washington Post*, 13 July 2011. Note that the US Air Force had, for its own part, made a very similar declaration in 2005.
6. A comprehensive early survey of these sorts of exploits can be found in Verton (2003). For more recent instances, see Clarke and Knake (2010).
7. Thomas Aquinas, *Summa Theologica*, II-II, qu. 40, Article 1 (quoted from Aquinas 2008).
8. For example, see Arquilla (1999: 203-12).

9. As cyber-weapons grow in sophistication, it is possible that their effects may one day move beyond just disruption, becoming more lethal as well. This development would make the issue of non-combatant immunity in cyberwar even more complex.

10. As Thucydides (1951) notes in his *History of the Peloponnesian War*, its cause was the 'growth of power in Athens and the fear this inspired in Sparta' (Book I, 1.34 -[3]). For Francis Bacon (1908: 86), in his essay 'Of Empire,' the main point was that 'there is no question but a just fear of an imminent danger, though there be no blow given, is a lawful cause of war.'

REFERENCES

Aquinas, Thomas (2008) *Summa Theologica*, online edition (based on the translation by the Fathers of the English Dominican Province), II-II, qu. 40, accessed 18 March 2013, available at: http://www.newadvent.org/summa/3040.htm; Internet.

Arquilla, John (1999) Can Information Warfare Ever Be Just?, *Ethics and Information Technology*, 1(1), pp. 203–212.

Arquilla, John & Ronfeldt, David (1993) Cyberwar Is Coming!, *Comparative Strategy*, 12(2), pp. 141–165.

Arquilla, John & Ronfeldt, David (1997) *In Athena's Camp: Preparing for Conflict in the Information Age* (Washington, DC: RAND).

Associated Press (1967) *Lightning out of Israel: The Six-Day War in the Middle East* (New York: The Associated Press).

Bacon, Francis (1908) Of Empire, in: Mary Augusta Scott (Ed), *The Essays of Francis Bacon* (New York: Charles Scribner's Sons).

Blank, Stephen (2008) Web War I: Is Europe's First Information War a New Kind of War?, 27(3), pp. 227–247.

Clarke, Richard A. & Knake, Robert K. (2010) *Cyber War* (New York: HarperCollins Publishers).

Danks, David & Danks, Joseph H. (2013) The Moral Permissibility of Automated Responses during Cyberwarfare, *Journal of Military Ethics*, 12(1), pp. 18–33.

Dipert, Randall R. (2013) Other-than-Internet (OTI) Cyberwarfare: Challenges for Ethics, Law, and Policy, *Journal of Military Ethics*, 12(1), pp. 34–53.

Eberle, Christopher (2013) Just War and Cyberwar, *Journal of Military Ethics*, 12(1), pp. 54–67.

Fidler, David P. (2012) *Inter arma silent leges Redux?* The Law of Armed Conflict and Cyber Conflict, in: Derek S. Reveron (Ed), *Cyberspace and National Security*, pp. 71–87 (Washington, DC: Georgetown University Press).

Gray, Chris Hables (1997) *Postmodern War* (London: The Guilford Press).

Jenkins, Ryan (2013) Is Stuxnet Physical? Does It Matter? *Journal of Military Ethics*, 12(1), pp. 68–79.

Pape, Robert (1996) *Bombing to Win: Air Power and Coercion in War* (Ithaca, NY: Cornell University Press).

Rattray, Gregory (2001) *Strategic Warfare in Cyberspace* (Cambridge, MA: The MIT Press).

Rid, Thomas (2012) Cyberwar: Think Again, *Foreign Policy Magazine*, pp. 58–61. 26 February.

Schelling, Thomas C. (1966) *Arms and Influence* (New Haven, CT: Yale University Press).

Thucydides (1951) *The Peloponnesian War*, trans. Richard Crawley (New York: The Modern Library).

Verton, Dan (2003) *Black Ice: The Invisible Threat of Cyber-Terrorism* (New York: McGraw-Hill).

Walzer, Michael (1977) *Just and Unjust Wars: A Moral Argument with Historical Illustrations* (New York: Basic Books).

John Arquilla is professor and chair in the Department of Defense Analysis at the United States Naval Postgraduate School. He earned his degrees in political science at Rosary College (BA 1975) and Stanford University (MA 1989, PhD 1991). He is the author of several books and over 100 articles on a wide range of topics in military and security affairs. He is best known for his collaborative studies with David Ronfeldt on cyberwar, netwar, and swarm tactics, concepts that they pioneered together. Overall, Dr. Arquilla's body of work remains primarily focused on developing an understanding of the strategic implications of the information revolution for society and security. *Correspondence Address*: Department of Defense Analysis, 589 Dyer Road, Room 214, Naval Postgraduate School, Monterey, CA 93943, USA. *Email Address*: jarquilla@nps.edu

Epilogue
"Beyond Clausewitz" – Military Ethics for Postmodern War

George R. Lucas, Jr.

1. Postmodern War

At the beginning of the first Gulf War in 1991, Umberto Eco wrote a short essay for an Italian magazine, entitled "Reflections on War." He contrasted war as it was then being fought in the deserts of Kuwait with "modern" war, as described by Karl von Clausewitz. According to Eco, the Clausewitzian conception of war is a thoroughly conventional, modernistic, state-centered enterprise, a chess game in which the object is, not simply to take the opponent's pieces, but ultimately to attain complete domination, or "checkmate." Indeed, it strengthens the contrast that Eco proceeds to draw by first recalling that Clausewitz himself used explicitly Newtonian metaphors, drawn from classical (early modern) physics, to describe the contest of political wills between nation-states as analogous to "physical forces" acting upon a "center of gravity," seeking to move that center to a more favorable position by breaking the political will and overcoming the military forces of one's adversary.

After two world wars and the Cold War, by contrast, Umberto Eco believed war could no longer be characterized in this modernist, Clausewitzian fashion, in terms of the straightforward linear vectors of force operating between clearly-defined rival centers of power (any more, we might add, than can classical physics handle the anomalies of relativity or quantum mechanics). In his interesting essay, he instead quoted his fellow postmodernist, Michel Foucault, to the effect that "power is no longer monolithic and monocephalous: it is diffused, packeted, made of the continuous agglomeration and breaking down of consensus." War, Eco went on to observe, pits a multiplicity of powers in competition with one another: no longer simply "two opposing states," but the controlling governments of states versus their own internal, rival political parties and religious factions; the media, embedded and reporting from behind "enemy lines;" Wall Street and the financial sector, heavily invested in hope of profit, but with no clear strategic goal or financial objective, just (especially in the case of the stock market) "oscillations in the play of powers."

Brilliantly and presciently, Eco proceeded to characterize postmodern war in terms of a technological contrast that had only emerged into public consciousness at that time: an analogy between serial computing and parallel processing. War is and will henceforth be of the latter sort, he claimed: no longer a simple serial sequence of events, but all sorts of things going on at once, what he termed a "neoconnectionist" or neural network system, with diffuse, discrete centers of power and influence, "no longer a phenomenon in which the calculations and intentions of the protagonists have any value" (Eco 1991: 13), or in which the desired outcome is clearly defined in terms of competing interests. War is now, he stated, more like a chess game in which every antagonist takes pieces of the same color, and no one can any longer clearly state just what the strategic goal is, or precisely who the real enemies are.

Baron von Clausewitz also, of course, cautioned that "every age had its own kind of war, its own limiting conditions and its own peculiar preconceptions."[1] Ours is the era of "irregular," "hybrid," or unconventional war, together with the so-called "revolution in military affairs" (RMA), including emerging new military technologies that accompany it. And indeed, as Eco foretold, the broad conception of "net-centric" warfare (NCW) is now central to this so-called "revolution in military affairs" that several of the essays in this volume purport to encompass, though this is not thought by its advocates to be altogether a bad thing. Originally the brainchild of military futurist and strategist, Vice Admiral Arthur K. Cebrowski, the driving ideal of NCW was to move beyond conventional or classical Clausewitzian conflict by linking up a small number of highly-trained human warriors with equally small, fast, agile weapons systems and mechanized support, all linked in turn to one another via GPS and SATCOMs into an intricate, interconnected, networked system, in which the behavior of components would be mutually enhanced and coordinated by constant exchange of real-time battlefield information – the very "neural network" that Eco had foretold.

Admiral Cebrowski's final tour of duty in the US Navy was as president of the Naval War College (Newport, RI), where he advocated forcefully for this new vision of military power. Upon retirement, he subsequently served under then US Secretary of Defense, Donald Rumsfeld, in what, at the time, was the Pentagon's new "Office of Force Transformation," leading the charge toward what Rumsfeld himself championed at the time as a new technological RMA. The Cebrowski Institute at the Naval Postgraduate School Monterey, CA) still carries on this integrated approach to operations research in engineering and military tactics today. The reining conception continues in the broad sense toward replacing huge forces of personnel and massive weapons platforms, operating in the Clausewitzian mode of classical physics, with fewer individuals doing many nuanced things at once, with enhanced power, presumably involving less risk to (at least some) combatants, and with the agility and flexible response attained through computerized information exchange on the parallel processing model. Cebrowski and his successors, we might say, have attempted to put a positive spin on what, in Eco's original vision, constituted a sinister nightmare.

The immediate legacy of Cebrowski's own vision of "net-centric warfare," along with Rumsfeld's own proposed "revolution," have, of course, long since been swept aside, or else assimilated within a radically-altered context. The US Army's companion technological focus, for example, early in the first decade of the twenty-first century, on the development of "Future Combat Systems" (FCS) was relatively quickly abandoned as expensive, wasteful, and ill-advised. But these subsequent developments merely constitute illustrations of what P.W. Singer, in his opening essay in this volume, describes as the operation of "Moore's Law." The pace of technological change and transformation itself is exponential, doubling overall technological capacity every two years (and now, more likely, every 18 months!). That pace of technological transformation itself quickly overwhelms, supplants, and renders obsolete what seemed exotic and visionary only a few years, or even months, before. Momentary policy fads and foci like RMA, FCS, and NCW may quickly fade, but they come and go against a backdrop of relentless technological development and transformation that continues unabated. The challenge is for policy-makers, including ethicists and philosophers concerned with military ethics, to keep abreast of this "technological tsunami" that defines Eco's original vision of "postmodern war."

That challenge was taken up late in the last decade by a significant new group of collaborating scholars and institutions, the "Consortium on Emerging Technologies, Military Operations, and National Security" (CETMONS). Beginning in 2008, the consortium began to convene experts in international relations, international law, and military ethics to meet with leading scientists, social scientists, and most especially engineers involved in the development of "new warriors" and "new weapons," for the larger purpose of collaborating more effectively in the assessment of ethical and legal ramifications of key elements of this emerging, ongoing, technologically-driven revolution in the nature of military conflict. Robots and unmanned (or remotely operated) air vehicles, the concomitant advent of increasingly discriminate and "riskless" war, the biological, mechanical, and even psychological enhancement of human warriors, novel developments in military uses of nanotechnology and the advent of three-dimensional printing as a new form of manufacturing, the increasing use of "non-lethal" weapons, and the advent of new forms of "virtual" war in cyberspace (occasionally accompanied by chilling kinetic implications in the real world) are all part of this emerging – albeit complex and confusing – picture.

As a founding institutional member of that consortium, the Stockdale Center for Ethical Leadership at the US Naval Academy embarked upon an ambitious program to accelerate the level of collaborative research in these key areas, in order to bring the level of ethical and legal discourse into closer synchronization with the pace of technological change itself. P.W. Singer inaugurated this "ethics surge" immediately upon the release of his influential and widely-read study of ethics and military technologies, *Wired for War* (Singer 2009), addressing a convocation of midshipmen, engineers, and military ethicists at the US Naval Academy on the urgent moral challenges presented by these new technologies in March, 2009, and subsequently at the Naval Postgraduate School in February of 2010. Meanwhile, scholars from CETMONS' supporting institutions were appointed fellows in the Stockdale Center during the 2009–10 academic year, during which time they met weekly for discussion of several of the areas of military technology cited above, and again in the ensuing academic year, to focus exclusively upon the growing crisis of cyber conflict. Those fellows, in turn, played leading roles in organizing and presenting their research at two major national workshops on ethics and military technologies organized by CETMONS at Case-Western University and Arizona State University in the fall of 2009 and spring of 2010, respectively.

Simultaneously, the 10th and 12th Annual McCain Conferences of US Service Academies and War Colleges[2] were dedicated to these topics, convening nationally and internationally recognized engineers and scientists working in these various subfields of the new military technologies with the Stockdale Fellows and other experts in law, policy and international relations. These annual gatherings of key scholars and educators in ethics and leadership, drawn from the nation's military academies and senior war colleges, intended in their turn to examine, and to begin to acquaint military personnel themselves more thoroughly with, the ethical ramifications of these emerging new military technologies.

2. Ethics and Unmanned Systems

Beginning with P.W. Singer's 2010 keynote address for the first of these two conferences, the essays reprinted in this volume stemmed predominantly (although not exclusively) from two special issues of the *Journal of Military Ethics* drawn largely, in turn, from the deliberations and presentations at these two McCain Conferences, and also from follow-on conferences for European and NATO allies on ethics, robotics, and cyber conflict sponsored by the Center for Ethics Research (CREC) at the French Military Academy in Saint-Cyr, and held at the *Ecole Militaire* in Paris in June, 2010, November 2011, and (on cyber conflict) in October 2013. Two essays which appear in the first part of this new volume, by Singer and Patrick Lin, respectively, are primarily descriptive, offering accounts of the varieties of new technologies presently under development, of their proposed military uses, and raising questions about the challenges they will present, or the obstacles we will need to overcome in order to integrate them effectively into our fighting forces.

Singer's book, *Wired for War* (2009) itself constituted one of the first widely-heralded and broadly accessible accounts of the impact of emerging military technologies on the conduct of war, and on social policy more generally. "The Ethics of Killer Applications" updates and continues this work, underscoring some of the key points from the earlier book concerning the blistering exponential pace of technological innovation (Moore's Law) and its sometimes devastating and irreparable impact on the social order. He adds some key new considerations, such as the hidden financial interests driving technological developments, and conflicts of interest, both financial and political, that may underlie policies and attitudes towards the risks posed by new military technologies, as well as induce scientists and engineers themselves to undertake projects which they may personally harbor deep moral reservations or misgivings.

Patrick Lin, a key CETMONS participant, perennial Stockdale Fellow, and co-author of an influential and widely-studied research report on legal and ethical risks associated with military robotics for the US Office of Naval Research (Lin et al. 2008), continues this discussion by comparing robotics and human enhancement as two different approaches to the common problem of enhancing war-fighting competence. With a particular emphasis on the biological and psychological enhancement of human soldiers, he describes the unintended impact that the diffusion of such technologies (or of their effects) may have in the civilian sector, both in terms of new civilian applications with ramifications that may prove surprising and sometimes decidedly unwelcome, as well as the prospects that these new technologies may finally make it all too easy for political authorities to resort to war as other than a last resort for conflict resolution. Like Singer himself, Lin is now recognized as one of the leading authorities on these technological innovations in the world today, especially on attempts to apply human enhancement technologies to what the Defense Advanced Research Projects agency (DARPA) terms "peak soldier performance."

While neither of the foregoing authors advocates suspension of research and development, both express what might best be termed "generalized moral anxiety" over the unreflective development and deployment of the various military technologies they describe. By contrast, when we turn to robotics and unmanned systems specifically, the next two authors have, it seems to me, argued for the moral superiority of the military technologies they advocate, suggesting that their use would *lessen* the occurrence of noncombatant fatalities and thus ease the humanitarian tragedies that have conventionally been thought an irreducible feature of warfare itself. That is certainly a startling conclusion, at odds with the public anxiety over drones and "targeted killing," for example, and it bears closer scrutiny.

Ronald C. Arkin, a noted computer scientist and roboticist, argues that lethal autonomous systems ("killer robots") would prove superior to their human warrior counterparts in faithful adherence to the provisions of international humanitarian law (Law of Armed Conflict). Because they are utterly immune to hatred, prejudice, or fear, "robot warriors" are far less likely to engage in deliberate violations of the laws of war, he suggests, than are the human beings who are especially prone to such sentiments in the midst of armed conflict. Developing and deploying such systems, he concludes, would serve to lessen the inhumanity of war itself. Philosopher and former US Air Force officer, B.J. Strawser, offers a further defense and moral justification of the use of unmanned aerial vehicles like the "Predator," on the grounds that agents engaged in otherwise justifiable acts (i.e., morally justified combatants) are entitled to as much protection from harm in the conduct of such justified actions as can be afforded, so long as such protection does not impede the agent's ability to act justly. This principle he labels "The Principle of Unnecessary Risk," and it has since become a central feature in the debate about the overall morality of building and using unmanned systems. If Stawser's argument is valid, for example, then, contrary to critics of the enterprise of military robotics, the development and use of technologies like the Predator and Raptor is not simply permissible, it is obligatory (provided, of course, that the conflicts in which these devices are deployed are indeed morally justified otherwise).

Computer scientist Noel Sharkey of the University of Sheffield, UK, however, is a well-known critic of such views, and, over the past several years, has been a frequent opponent of Ron Arkin in public debates over the moral dilemmas attendant on the development and use of autonomous robots armed with lethal force. His essay in this volume offers an eloquent and succinct summary of points he has repeatedly made in opposition to Arkin, in particular, concerning the wisdom and moral probity of the proposed military uses of robots in combat (e.g., Sharkey 2007, 2008), and the need for international governance measures to ensure that what he describes as a nightmarish science fiction scenario does not in fact come to pass. It is a particularly exasperating error, he finds, to maintain that robots themselves can be designed to be "moral," since that requires dimensions of interiority and self-regard in reasoning that machines, by definition, do not have. The best that might be hoped is that robots could be designed to comply with applicable legal restrictions on their operations in combat, which is itself a far cry from endowing them with true "morality."

It bears mention that Ron Arkin, along with Patrick Lin, is a key member and participant in the collaborative work of CETMONS, and that he is also a member, while Sharkey is a co-founder (with American philosopher, Peter Asaro), of ICRAC – the International Committee for Robot Arms Control – which advocates publicly for new treaties and improved legal oversight of the future development and use of robots in armed conflict.

3. Ethics and Cyber Conflict

The final essays in this special volume turn in some detail to perhaps the most vexed and ominous of the ongoing ethics debates over military technology, the prospects for cyber warfare. I characterize it thus because, of all the technologies surveyed, the cyber warfare prospects are perhaps the most widespread and devastating on a human scale, the most representative of the common themes and ethical concerns pervading all the other technologies considered, and yet, of all of these, the least thoroughly examined from the standpoint of ethics and governance. That is, cyberwar and cyber weapons, just like all the other innovations in military technology considered here, invoke, in various ways, what we might label the "threshold" problem, the "accountability" problem, and the "discrimination/proportionality" problem. In so doing, all, including cyberwar, contribute to desensitizing the public to the true costs of war.

So, for example, the first thread of concern running through all of these discussions has been the "threshold" question: will the technology in question, including resort to cyberwar, lower the threshold for resorting to war of any sort, traditionally consigned to being the last (rather than the earliest) resort to conflict resolution with adversaries or competitors. Any technology, weapon, or tactic that makes it inherently easier to resort to destructive uses of force in order to resolve disputes automatically constitutes a ground for concern on this criterion. Using robots cuts down on human casualties and costs, for example, while cyberwar is "virtual" war, and may seem more like a game than reality. Thus, in both instances, we might more readily resort to war using these technologies when we should instead refrain or at least forebear in favor of alternative, non-militaristic approaches to conflict resolution.

Secondly there is what we might term the *jus in bello*/LOAC question: will the technology in question present increased risks of harm to civilians, or otherwise threaten disproportionate destruction and collateral damage in war? Is the very development of the technology itself a violation of international humanitarian law? With armed autonomous robots, for example, a principal concern first raised by Australian philosopher Rob Sparrow, is with lack of meaningful accountability (Sparrow 2007). One may interpret the so-called "Martens Clause" in international law as prohibiting the deliberate development of any weapons system for whose use or misuse military personnel or their governments cannot be held reasonably accountable under LOAC, since such accountability is, in all other contexts, a feature of custom and law as these pertain to existing weapons systems. It is difficult, if not impossible, Sparrow argued, to conceive of how an autonomous lethal machine can be meaningfully held accountable for its actions, and so the prohibition entailed in the Martens Clause prohibits the development or use of these weapons.

With cyber warfare, the principal LOAC concern has been slightly different. Accountability for the use of cyber weapons is an enormous challenge in its own right, to be sure, owing to the still-formidable difficulty of being certain beyond reasonable doubt of who has launched an attack, especially when the suspected attacker denies culpability. The moral debate over cyber strategy, however, has focused even more upon the indiscriminate (and sometimes also wildly disproportionate) properties of cyber weapons and tactics that are sometimes proposed. Computer science professor Neil Rowe at the Naval Postgraduate School, for example, has published several papers raising an alarm that the weapons and tactics frequently and (from the standpoint of ethics, at least) unreflectively envisioned as part of any routine preparation to wage or defend against cyberwar are aimed at civilians, and that their use would cause widespread destruction of lives and property, and otherwise inflict surprisingly massive and terrible suffering among the civilian population of the target state (Rowe 2007, 2008, 2010). Such cyber-strategy is thus, he argues, inherently a violation of LOAC.

Philosopher Randall R. Dipert (University of Buffalo) cites this work in the first of his two essays in this volume, and gives careful consideration to Rowe's concerns in the course of offering what remains (in my estimation) the most technically detailed, precise, and comprehensive account of warfare, weapons, strategy, and tactics in cyber space – all part of what Dipert sees as the puzzling and paradoxical "ontology of cyber space." More broadly than Rowe's *jus in bello* analysis, however, Dipert attempts to evaluate both the strategic justification for cyber warfare, and the morality and legality of the tactics proposed for the conduct of such warfare, through the lens of classical just war doctrine. His earliest conclusion in 2010 was that both international law and just war doctrine are woefully deficient in providing reliable guidance for this novel and unique kind of futuristic warfare, and, in this path-breaking essay, he originally proposed some guidelines that might help remedy these deficiencies in governance as the prospects for the conduct of such warfare grow ever more imminent. His specific conclusions on the inapplicability of existing law and extant understandings of just war doctrine, however, are views against which I have subsequently argued strongly (e.g., Lucas 2011, 2013a, 2013b)

Col. James Cook, a philosopher at the US Air Force Academy, however, argued in response that this appearance of deficiency is illusory. Cook specifically rejected what he terms "the argument from disanalogy" in Dipert's account, and attempts to demonstrate in brief compass that the new dimensions of cyberwar and cyber weapons can, or can be made to, fit well within the framework of guidance provided by conventional just war doctrine. In an intriguing Heideggerian turn, Cook defined "cyberation" by analogy with the earlier derivation of "aviation" from "navigation," as the collective noun meant to encompass the "goings-on" of cyber "things" and "events" within cyberspace. Like aviation applied to the air, or navigation to the sea, cyberation includes a central, but by no means exclusive focus on military operations and considerations, alongside other operations and considerations that are decidedly non-military, with no obvious and immediate demarcation between or among them. In the end, and despite the confusing novelty of the medium and its activities, it is no less possible to encompass the military operations in cyberation within the framework of just war theory and international law than it is in the realms of aviation and navigation, respectively. Rather, it takes a considerable degree of patience and judgment to sort out which things are which. This intriguing and original analogy has gained widespread currency, and has proved useful in helping situate our confusion over activities in the cyber domain in historical context, against which backdrop these current confusions are merely the latest episode in an ongoing drama of post facto social adjustment to technological transformations.

Many of these basic concerns are reiterated in the remaining essays on cyber warfare which, unlike the two preceding pieces, were drawn from the second special issue of *JME*. While some of these stemmed from the second (2012) McCain Conference, entirely devoted to cyberwar, surprisingly, a good many of the seminal pieces by well-known cyber experts (e.g., Richard Clarke) were omitted, while others gathered from a subsequent call for papers are included. This makes for a rather uneven treatment of the topic, with a good deal of overlap and repetition of basic material on that part of some of the authors, who are relative newcomers to this field.

The publication of this special issue itself testifies to new awakening of interest in the topic of cyber conflict within the field of ethics, very much in the manner that two lengthy wars in the last decade re-awakened scholarly interest in just war doctrine itself, which had languished as a mainstream interest for decades after the Vietnam War. Prior to the appearance of this special issue, the topic of cyber conflict had received only scattered treatment by subject matter experts in ethics, especially in comparison with the voluminous body of interpretive scholarship generated by international lawyers in the aftermath of the Estonian conflict. Unfortunately, it is also invariably the case that, following in the wake of some massive technological upheaval (of the sort that cyber conflict has come to represent over the past decade), an initial period of embarrassed silence or woefully inadequate response is quickly followed by a somewhat uncritical overabundance of eager, enthusiastic, but sometimes not particularly well-researched contributions, intended to fill that initial conceptual vacuum before accurately assessing just how deep it is – a phenomenon that the late British philosopher, Gilbert Ryle, while editing the prestigious journal, *Mind*, used to refer to disparagingly as "philosophy without footnotes."

There are two very substantial and original exceptions to that rule in this collection, written once again by Randall R. Dipert, with a closing summative essay by John Arquilla. Despite having disagreed with some of his findings regarding the relevance of just war doctrine, there is no question that Dipert is the single most knowledgeable and substantial contributor to the current, relatively nascent cyber-ethics debate. John Arquilla, for his part, is in many respects the founder or father of the entire field of cyber-ethics, as his thoughtful reflections on some twenty years of this discussion (that often included acknowledgement of his distinguished colleague, computer scientist, Dorothy Denning) gently reminds us.

In his second essay, Dipert reminds us that the cyber domain is not restricted to the internet itself, but includes a number of other important facets (what he terms "other-than-internet," or OTI attacks) that offer their own ethical challenges apart from those considered by the remaining contributors (who seem entirely unaware of these additional dimensions of cyber operations, and seem to equate "cyber space" itself wholly with what transpires on the internet). While Dipert essay focuses on techniques like infecting thumb-drives (the vector through which the Stuxnet worm may have been introduced to the "air-gapped" Iranian military internet system) and interfering with buried optical cables (that he surmises as the most likely manner of sabotaging Syrian air defense systems in advance of the Israeli bombing attack on the nuclear reactor under construction at Dayr-al Zawr), his account would have been much strengthened by considering the prospects revealed in the Snowden accounts of cell phone network hacking and even tapping into the GPS system, both of which are connected to, interact with, but physically distinct from the "internet" *per se* (far more so than thumb drives, which, like many of Dipert's other OTI examples, must eventually be introduced into the internet itself to take effect).

It is a helpful illustration to recall that the ubiquitous social-networking sight, Facebook, is an internet-based phenomenon that is struggling to gain a foothold in the cellular world, while its cellular counterpart, Twitter, was explicitly designed with the original limitations of the cellular network in mind, and is only now beginning to make a foray onto the internet. GPS navigational devices operate on a frequency entirely separate from both. All are part of the cyber domain, and the now-infamous "Enterprise Knowledge System" laid bare in Edward Snowden's release of NSA Management Directive #424, tracks and correlates meta-data from these distinct sources, along with conventional telephone (both copper and fiber optic cable) local usage details ("luds") as well. It is also the case that devices, like the unmanned systems discussed above, or like the elder care robots and robotic cars that loom on the near horizon, make use of a variety of these different dimensions of the cyber domain, and thus are, perforce, part of that domain as well. Notwithstanding this technical quibble over the great diversity of forms a genuine OTI attack might assume, however, Dipert's account of the severity and ethical conundrums of OTI-sourced attacks helpfully broadens the discussion to include threats simply not envisioned in the remaining, exclusively internet-focused analyses (to which I will shortly turn in conclusion, below).

Arquilla's most recent critic and cyber debating opponent has become the British international relations scholar, Thomas Rid, who, contrary to all Arquilla's assertions and concerns regarding the threat of cyber conflict over the past two decades, has forcefully denied that genuine cyber "warfare," involving death, harm, and destruction equal to that of conventional warfare, could ever truly occur (e.g., Rid, 2011, 2012). This (as I have repeatedly argued in Arquilla's defense: Lucas 2013a, 2013b) is far too stringent and unrealistic a definition of "armed attack" or its equivalent, and Arquilla is right to continue to resist such an iconoclastic characterization of warfare.

Arquilla has made his own reputation through being something of a strategic contrarian, taking sharp issue with the received views of the defense establishment – for example, by advocating, nearly two decades ago, increased reliance on special forces in response to the growth of humanitarian intervention and counterinsurgency "small wars." The military establishment during most of that period was instead bent on investing personnel and scarce resources to build overwhelming (and exorbitantly expensive) conventional superiority. In the midst of this sometimes-acrimonious debate, Arquilla quite literally (with colleague, David Ronfeldt) coined the term "cyberwar" to designate a form of conflict that, at the time (from 1992 until at least 1999) few took seriously.

Time, we might observe, has proven Arquilla a remarkably prescient prognosticator. Yet he does err when he challenges Dipert's assertion that there are few, if any, historical precedents of military non-attribution that can be cited. Arquilla responds with earlier examples of unattributed *cyber* attacks, when what Dipert clearly intends instead is that there is a dearth of historical examples of unattributed *conventional* attacks, from which to draw conclusions by analogy on the limits of acceptable responses to the cyber cases (including those that Arquilla cites in response).

This is an unfortunate confusion, since a well-informed historian like Arquilla could have easily produced interesting count-examples of instances in conventional warfare, such as the sinking of British merchant convey ships (allegedly) by Italian submarines early in World War II, and (perhaps even more apt to contemporary cyber attribution and response) the sinking of the US battleship *Maine*, just prior to the Spanish-American war of 1898. To this day, in the latter example, no one has been able to conclude definitively beyond a reasonable doubt who or what actually sunk the *Maine*, although suspicion, even from shortly after the event itself, was that it surely wasn't the Spanish government (which was nonetheless blamed for it). This instance provides a perfect analogue for the attribution problem in cyber, and refutes Dipert's contention that we don't have solid ground to stand on in evaluating how to proceed in the event of uncertainty. And the WW II instance seems even more definitive, inasmuch as, following Churchill's threats to "bomb Italy into the Stone Age," the mysterious submarine attacks on British conveys in the Mediterranean ceased entirely.

Both instances, and others like them, are instructive for cyber conflict in different ways. The sinking of the *Maine* produced what was likely a deliberately-provoked and unjust war, lacking a genuine just cause, let alone right intention, on the part of the US. In contrast, the cessation of submarine attacks following a threat of use of force against Italy strongly suggested, *post facto*, that the Italian suspects were indeed the guilty party (in which case, retaliation by the British for any subsequent attacks would have been morally justified). Following the leads of "probable cause" and focusing suspicions on the most obvious beneficiaries of an "unattributable" cyber attack, as Dipert has alluded to in his earlier essays, is likely "close enough for government work" – also suggesting, as he has also helpfully implied, that the so-called "attribution problem," while significant, need not be seen as an absolute obstacle to the prosecution of acceptable defense and retaliation against cyber attacks.

The remaining essays on cyber conflict seem wholly occupied with the now well-studied properties for internet mischief, ignorant of the other forms of cyber vulnerability of which Dipert writes, even while they collectively fail in addition to distinguish carefully the important legal and moral boundaries between crime, vandalism, activism or "hacktivism, corporate (criminal) espionage, state-sponsored espionage, covert action, sabotage, and the as-yet very few instances of outright cyber equivalents of kinetic armed attacks. Aside from this frustrating and systematic equivocation, however, these pieces raise a number of important and disturbing points.

Perhaps most disturbingly, David and Joseph Danks offer a poignant analysis of the manner in which automated cyber response systems can go awry: for example, the "death spiral" of cyber violence that might result from retaliation in the absence of definitive attribution, especially if autonomous retaliatory defense mechanisms are put in place, in a manner analogous to the fears of such unintended escalation during the age of "nuclear deterrence." This is indeed a grave concern, especially given the rapidity with which cyber events may unfold in time, as illustrated by the authors in the highly unstable computer securities trading systems, which have caused extreme oscillations and even unintended but grave "meltdowns" in securities markets in a matter of seconds. Automation, they aver, can produce some other unintended consequences, such as when a DoS (denial of service) attack provokes a blackout of the offending source. If it turns out the "zombie" is an unprotected hospital computer system, the counter-attack might produce unacceptable collateral damage in the form of denial of access to medical records needed in life-saving treatment. The problem's complexity is multiplied exponentially if an automated defense system is responding to a "distributed" DoS (DDoS) attack, as in Estonia, but instead launched from a botnet incorporating perhaps thousands of unsuspecting users or organizational systems who themselves have done nothing wrong, and are not legitimately the object of the counter-attack. Combined with the well-studied attribution problem, these authors conclude that factors in increasingly necessary automated defense systems have not been sufficiently studied, and the moral impact of their use remains unclear, especially if the best automated "defense" becomes an offensive or "preemptive" system.

Ryan Jenkins' article on Stuxnet, attempts to bridge the gap between the virtual and the physical, and determine whether the use of a weapon of this sort might (as his opening quote from General Michael Hayden suggests) cross a line of significant demarcation between intention in the virtual world and genuine harm in the real world. Jenkins does not engage a questionable view, set forth in a companion essay by Christopher J. Eberle, that "mere" destruction of property without loss of life can never constitute a just cause for war. In this case, the destruction of a military asset, just as in the case of the Israeli attacks on nuclear facilities in Iraq and Syria, regardless of loss of life, constitutes a use of armed force that would provide a just cause for war, should the victim of the attack choose to regard it as such. Jenkins' account of the ontology of cyberspace, of effects-based analysis of cyber weapons in determining the justification for a response, and in the details of Stuxnet in particular, regarding its extraordinary degree of discrimination, are all on point, and would provide a telling summary of this event, had every element of this account not already been exhaustively discussed in similar fashion in a rather considerable body of literature that the author does not cite (e.g., the Symantec security analysis of the structure and risks of Stuxnet in 2011). Still, this is a useful, accurate, and informative account for those who may not have followed those earlier developments.

In distinguishing between types of cyber attacks that do, do not, and might (or might not) constitute just causes for war, Christopher Eberle, in a subsequent essay, helpfully suggests that "right intention," in terms of the plan or purpose envisioned by the attacker in launching such an attack, plays a deciding role in such retaliation. This seems intuitively correct: before launching some sort of retaliation, especially kinetic retaliation, as the US Department of Defense cyber strategy (Dod 2012) threatens, one might wish to know not only who launched the cyber attack, but why, with what "purpose?" An accident, let alone the actions of cyber "rogues" or "patriotic hacktivists" within a nation's sovereign borders (as the Russian government characterized the unattributed attacks on Estonia launched from within its borders in 2007), might call for a different kind of retaliation than a deliberate attempt to do harm for reasons of state. In any event, Eberle finds that the very recourse to cyber weapons, which generally do less or even no "real" harm, would never provide a moral justification for the kind of kinetic retaliation that the US has threatened (presumably as a deterrent to suffering such attacks). This seems unresolvable, however, absent a specific instance or context to analyze. I would think Eberle's constraint would hold (as it did, finally, hold) in the Estonian case, where virtual cyber mischief, even on a heretofore unprecedented scale, did not inflict the kind of permanent damage or harm that would justify kinetic retaliation. The "cyber-Armageddon" or "Pearl Harbor" that pundits like Richard Clarke predict, as Eberle acknowledges, would constitute a different matter altogether, were such an attack to be launched. The harm would perforce be substantial, and fully equivalent to the physical effects of a conventional attack. The recourse to cyber weapons, moreover, would not automatically signify, as Eberle seems to think, a desire to do little or no actual physical harm, but instead would simply represent the inability of the attacking state to resort effectively to a kinetic attack to achieve the damage easily wrought from afar by the use of cyber weapons.

Here Eberle takes up one of the most recalcitrant of cyber threats, the planting of "trap doors" and "logic bombs" by adversaries within a nation's vital infrastructure, presumably in anticipation of a future cyber attack. He likens these to nuclear missiles, aimed at a nation from its adversary's soil. Inasmuch as the existence and even the positioning of the latter do not constitute just cause for war, neither, he suggests, would this form of systematic sabotage (since no actual harm has been inflicted). The nuclear analogy, however, is tricky in this instance and can cut both ways (including against the point that Eberle is at pains to establish regarding the illegitimacy of preventive self defense). The development of nuclear weapons capabilities by adversary governments, combined with the thinly-veiled threat to use them, *was* indeed thought by a great many leaders in many countries to constitute a just cause for preventive war on several occasions: against the mounting Soviet nuclear threat in the 1950s, against the introduction of those same missiles in Cuba in the 1960s, and in the present day, in the nuclear weapons programs and accompanying public threats to use them, first by Iraq in 1981, then by Syria prior to 2007, and most recently by Iran. These are hardly uncontested examples of "just cause," but the controversy over them indicates that the matter is not easily resolved.

Independently, there seems something oddly *dis-analogous* between the nuclear and the cyber case. A better analogy would be the discovery of a group of enemy operatives systematically planting real bombs on real infrastructure, like highways, railroads, and bridges. When noticed, it would seem a bit disingenuous for the organizing adversary to claim that these were merely preventive measures, that no *actual* harm had been done, and thus that no retaliation for these acts of (potential) sabotage was justified – at least until the devices were detonated. Obviously, that is not a satisfactory threat assessment. At very least, mining an adversary's civilian infrastructure with logic bombs constitutes a massive crime, for which the host nation could be expected to cooperate in interdicting the guilty parties, and if not, under the Cybercrime Convention of 2001, the nation itself could be held accountable (whatever that would mean: Lucas 2011, 2013b) for failing to put a halt to this cyber mischief.

This brings us full circle to Eberle's underlying concern with the adversary's plans, purposes, and the just war criterion of "right intention." In all these instances, both nuclear and cyber, it would help immensely to know what the intentions of the various actors truly were. Yet, whether discussing logic bombs planted by the Chinese cyber army or the development of nuclear capabilities by Saddam Hussein, Bashir al-Assad, or the inscrutable Iranian leadership, we can never fully gauge intentionality apart from concrete actions. If the actions appear hostile and threatening, especially if accompanied by bellicose public rhetoric, we can hardly assume the underlying intentions, whatever they may be, are benign. In the State of Texas, unlike the UK, we do not pause, other than as fair warning of our own intent, to discuss with masked intruders we discover invading our home at midnight, what is their purpose in being there, whether they intend merely to rob us, or also to kill us and burn the house down? We simply shoot them, and permit their surviving relatives to explain to our supportive public the variety of illicit and morally unjustified purposes that might have brought them hence.

This crude and ineloquent summary of justification for the use of force would appear to dovetail nicely with the vastly more diplomatic statement of this principle by Col. Edward Barrett, the original driving vision behind this entire project on the ethics of cyber warfare. And yet, somewhat akin to Eberle, he appears to waiver when the time for decisiveness arises, stating (in tepid defense of my point about anticipatory self defense in the previous paragraph), that the acquisition and deployment of nuclear weapons or logic bombs is not itself a cause for war, and that any forcible response designed to pre-empt an imminent attack could only occur if we, the potential victims of attack, "possessed absolute clarity about an adversary's capabilities, intentions, preparations, and especially identity."

Poppycock! Horse-feathers! This limp-wristed waffling by both authors constitutes entirely too much anticipatory pussy-footing for my taste, and their views place the entire burden of proof upon the victim of an attack, without any residual accountability for the perpetrator. Granted, the principle of proportionality, in both war and individual self-defense, does place a reasonable burden of proof on anyone who would employ deadly force to refrain from doing so if the situation (as judged by reasonably impartial third persons) would not clearly warrant it. The "Texan" is not wholly free, even in Texas or Florida, to simply shoot first and ask questions later. A warning must be issued, an opportunity afforded for the invaders to surrender or flee. But it seems patently absurd, after a point, to deny the victim the right of defense of life and property if these conditions are met, even absent knowledge of who the intruders are, or what they intend (since their initial trespass, coupled with their unwillingness to surrender or attempt to flee when warned or ordered to do so, seems to constitute reasonable doubt about the goodness of their intentions).

In like manner, if an adversary nation is busily engaged in stealing everything I own, and festooning my infrastructure with devices of sabotage whose intent, if used, cannot but be devastating to life and limb, exactly how certain do I need to be before putting a stop to the mischief? Forbearance in the face of such circumstances would unquestionably constitute a virtue, but one of supererogation, rather than a fundamental claim of morality. None of this mitigates the entirely separate concern for unintended spiralling escalation and harm that Danks and Danks caution against, in the case of wholly automated and unsupervised systems of cyber self-defense. But it does pose a limit of sound judgment upon the anxieties about discriminate and proportionate response to cyber attack. I argued a few years ago (Lucas 2011) that "Stuxnet," on any reasonable account, represented a legal and morally justifiable response to criminal conspiracy in violation of international laws and regulations, especially because (as Eberle and Jenkins note) it managed to accomplish this without so much as "mussing the hair" of a bystander or conspirator. And this attack, to date, was the most physically destructive we have witnessed. In general, we are discussing forms of virtual, not real harm, mighty inconvenience and loss, rather than death, destruction, and devastation. If so, perhaps we are being too precious about the measures we may reasonably undertake regarding cyber defense. If, as Eberle in particular seems to feel, no one and no thing is actually harmed or damaged in a cyber attack, then the reverse would also be true as we, the victim, responded in kind, presuming we did only that, and not something more dire. In that sense, I would defend the morality not only of Stuxnet, but also, properly understood, the morality of anticipatory self-defense inherent in NSA's now much-maligned "Enterprise Knowledge System." The unjustifiable component was the secrecy and deception, not the institution of the wholly defensive measures. But that is an argument I have made at great length elsewhere (e.g., Lucas 2013c, 2013d, 2014), and in any case, would take us far afield from the limited purview of these interesting essays.

I return in conclusion of Col. James Cook's earlier astute observations in response to Dipert, which offer a helpful caution and corrective to our examination of the ethics of military technology generally. We may sometimes need to exercise caution about overdramatizing the "unprecedented" nature of technological transformation or of the moral challenges such transformations pose. Engineers and pundits sometimes tend to terrify the public inappropriately with their sweeping and occasionally undocumented claims regarding the near-term future of military technology, while opponents, fearful of these developments, tend to exaggerate the risk (Lucas 2010). Patrick Lin, in his essay toward the beginning of this volume, instinctively may have gotten the balance just about right when contrasting what I would explicitly call the *changing technological context* of the war-*fighter (*Lin's catapults and cross-bows, pistols and air war), with the impact of those technologies (like Predators or autonomous robots) that aim finally to *replace* the war-fighter altogether. The latter, in turn, contribute to what we might likewise term the *de-valorizing* of war, as well as perhaps the ease with which we engage in it.

On the other hand, this difference of opinion between Dipert and Cook over the applicability of just war doctrine to cyberwar (especially to its proposed weapons and tactics) raises a deeply troubling concern, even while it invites further comparisons with robotics and the other military technologies discussed. We see in Dipert's account, as well as in Rowe's earlier criticisms of prevailing cyber tactics, that the purposive targeting of civilians, civilian property (e.g., financial accounts), or public infrastructure (electric power grids, dams, air traffic control systems) represent common features (though by no means the exclusive focus) of the tactics of cyber warfare.

This last parallel offers some intriguing conclusions regarding the scope of this entire volume. First, roboticists like Ronald Arkin are part of the overall defense industry's research and development sector, that reports finally to the military, delivering hard platforms and kinetic weapons (even when these are controlled by microcomputers and innovative software). The operational or combat commands are the customers, and those customers think instinctively like warriors: their job, they say, is to "kill people and break things," but they instinctively operate within known constraints of military ethics and the law when considering whom to kill and what to break.

By contrast, cyber warfare experts deal almost exclusively in software: their weapons are "virtual" rather than kinetic (although some can be made to do a great deal of physical damage). Theft of data, denial of access, and most of all, deception and psych-ops to sow confusion and demoralization – all these are tools of the intelligence and espionage community, rather than of the combat warrior. By custom and conventional practice, intelligence gathering, espionage, and even covert operations are not thought to be so rigidly governed or constrained by LOAC, while "intelligence ethics" is a decidedly nascent field (Goldman 2006; Perry 2009). In any case, "targeting civilians" for information, access to data, or to sow confusion and demoralize through deception, is all "standard operating procedure." This difference in the moral and legal background assumptions of the two distinct communities helps, I think, account for why cyberwar strategists have been more ready to aim their weapons at civilian targets than have their counterparts in conventional military combat. As I have taken frequent occasion to observe, the still unrecognized facet of cyber warfare, in contrast to all other, more conventional forms of conflict, is that it *is unrestricted warfare carried out by spies and espionage agents who do not think themselves bound by any legal restraints, rather than by conventional combatants trained in the Law of Armed Conflict. Unrestricted warfare is NOT legally permissible or morally justifiable in the conventional case, but it is routine practice among agents of espionage* (Lucas 2014).

Finally, and notwithstanding the very real physical damage and genuine suffering that denial of access or theft of data or property can inflict on the target victims, there is a background assumption that majority of such destruction or suffering is "virtual" rather than "real," and can even be readily reversed. Rowe, for example, acknowledges this in his analysis, and even cautiously commends *some* cyber tactics as morally (and legally) superior to conventional, kinetic counterparts because the damage is momentary, easily contained, and easily reversed. One can, upon resolution of the conflict, for example, restore electrical power or access to financial accounts by supplying the code or password that is presently locking users out far more easily than one can rebuild actual, physical banks, dams, or power plants that were destroyed in a conventional attack (as NATO forces learned in Kosovo).

We thus have a cyber-counterpart to roboticist Ron Arkin's advocacy of the superior morality of robotic warriors: namely, that cyber warfare, too, if rightly handled, could end up being more discriminate, more proportional, and thus more in compliance with the statutes of LOAC and the moral principles of *jus in bello*, than any conventional counterpart. That would constitute an interesting irony, to say the least.

Notes

1 *On War* (1832: Book VIII, Ch 3: 593).

2 Funded by a bequest from Cindy McCain in honor of her husband, Arizona Senator John McCain (US Naval Academy Class of 1958), the McCain Conference annually convenes scholars and educators in ethics and leadership at the US Naval Academy (Annapolis MD), drawn from US and allied-nation military service academies, Staff and Command Colleges, and senior War Colleges and Defence Academies to examine, and to improve ethics and leadership education, in areas of emerging ethical concern. Past conferences have dealt with private military and security contractors, the rise of military anthropology, Islam and liberal democracy, and new developments in just war doctrine. The conference annually produces an Executive Summary of findings and recommendations on the conference theme for use by senior military leaders and members of Congress.

References

Arkin, Ronald C. (2007) *Governing Lethal Behavior: Embedding Ethics in a Hybrid Deliberative/ Reactive Robot Architecture*, Report GIT-GVU-07-11, Atlanta, GA: Georgia Institute of Technology's GVU Center: http://www.cc.gatech.edu/ai/robot-lab/online-publications/formalizationv35.pdf

Bekey, George (2005) Autonomous Robots: From Biological Inspiration to Implementation and Control, (Cambridge, MA: MIT Press).

Canning, John, Riggs, G.W., Holland, Thomas O., Blakelock, Carolyn (2004) A Concept for the Operation of Armed Autonomous Systems on the Battlefield, *Proceedings of Association for Unmanned Vehicle Systems International's (AUVSI) Unmanned Systems North America*, August 3-5, Anaheim, CA.

Canning, John (2008) Weaponized Unmanned Systems: A Transformational Warfighting Opportunity, Government Roles in Making it Happen, *Proceedings of Engineering the Total Ship (ETS)*, September 23-25, Falls Church, VA.

DoD, (2012) *US Department of Defense Strategy for Operating in Cyber Space.* (Washington, DC: US Dept. of Defense, 2011): http://www.defense.gov/news/d20110714cyber.pdf.

Eco, Umberto (1991) Reflections on War, *La Rivista dei libri* (1 April 1991); reprinted in: Umberto Eco, *Five Moral Pieces*, trans. Alastair McEwen, pp. 1-17 (New York: Harcourt, Inc., 1997).

Goldman, Jan (Ed) (2006) *The Ethics of Spying* (Lanham, MD: The Scarecrow Press).

Krishnan, Armin (2009) *Killer Robots: Legality and Ethicality of Autonomous Weapons* (London: Ashgate Press).

Lin, Patrick, Bekey, George, Abney, Keith (2008) *Autonomous Military Robotics: Risk, Ethics, and Design.* US Department of the Navy, Office of Naval Research (20 December 2008), pp. 112.

Lucas, George R. Jr, (2010) Nerds Gone Wild: Can Moore's Law Remain Valid Indefinitely?, *International Journal of Applied Ethics*, 24(1), pp. 71-80.

Lucas, George R. Jr, (2011) *Permissible Preventive Cyberwar: Restricting Cyber Conflict to Justified Military Targets*, opening address for a UNESCO-sponsored conference on Cyber War and Ethics at the University of Hertfordshire (July 1, 2011): http://www.elac.ox.ac.uk/downloads/Permissible%20Preventive%20Cyberwar%20UNESCO%202011.pdf. Luciano Floridi and Mariarosario Taddeus (Eds). (The Hague: Springer Verlag, forthcoming 2014), Part II, Ch. 5.

Lucas, George R. Jr, (2012) *Just War and Cyber Conflict*, Annual Stutt Lecture on Ethics, US Naval Academy (9 April 2012): http://www.youtube.com/watch?v=hCj2ra6yzl0. Text at: http://www.usna.edu/Ethics/_files/documents/Just%20War%20and%20Cyber%20War%20GR%20Lucas.pdf.

Lucas, George R. Jr, (2013a) Can there Be an Ethical Cyberwar?, Panayotis A Yannakogeorgos and Adam B Lowther, (Eds) *Conflict and Cooperation in Cyberspace*, pp. 195-210 (Boca Raton, FL: CRC Press/Taylor and Francis Publishers)

Lucas, George R. Jr, (2013b) *Jus in silico*: moral restrictions on the use of cyber warfare, in: Allhoff, Evans, & Henschke (Eds) *The Routledge Handbook of War and Ethics*, pp. 367-380 (Oxford: Routledge).

Lucas, George R. Jr, (2013c) *Emerging Norms for Cyber Warfare*. Keynote Address, "Conference on Ethics & Cyber Security," Center for Applied Philosophy & Practical Ethics, Australian National and Charles Sturt Universities (Canberra, AU: 5-6 August): http://philevents.org/event/show/11114.

Lucas, George R. Jr, (2013d) *Navigation, Aviation, 'Cyberation': Developing New Rules of the Road for the Cyber Domain*. Inaugural Public Security Lecture, National Security College, Australian National University (Canberra, AU: 7 August): http://www.youtube.com/watch?v=cs7RXAPzG84.

Lucas, George R. Jr, (2014) NSA Management Directive # 424: Secrecy and Privacy in the Aftermath of Snowden, *Ethics and International Affairs*.

OSD (2009) Office of the Secretary of Defense. FY 2009-2034 Unmanned Systems Integrated Roadmap, Second Edition. http://www.acq.osd.mil/psa/docs/UMSIntegratedRoadmap2009.pdf

Perry, David (2009) *Partly Cloudy: Ethics in War, Espionage, Covert Action and Interrogation* (Lanham, MD: The Scarecrow Press).

Rid, Thomas (2011) Cyber War Will Not Take Place, *The Journal of Strategic Studies*, 35(1), (February 2012), pp. 5-32.

Rid, Thomas C. (2012) Cyber War: Think Again, *Foreign Policy Magazine* (26 February 2012), pp. 58-61.

Rowe, Neil C. (2007) War Crimes from Cyberweapons, *Journal of Information Warfare*, 6(3), pp. 15-25.

Rowe, Neil C. (2008) Ethics of Cyber War Attacks, in: Lech J. Janczewski & Andrew M. Colarik (Eds), *Cyber Warfare and Cyber Terrorism* (Hershey, PA: Information Science Reference).

Rowe, Neil C. (2010) The Ethics of Cyberweapons in Warfare, *International Journal of Cyberethics*, 1(1), pp. 20-34.

Sharkey, Noel (2007a) Robot Wars are a Reality, *The Guardian* (UK), August 18, 2007, p. 29. http://www.guardian.co.uk/commentisfree/2007/aug/18/comment.military

Sharkey, Noel (2007b) Automated Killers and the Computing Profession, *Computer*, 40, pp. 122-124.

Sharkey, Noel (2008a) Cassandra or False Prophet of Doom: AI Robots and War, *IEEE Intelligent Systems*, July/August 2008, pp. 14-17.

Sharkey, Noel (2008b) Grounds for Discrimination: Autonomous Robot Weapons, *RUSI Defence Systems*, 11(2), pp. 86-89.

Singer, P.W. (2009) *Wired for War* (New York: Penguin Press).

Sparrow, Robert (2007) Killer Robots, *Journal of Applied Philosophy*, 24(1), pp. 62-77.

Sparrow, Robert (2008) Building a Better WarBot: Ethical Issues in the Design of Unmanned Systems for Military Applications, *Science and Engineering Ethics* 15(2), pp. 169-187.

Sparrow, Robert (2009) Predators or Plowshares? Arms Control of Robotic Weapons, *IEEE Technology and Society Magazine*, 28(1), pp. 25-29.

Symantec (2011) *W32.Stuxnet Dossier*, Nicholas Falliere, Liam O. Murchu, and Eric Chien (February 2011): http://www.symantec.com/content/en/us/enterprise/media/security_response/whitepapers/w32_stuxnet_dossier.pdf.

Recommended Reading

The literature on emerging technologies is growing rapidly. The works cited below are provided in order to direct readers to readings that are beneficial beyond those referenced at the end of each chapter. The entries are a starting point. In the fast-moving and multi-disciplined world where military ethics and emerging technologies bisect, attempts for a comprehensive bibliography would be futile. However, in pursuing these and other writings, readers will find valuable information for further study regarding the importance and ethics of emerging technologies.

Altmann, Jürgen (2006) *Military Nanotechnology: Potential Applications and Preventive Arms Control* (New York: Routledge).

Anderson, Michael & Anderson, Susan Leigh (Ed) (2011) *Machine Ethics* (Cambridge: Cambridge University Press).

Arkin, Ronald C. (2009) *Governing Lethal Behavior in Autonomous Robots* (Boca Raton, FL: CRC Press).

Asaro, Peter M. (2008) How Just Could a Robot War Be? In Philip Brey, Adam Briggle & Katrina Waelbers (Eds), *Current Issues in Computing and Philosophy*, pp. 50–64 (Amsterdam: IOS Press).

Asaro, Peter M. (2006) What Should We Want From a Robot Ethic? *International Review of Information Ethics* 6, pp. 9–16.

Austin, Reg. (2010) *Unmanned Aircraft Systems: UAVS Design, Development and Deployment* (West Sussex, United Kingdom: John Wiley and Sons).

Axe, David (2008) *Warbots* (Ann Arbor, MI: Nimble Books LLC).

Barnes, Michael. (2010) *Human-Robot Interactions in Future Military Operations* (Surrey, UK & Burlington, VT: Ashgate Publishing).

Bekey, George (2008) *Robotics: State of the Art and Future Challenges* (London, England: Imperial College Press).

Benjamin, Medea. (2012) *Drone Warfare-Killing by Remote Control* (New York: OR Books).

Button, Robert (2009) *A Survey of Missions for Unmanned Undersea Vehicles* (Santa Monica, CA: RAND Corporation).

Davison, Neil (2009) *Non-Lethal Weapons* (New York: Palgrave Macmillan).

Freidman, Norman (2010) *Unmanned Combat Air Systems* (Annapolis: Naval Institute Press).

Giordano, James (2014) *Neurotechnology in National Security and Defense: Practical Consideration, Neuroethical Concerns* (Abingdon: CRC Press).

Giordano, J., Forsythe, C. & Olds, J. (2010) Neuroscience, Neurotechnology, and National Security: The Need for Preparedness and an Ethics of Responsible Action, *AJOB-Neuroscience*, 1(2), pp. 35–36.

Hickey, James E. (2012) *Precision-Guided Munitions and Human Suffering in War* (Surrey, UK & Burlington, VT: Ashgate Publishing).

Koplow, David A. (2006) *The Law and Policy of Revolutionary Technologies for the Military and Law Enforcement* (Cambridge: Cambridge University Press).

Krishnan, Armin (2009) *Killer Robots: Legality and Ethicality of Autonomous Weapons* (Surrey, UK & Burlington, VT: Ashgate Publishing).

Lele, Ajey (2009) *Strategic Technologies for the Military: Breaking New Frontiers* (Los Angeles: Sage Publications).

Lin, Patrick, Abney, Keith & Bekey, George A. (2012) *Robot Ethics: The Ethical and Social Implications of Robotics* (Cambridge, MA: The MIT Press).

Mets, David R. (2009) *Airpower and Technology: Smart and Unmanned Weapons* (Westport, CT: Praeger Security International).

Moreno, Jonathan D. (2013) *Mind Wars: Brain Science and the Military in the 21st Century*, 2nd ed. (New York: Bellevue Literary Press).

O'Donovan, Oliver (2003) *The Just War Revisited* (New York: Cambridge University Press).

Orend, Brian (2006) *The Morality of War* (Toronto: Braodview Press).

Rappert, Brian (2003) *Non-Lethal Weapons as Legitimizing Forces?: Technology, Politics and the Management of Conflict* (London: Frank Cass).

Riza, Shane M. (2013) *Killing Without Heart* (Dulles, VA: Potomac Books).

Schmitt, Michael (2013) Extraterritorial Lethal Targeting: Deconstructing the Logic of International Law, *Columbia Journal of Transnational Law*, Vol. 52.

Singer, P. W. (2009) *Wired for War: The Robotics Revolution and Conflict in the 21st Century* (New York: Penguin Press).

Singer, P. W. & Friedman, Allan (2014) *Cybersecurity and Cyberwar: What Everyone Needs to Know* (New York: Oxford University Press).

Sparrow, Robert (2009) Building a Better Warbot: Ethical Issues in the Design of Unmanned Systems for Military Applications, *Science and Engineering Ethics*, 15, pp. 169–87.

Sparrow, Robert (2007) Killer Robots, *Journal of Applied Philosophy*, 24(1), pp. 62–77.

Sparrow, Robert (2011) Robotic Weapons and the Future of War in: Jessica Wolfendale & Paolo Tripodi (Eds), *New Wars and New Soldiers: Military Ethics in the Contemporary World*, pp. 117–33 (Surrey, UK & Burlington, VT: Ashgate).

Strawser, Bradley J. (Ed) (2013) *Killing by Remote Control: The Ethics of an Unmanned Military* (New York: Oxford University Press).

Sullins, John (2006) When is a Robot a Moral Agent?, *International Review of Information Ethics*, 6, pp. 23–30.

Turse, Nick & Engelhardt, Tom (2012) *Terminator Planet: The First History of Drone Warfare, 2001–2050* (Lexington, KY: Dispatch Books).

US Department of Defense (2011) *FY 2011–2036 Unmanned Systems Integrated Roadmap.*

Vaseashta, A., Braman, E. & Sussman, P. (Eds) (2012) *Technological Innovation in Sensing and Detecting Chemical, Biological, Radiological, Nuclear Threats and Ecological Terrorism,* NATO Science for Peace and Security Series (NY: Springer).

Wallach, Wendell and Allen, Colin (2009) *Moral Machines: Teaching Robots Right from Wrong* (Oxford: Oxford University Press).

Index

Note:
Page numbers in **bold** type refer to figures
Page numbers in *italic* type refer to tables
Page numbers followed by 'n' refer to notes

FSC
www.fsc.org
MIX
Papier | Fördert gute Waldnutzung
FSC® C083411

Kolibri 360 GmbH
Ferdinand-Jühlke-Straße 7
99095 Erfurt, Deutschland
produktsicherheit@kolibri360.de